Rainforest Cities

Rainforest Cities

Urbanization, Development, and Globalization of the Brazilian Amazon

John O. Browder and Brian J. Godfrey

COLUMBIA UNIVERSITY PRESS

New York

Columbia University Press
Publishers Since 1893
New York   Chichester, West Sussex

Copyright © 1997 Columbia University Press
All rights reserved

Library of Congress Cataloging-in-Publication Data
Browder, John O.
Rainforest Cities : urbanization, development, and globalization
of the Brazilian Amazon / John O. Browder and Brian J. Godfrey.
        p.   cm
Includes bibliographical references and index.
ISBN 978-0-231-10655-9 (pbk.)
1. City planning—Amazon River Region. 2. Regional Development—Amazon River
Region. 3. Land use, Rural—Amazon River Region. 4.Land settlement patterns—
Amazon River Region. 5. Amazon River Region—Urbanization. I. Godfrey, Brian J. II.
Title.
HT169.A5B76   1997
307.1'216'09811—dc20   96-33481
                                       CIP

Casebound editions of Columbia University Press books
are printed on permanent and durable acid-free paper.

Printed in the United States of America

To Jayna Madeleine

(JOB)

In Memory of Antonio Aparecido de Souza (1953—1994)

(BJG)

and

A nossos amigos brasileiros

(BJG & JOB)

Contents

❧᯼❧

Figures

❧..❧

Tables

❧❦

Plates

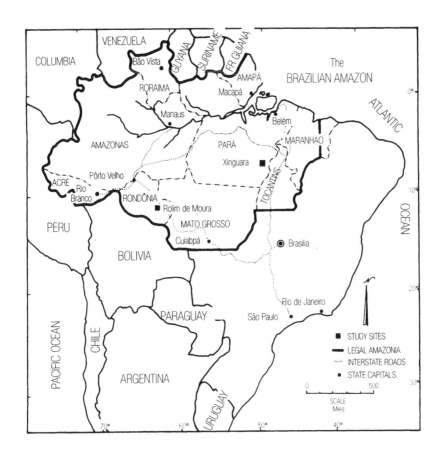

Map of Brazilian Amazonia and Study Sites

Preface

❧⸎❧

The idea for this book on the rainforest cities of Amazonia arose in an unlikely place, the desert city of Phoenix, Arizona. We first met there in April of 1988, at the annual meeting of the Association of American Geographers (AAG), where we both presented papers on the Brazilian Amazon. We discovered mutual interests in a widespread but still largely unrecognized regional trend: the rapid urbanization of the Amazon's extensive settlement frontiers. Since our research areas covered two of the most dynamic expansion fronts in eastern and western Amazonia—southern Pará and Rondônia—the idea of a collaborative research project immediately arose. At the time Amazonia's burgeoning boomtowns had scarcely been studied by scholars, and the few exceptions were generally self-contained case studies. Shortly after the 1988 AAG meeting, we resolved to seek joint funding for a comparative study of Amazon frontier urbanization.

As an urban planner and an urban geographer, we approached the subject of Amazon urbanization from the major traditions of our academic fields: urban morphology and city-systems dynamics, as framed by spatial economics, central place theory, and historical geography. Our initial conceptualization of the frontier urbanization process took the form of a linear transition model, hypothesizing a general historical-geographical sequence of stages from pioneer agriculture to urban primacy, couched within the lineaments of world systems theory (Browder and Godfrey 1990). Empirical studies in the Amazon Basin sometimes supported the transition model (Brown et al. 1994) and sometimes found local exceptions (Roberts 1992). In our own empirical work the urban-transition model turned out to be largely valid in describing urban system dynamics in some parts of the Amazon (Rondônia) but not in others (southern Pará).

This empirical variability led us to rethink conventional conceptualizations of the frontier as a homogeneous regional type. Our subsequent comparative studies have convinced us that Amazonia is a highly fragmented and fluid social space; patterns of regional development are variable, both geographically and historically. The dynamics of socioeconomic change on the frontier are contingent and contextual, specific to particular localities, where key sectors are differentially articulated to regional, national, and global circuits of capital, information, technology, and political influence. These conditions produce polymorphous urban systems whose functions are differentially articulated to different levels of the global economy.

This book explores the theoretical implications of Amazonia's urban and regional diversity. We argue that the Amazon's range of settlement patterns does not favor hegemonic master theories of explanation, which have dominated development debates for the last three decades. Rather, we find in Amazonia fertile ground for a new conceptual framework of spatial analysis, grounded in what we call "ecological pluralism." We have not come to this conclusion cavalierly. Some readers may find our comparative review of the conventional conceptual models (chapter 2) and our pluralistic theory of disarticulated urbanization (chapter 4) as useful as our interpretation of empirical field data (chapters 5–10). So that interested readers may understand fully how we conducted our empirical research, we include a brief technical appendix delineating our methodologies. Throughout the text, all translations from the Portuguese are ours, except where otherwise stated.

Our purpose in writing this book ultimately reaches beyond providing what we believe to be an accurate analysis of urbanization patterns in Brazil's rainforest frontier. Additionally, we hope that our study contributes to an intellectual movement afoot in several disciplines, which seeks new and integrated ways of constructing knowledge in a *post*-postmodernist era. May this book encourage others to press forward with this ongoing task of intellectual renewal.

Acknowledgments

Working effectively in Amazonia, confronting the region's daunting distances, annoying discomforts, and unexpected obstacles can never be an individual effort. It is simply impossible to do any significant regional research without the considerable help of others. We owe an enormous debt to colleagues and coworkers, both in Brazil and the United States, who provided intellectual inspiration, gave generously of their time, and lent essential logistical support and moral encouragement to what, in retrospect, was the most complex undertaking either of the authors ever embarked upon.

First, we acknowledge the vital support we received from our families and close friends: Diane Browder, Olin and Olive Browder, Ann Sorensen, and Catherine Browder-Morris.

Similarly, Brian Godfrey's family and friends made feasible both the fieldwork in Brazil and the subsequent writing of this book. In the U.S., special thanks go to Thies M. Grimes, John H. Godfrey, Richard Juhl, Eileen Donahue, James Lambert, Jack Nix, James Chappell, and Ruth Levine. Close Brazilian friends made living abroad feel like home: in São Paulo, João Carlos Attarian and the late Antonio Aparecido de Souza; in Rio de Janeiro, Antonio J. C. Pereira and Jurandyr Florentino Miguez; and in Belém, Gino Pina and Antonio Fadul. *Muito obrigado a todos!*

We are intellectually indebted to Bertha Becker and Donald Sawyer, two long-time friends and mentors, and to our colleagues Marianne Schmink, Charles Wood, Roberto Samanez Mercado, and Marília Andrade. They engaged us in long hours of discussions on the substantive theoretical questions at stake in our research and provided logistical support and helpful suggestions. Their own pathbreaking works on frontier expansion have served as a continuing inspiration for us.

This project is really the triumph of twenty-five Brazil-based students and young researchers who took time from their everyday lives to join our research teams as interviewers. We are grateful to our Pará research team, which included Alberto Carlos Lourenço Pereira, Rebecca Abers, Domingos M. S. Macedo, Ana Luiza Nabuco Palhano, Glaúcia Moreira Pinto, Moisés Machado, Oderval Esteves Duarte Filho, Ricardo "Trescentos" Rodriguez Dutra, and Frederico Poley Martins Ferreira, Steve McCracken, and David Wasserman. In Rondônia, the 1990 field research consisted of Igor Mousasticoshvily, Jr., Fernando Fernandes, Humberto Paiva Isidoro, José Irineu Rangel Rigotti, Lorena Ribeiro Fonseca, Dilse Rocha do Amaral, Adnilson de Almeida Silva, Valdineuza Maria do Nascimento Borges, Paul Sergio Tavares da Silva, and Maria Helena de Bezerra. The peri-urban surveys in Porto Velho were undertaken in 1993 with the dedicated assistance of the Centro de Educação e Alfabização Popular (CEAP): Gilvandro Barros Pinheiro, Joana Darc Moura Silva do Amaral, Benedita do Nascimento Pereira, and Carlos Macedo Dias. All of these researchers battled heat, hunger, and fatigue to successfully complete the nearly two thousand household interviews that form much of the empirical base of this book. To them, along with our other *amigos brasileiros*, this book is dedicated.

The empirical core of this book, our field surveys, was made possible by the receipt of a research grant from the Geography and Regional Science section of the National Science Foundation (Grant No. SES-8910972) entitled "Frontier urbanization in the Brazilian Amazon: Implications for regional development." In Brazil, survey work became feasible through the logistical support of several important research institutions. At the Instituto Sociedade, População, and Natureza (ISPN) in Brasília, Donald Sawyer took an immediate interest in our project and helped us in innumerable ways, ranging from professional contacts to advice on the questionnaire design. The Centro de Desenvolvimento e Planejamento Regional (CEDEPLAR) at the Federal University of Minas Gerais welcomed us as Research Associates (Pesquisadores Associados), duplicated our survey questionnaires, and recruited graduate students to serve as field interviewers. We are particularly indebted to the CEDEPLAR director at the time, Prof. João Antônio de Paula, for his generous institutional assistance. The Rondônia project would not have been possible without the generous logistical support of several Rondônia State and município government officials, especially Maria Emilia da Silva (Secretaria de Planejamento do Estado de Rondônia), Edson Mugrave (formerly Instituto Estadual de Florestas de Rondônia), Pedro Wilson Dias Pinheiro (Secretaria de Agricultura do Estado de Rondônia), Eraldo Matricardi (Secretaria de Desenvolvimento Ambiental do Estado de Rondônia), and the Prefeitura of Rolim de Moura. A special word of gratitude is offered to João Baptista Lopes, founder of Rolim de Moura, for his hospitality and guidance in getting around the town since 1984. Research in Belém, a frequent place of residence for both authors, counted on the frequent assistance of the staffs at the Instituto do Desenvolvimento Econômico-Social do Pará (IDESP), the Instituto Brasileiro de Geografia e Estatística (IBGE), the Museu Paraense "Emilio Goeldi," the Superintendência da Campanhia contra a Malária (SUCAM), and the Superintendência do Desenvolvimento da Amazônia (SUDAM). The Núcleo de Altos Estudos Amazônicos (NAEA) at the Federal University of Pará constituted an agreeable environment to consult the library and exchange ideas with colleagues. Chris and Nancy Uhl, along with colleagues at IMAZON, also provided assistance in Belém.

Our home institutions also played a pivotal role in mobilizing resources for this research. At Virginia Tech, John Browder received a Creative Match Grant that financed travel to Brazil in 1989 to lay the groundwork for the field surveys the following year. A separate small grant was received from the Agency for International Development through Virginia Tech's Office of International Research and Development that financed the farm surveys in Rondônia. Two graduate research assistants in the College of Architecture and Urban Studies, Marcos Pedlowski and Lyle Boelens, helped in data analysis at Virginia Tech and went on to produce a doctoral dissertation and master's thesis, respectively, on development issues in Rondônia. John Browder is particularly grateful to Professors James Bohland (former chair of the Urban Affairs and Planning Program) and to Patricia Edwards (Dean of the College of Architecture and Urban Studies) for assisting him in securing a yearlong research leave in 1994–1995, during which the book manuscript was completed. While on leave, John Browder was welcomed and supported by colleagues in the School of Natural Resources and the Environment at the University of Michigan, especially by Patrick West, and in the Center for Latin American Studies and the Tropical Conservation and Development Program at the University of Florida, namely Terry McCoy, Steven Sanderson, and Peter Polshek. My gratitude goes to them as well.

At Vassar College, Brian Godfrey received several awards for travels and materials from the Committee on Research. A Mellon Grant for non-Western Cultures, administered by the International Studies Program, provided additional generous assistance for field research during his sabbatical year in 1990–1991. The Department of Geology and Geography allowed the use of its research facilities, secretarial time, and student research assistance throughout this project. Besides the ongoing support of secretary Sandy Ponte, several excellent student research assistants deserve special mention for their work on the maps and figures: Adriane Fowler (Vassar class of 1994), Rebecca Penniman (class of 1995), and Doug Ornstein (class of 1996). While a Fulbright scholar at the Federal University of Santa Catarina during the latter half of 1988, Brian Godfrey also benefited from the professional contacts and friendships of Harrysson L. Silva, E. Jean Langdon, and Dennis Werner.

We have also appreciated working with many U.S. colleagues who in different ways have generously shared their insights with us: J. Timmons Roberts at Tulane University; Stephen Bunker at the University of Wisconsin; Paul Knox at Virginia Tech; Larry Brown at Ohio State University; Douglas McManis, long-time editor of the *Geographical Review;* Marie Price of George Washington University; Candace Slater of the University of California, Berkeley; James E. Vance, Jr., James J. Parsons, and Hilgard O'Reilly Sternberg, now professors emeriti at Berkeley's Department of Geography. Needless to say, we hold none of the above-mentioned individuals and institutions accountable for our research and conclusions, for which we, the authors, alone are responsible.

Finally, we gratefully acknowledge the generous assistance of countless local residents in the Amazon towns studied. These individuals are too numerous to mention, but our study would not have been possible without them. Our local friends and contacts made field research a never-ending source of pleasure, surprise, and revelation. If this study serves to increase public awareness and effective policy responses to the many hardships of daily life in urban Amazonia, our efforts will have been rewarded.

Glossary of Acronyms

❧⸎❧

BASA. Banco da Amazônia, Sociedade Anônima (Amazon Bank, Inc.).

CAPEMI. Caixa de Pecúlio dos Militares (Military Pension Fund).

CONSAG. Construtora Andrade Gutierrez (Andrade Gutierrez Construction Company).

CVRD. Companhia Vale do Rio Doce (Valley of Rio Doce Company).

ELETRONORTE. Centrais Elétricas do Norte do Brasil (North Brasil Electricity Board; the Amazon's regional affiliate of Eletrobrás, the national electrical power agency).

FUNAI. Fundação Nacional do Indio (National Indian Foundation).

GETAT. Grupo Executivo de Terras do Araguaia-Tocantins (Executive Group for the Araguaia-Tocantins Lands).

IDESP. Instituto de Desenvolvimento Econômico-Social do Pará (Pará State Institute for Socioeconomic Development).

INCRA. Instituto Nacional de Colonização e Reforma Agrária (National Institute of Colonization and Agrarian Reform).

ISI. Import Substitution Industrialization.

ITERPA. Instituto de Terras do Pará (Land Institute of Pará).

MINTER. Ministério de Interior (Ministry of the Interior).

PIN. Programa de Integração Nacional (National Integration Program).

POLAMAZONIA. Programa de Polos Agropecuários e Agrominerais da Amazônia (Program of Agricultural and Mineral Development Poles in the Amazon).

POLONOROESTE. Programa do Desenvolvimento Regional do Noroeste (Northwest Regional Development Program).

SPVEA. Superintendência do Plano de Valorização Econômico da Amazônia (Superintendency for the Economic Valorization Plan of Amazonia).

SUCAM. Superintendência da Campanha Contra a Malária (Superintendency for the Campaign Against Malaria).

SUDAM. Superintendência do Desenvolvimento da Amazônia (Superintendency for Amazon Development).

SUDECO. Superintendência do Desenvolvimento do Centro-oeste (Superintendency for the Development of the Center-West).

SUFRAMA. Superintendência da Zona Franca de Manaus (Superintendency for the Manaus Free Trade Zone).

ZFM. Zona Franca de Manaus (Manaus Free Trade Zone).

Rainforest Cities

1

Amazon Town Revisited: New Urban Realities on Brazil's Rainforest Frontier

The Amazon Basin strikes the core of our imagination. The continental scale of its tropical forests, the mysteries enshrouding its inestimable life forms, and the legends of its untold natural treasures have excited imperial ambitions and fed frustrations for five hundred years. We have been inundated by a recent outpouring of popular and academic books, artistic and literary works, and television documentaries and commercial films dramatizing the destruction of the rainforest and social conflict on the planet's final great frontier. The metaphors describing Amazonia are typically monumentalist and misleading: El Dorado, Second Eden, Green Hell, Earth's Lung, and the Last Frontier (Godfrey 1993; Hecht and Cockburn 1989). Such entrenched regional images evoke essentially *rural* environments, sparsely populated by indigenous forest peoples, rubber tappers, peasant farmers, cattle ranchers, *caboclos*, and other country folk.

Lost in the popularized vision of this troubled rainforest realm is the paradox of the Amazon's current urbanization. Despite the region's long-standing image as a rural environment of receding rainforests, Amazonia has been predominantly urbanized since at least 1980. Most inhabitants of the Brazilian Amazon, along with increasing proportions of the Amazonian populations of the surrounding Spanish American countries, now reside in cities (Browder and Godfrey 1990; Brown et al. 1994).[1] While Brazil as a whole experienced an urban transition during the decade of the 1960s, according to the official decennial censuses, the six states and former territories of the Brazilian North Region became predominantly urbanized during the 1970s.[2] By 1991 the year of the most recent national census, the regional population of Amazonia was about 58 percent urban (table 1.1).[3]

Table 1.1 Population growth and urbanization in Brazil and the North Region, 1940–1991

| Year | BRAZIL | | NORTH REGION[A] | |
	Population	% Urban	Population	% Urban
1940	41,236,315	31.2%	1,462,420	27.7%
1950	51,944,397	36.2	1,844,655	31.5
1960	70,070,457	44.7	2,561,782	37.5
1970	93,139,037	55.9	3,603,866	45.1
1980	119,002,706	67.6	5,880,268	51.7
1991	150,367,800[b]	75.0[b]	9,337,150	57.8

Source: Instituto Brasileiro de Geografia e Estatística, various years.
a Includes the states and former territories of Acre, Amapá, Amazonas, Pará, Rondônia, and Roraima.
b Preliminary figures, according to the IBGE (1991).

Far from the lethargic and insular Amazon river town depicted by anthropologists only a generation ago, the current focal point of regional settlement is the volatile boomtown, seething with activity, deep in the interior of the Amazonian upland rainforests. New urban centers have proliferated across the landscape in previously inaccessible *terra firme* forest areas. Older towns and cities have become enveloped in a sea of periurban shantytowns. Unlike their remote counterparts on the nineteenth-century North American frontier, which were plagued by uncertain transportation and communications connections, the contemporary rainforest cities are permeated by global currents of information, trade, and politics. Still, some virtually vaporize into ghost towns, after riding brief boom-bust waves of the extractive activity so typical of the region. Others grow to become permanent settlements of continuing economic importance. Yet, most commentators have overlooked Amazonia's rapid urbanization, preferring instead to emphasize the forest frontier and other more familiar rural manifestations of the Amazon. Brazilian geographer Bertha Becker (1990:44) was one of the first scholars to recognize the novelty of

Amazonia's urban character: "Urbanization there is not a consequence of agricultural expansion: the frontier is born already urbanized, and it has a rhythm of urbanization faster than the rest of Brazil."[4]

This new urban reality remains at odds with both popular views and official regional development policies. Consequently, many pressing urban problems remain unaddressed in Amazonia, including deficient infrastructure, social and medical services, rapid shantytown growth, and pollution. Further complicating the formulation of appropriate urban policy in the region is the fact that the urbanization of the Amazon does not neatly replicate North American or European urban transition models. Rather, urban growth in the Amazon must be placed in locally specific contexts.

Unfortunately, the widespread ignorance about urban Amazonia is understandable, given the paucity of works on the subject. Only a handful of scholarly studies of Amazon cities has been carried out, and most of those, while often excellent works, have not gone beyond isolated case studies of urban life. Certainly, classic urban histories of Belém (Penteado 1968) and Manaus (Benchimol 1977) have effectively put metropolitan change in a wider regional context, while more recent work has focused on the rapid expansion of peri-urban shantytowns (Browder et al. 1994; Mitschein et al. 1989). Recognition of the urban transformation in the Amazon interior, however, began only in the late 1980s with seminal works by Becker (1985), Sawyer (1987), and a few others. The lament of Despres (1991:24) still stands: "The history of urban development for the Amazon region remains to be written." The time has come to fully acknowledge the Amazon as an urbanized region. In this book we explore the diverse processes of urban expansion and the role of Amazon towns and cities in shaping the larger regional economy. This introductory chapter examines the nature and magnitude of regional urban growth in the context of contemporary Brazilian national trends and presents our purpose in writing this book.

URBAN AMAZONIA: 1960 VERSUS 1991

A comparison of recent Brazilian censuses of population suggests the magnitude of changes in regional settlement structure. Amazonia's traditional dependence on economic boom-and-bust cycles in primary commodities long created a chasm separating the two large metropolises, Belém and Manaus, from the smaller settlements of the interior. The regional economy long relied on the exploitation and export of natural resources scattered widely in the rainforest, including diverse forest products (*drogas do sertão*), rubber latex, gold and diamonds, valuable timber, and so on. These commodities were commercialized through the primary river ports, mercantile centers from which foreign and national companies controlled the regional trade. From these strategic points in the Amazon's extensive network of rivers, long the primary means of regional transportation, elites from a few primary cities controlled a vast rural hinterland (fig. 1.1). This long-standing commercial geography resulted in a top-heavy urban system of two large metropolises in a scantily urbanized region.

As late as 1960, only twenty-two cities in the entire North Region of Brazil had populations exceeding 5,000. Just two of these cities (Belém and Manaus) had over 100,000 inhabitants. By 1991, however, the number of cities larger than 5,000 had grown to 133, eight of which exceeded

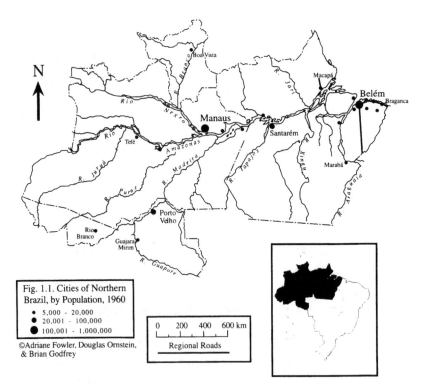

Fig. 1.1. Cities of Northern Brazil, by Population, 1960

- • 5,000 - 20,000
- • 20,001 - 100,000
- ● 100,001 - 1,000,000

©Adriane Fowler, Douglas Ornstein, & Brian Godfrey

0   200   400   600 km

Regional Roads

Figure 1.1. Cities of northern Brazil, by population, 1960

Map design: Adriane Fowler, Douglas Ornstein, and Brian Godfrey

Source: IBGE, 1960.

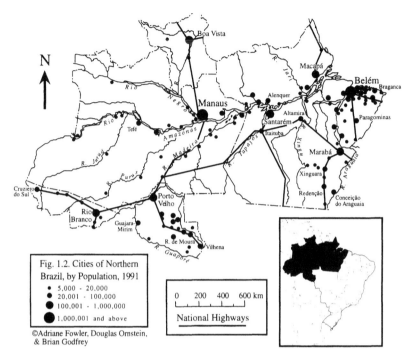

Figure 1.2. Cities of northern Brazil, by population, 1991
Map design: Adriane Fowler, Douglas Ornstein, and Brian Godfrey
Source: IBGE 1991.

100,000 residents by official estimates (table 1.2).[5] Although the regional economy has hardly shed its reliance upon boom-and-bust economic cycles, a more extensive and diversified urban system has emerged in recent years. Contemporary programs of economic integration, sponsored by the state in concert with private capital, have encouraged massive migrations to inland cities. These population movements have transformed the Amazon's traditional settlement structure by narrowing the formerly extreme gap between the upper and lower levels of the regional urban hierarchy. For the first time, the urban population tended to settle primarily in small- or medium-sized cities.

A regional map of contemporary urban centers reveals a concentration of cities in lower Amazonia, near Belém, and in the middle of the Amazon Basin, stretching between Manaus and Santarém. Indeed, the Amazon's recent urbanization in part reflects a continuing growth of the major cities of Belém and Manaus, each of which has become the center of an extensive metropolitan region radiating outward dozens of kilometers from the central city. Yet, the most impressive expansion of the

Table 1.2   Number of cities, classed by size of municipal urban populations, North Region of Brazil, 1960–1991

SIZE OF URBAN POPULATIONS	1960	1970	1980	1991
5,000–20,000	18	23	50	87
20,001–100,000	2	8	13	38
100,001–1,000,000	2	2	3	6
1,000,001 and above	0	0	1	2
Total cities with more than 5,000 population	22	33	67	133

Source: IBGE, Censos demográficos, various years.

Note: The North Region includes the states and former territories of Acre, Amapá, Amazonas, Pará, Rondônia, and Roraima. The state of Tocantins, created in 1988 out of northern Goiás, is not included for lack of comparative demographic data from earlier years. The urban population of Belém encompasses the municipalities of Belém and adjacent Ananindeua, which constitute the metropolitan area. The 1991 population of metropolitan Belém includes suburban and exurban "agglomerations of urban extension" in the rural zones.

region's network of cities and the fastest rates of urban population growth have occurred on the inland settlement frontiers of southern Pará (Marabá to Redenção) and Rondônia (Porto Velho to Vilhena). These two areas, previously isolated from the region's space economy, constitute the case study material of this book (fig. 1.2).[6]

CONTEMPORARY FRONTIER GROWTH: METROPOLITAN PRIMACY IN QUESTION?

The demographic consequences of Brazil's ambitious efforts to open up Amazonia have been dramatic, especially during the period of the military authoritarian regime (1964–1985). The total population of the North Region increased from roughly 2.5 million in 1960, just 37 percent of it urbanized, to more than 9.3 million in 1991, with nearly 58 percent urbanized. Projections of recent growth rates indicate that the population of this six-state region will increase to a total of more than 11.4 million by 2000, of which over 60 percent will live in urban centers. Particularly notable has been the rapid urban growth of the remote northern states: in 1991 the population of Amapá was 80.9 percent urbanized; that of Amazonas was 71.4 percent urbanized, and that of Roraima, the center of a contemporary gold rush, was 64.6 percent urbanized (table 1.3). In general, the urban growth rates in the Amazon region have been about double those of the nation as a whole (Godfrey 1988, 1990).

Amazonia, long dominated in political, economic, and demographic terms by Belém and Manaus, has suffered from an extreme case of metropolitan primacy. Few intermediate-sized cities challenged these metropolises, making the regional urban structure of Amazonia remarkably asymmetrical. In 1950 more than two-thirds of the urban population of the North Region resided in the municipalities of Belém and Manaus. Since that time, in large part due to frontier urbanization, the overall degree of regional urban concentration in these two metropolitan centers has declined steadily. The percentage of the regional urban population in Belém and Manaus combined fell from 67.9 percent in 1950 to 41.8 percent in 1991 (table 5.1). In other words, the last decennial census indicated for the first time that more than half of the region's urbanites resided outside the two largest metropolitan areas. Although Belém and Manaus now have reached unprecedented sizes, the regional trend toward metropolitan deconcentration has continued in recent years (see chapter 5).

Table 1.3   Population growth and urbanization in the Brazilian North Region, by state, 1960–1991

| STATES AND TERRITORIES | 1960 TOTAL | 1970 TOTAL | 1980 TOTAL | 1991 TOTAL |
	% Urban	% Urban	% Urban	% Urban
Acre	158,184	215,299	301,303	417,165
	20.7%	27.6%	43.8%	61.8%
Amapá	67,750	114,359	175,257	288,690
	51.4%	54.6%	59.2%	80.9%
Amazonas	708,459	955,235	1,430,089	2,102,901
	32.9%	42.5%	59.9%	71.4%
Pará	1,529,293	2,167,018	3,403,391	5,181,570
	40.2%	47.2%	49.0%	50.4%
Rondônia	69,792	111,064	491,069	1,130,874
	43.3%	53.6%	46.5%	58.2%
Roraima	28,304	40,885	79,159	215,950
	42.9%	42.8%	61.6%	64.6%
North Region	2,561,782	3,603,860	5,880,268	9,337,150
	37.5%	45.1%	51.7%	57.8%
Brazil	70,070,457	93,139,037	119,002,706	150,367,800[a]
	44.7%	55.9%	67.6%	75.0%[a]

Source: Instituto Brasileiro de Geografia e Estatística, various years.
[a]*Preliminary figures, according to the IBGE (1991).*

Despite the general decline in regional metropolitan primacy, much of Amazonia's recent urban expansion has occurred in the preexisting primate cities of Belém and Manaus and, to a lesser extent, in Porto Velho. These metropolitan centers have witnessed the creation of extensive peripheral shantytowns. Manaus, capital of the state of Amazonas, has grown spectacularly since being denominated a free trade zone by the Brazilian government in the late 1960s. The city has attracted many new businesses, factories, and migrants. Manaus now boasts a metropolitan population of about one million residents, nearly the size of its rival Belém in the lower Amazon Basin, a city that was almost twice its size in the early twentieth century. Manaus has gradually increased its share of the population of Amazonas, a state marked by few intermediate-sized cities. On the other hand, Belém's share of the population of the state of Pará has declined gradually with the growth of secondary cities, such as Santarém and Marabá.[7]

The emergence of these new urban realities in Amazonia marks a historic watershed and raises urgent new issues. Current local processes of frontier urbanization are complex, dynamic, and still inadequately understood in relation to larger issues of regional development, national policy, and the global economy. To better highlight the sociocultural changes resulting from contemporary regional urbanization, we revisit a classic ethnography of an Amazon river town. This community, often depicted as the prototype Amazon community, now stands in sharp contrast to present-day realities.

AMAZON TOWN: A TRADITIONAL COMMUNITY

Over three decades ago Charles Wagley (1953) prepared a now classic ethnography of a traditional Amazon riverine community, which went by the pseudonym "Itá"—later revealed to be the town of Gurupá in Pará. This town has provided the stereotypic image of a small town in Amazonia ever since. Secluded for centuries in "lethargy and backwardness" on a remote margin of the lower Amazon River, postwar Gurupá evolved over the generations into a tightly knit and insular community of roughly five hundred inhabitants, united by cohesive social ties and a long continuity of local traditions. Wagley's portrayal of Gurupá in many ways fits Tonnies's classic (1887) model of a premodern *Gemeinschaft* community: personal relations predominate over abstract market forces; people coalesce around shared values and ties of kinship, and the community's common experience results in a cohesive social order.

Gurupá began its recorded history as a strategic outpost during the colonial period. In 1623 the Portuguese captured the Dutch fort at this contested site and shortly thereafter declared Gurupá a *vila* or town. The local Indians gathered in Gurupá, gradually converted to Christianity, learned the rudiments of Portuguese, and intermarried with the coloniz-ers. The resulting miscegenated Portuguese-Indian inhabitants became the basis for the town's growing population. By 1800 the town contained 564 inhabitants. In 1850 a visitor estimated the population at 715: 482 were "white or mestizo" and 233 were Indian "slaves" (Wagley 1953).

Gurupá's history accompanied the region's boom-and-bust cycles in natural resources, especially the extraction of rubber latex (*Hevea brasiliensis*) in the late nineteenth and early twentieth centuries when the area reached its economic apogee. The population of the town grew steadily through in-migration and natural increase and became more dif-ferentiated in social and racial terms as Afro-Brazilians from the Northeast, Luso-Brazilian merchants, and others entered the area. New municipal infrastructure appeared—electric lights, new municipal build-ings, and so on. Following the collapse of the rubber boom in the early twentieth century, however, the town, like so many others, drifted into economic stagnation. After the bust the population fell to only about three hundred in 1920. Yet Gurupá retained some local importance as the *município-* or county-level governmental seat. This continuing political function ensured Gurupá's survival.[8]

By the middle of the twentieth century, Wagley observed that Gurupá's social structure had hardened into a virtual caste system of four clearly demarcated classes: a small upper class, mainly white, descended largely from wealthy merchants; a "second class" of local townspeople, predomi-nantly of mixed racial ancestry; farmers from the surrounding rural area; and the local rubber tappers and collectors of forest products. These classes were firmly delimited by race, class, and custom. Wagley's ethnog-raphy describes how family and marriage, festivals, religion, and other cul-tural aspects of the community functioned in a traditional Amazon town. Despite Gurupá's social stratification, Wagley portrays a community rela-tively free from strife, unified by a cohesive *caboclo* riverine culture.[9]

As if oblivious to the dramatic changes sweeping over the region, by 1980 Gurupá's urban population stood at only 2,315. Between 1980 and 1985 the municipality grew at a sluggish annual rate of 1.1 percent (half that of Brazil overall), as opposed to the explosive annual rates registered in areas of active frontier settlement. For example, in southern Pará, a zone of massive in-migration, the municipality of Conceição do Araguaia grew annually by 15.1 percent in the early 1980s and that of Marabá increased by 9.3 percent every year during this period. And in Rondônia

Prevailing theories of political economy emphasize the articulation of local dynamics to the penetration of capitalism into the countryside through cash cropping, advanced farming technologies, and the growing proletarianization of the peasantry displaced from the land to the cities. These theories draw upon both modes of production (de Janvry 1981) and world systems (Armstrong and McGee 1985) literatures. Yet, a convincing argument can be made that Amazonia has undergone at best only a partial capitalist transition, marked by such problematic developments as a persistence of traditional social relations, a burgeoning informal sector, a weakened central state, and blurred rural-urban distinctions (Cleary 1993).

How, then, can we best characterize and explain the polymorphous patterns of urbanization found in the Amazon today? To what extent do the urban systems of the Amazon support a process of autonomous regional development? What functions do rainforest cities provide to the processes of development at the national and global levels? How are rural and urban sectors of the Amazon linked? In addressing these questions we present a pluralistic theory of *disarticulated urbanization*. A theory is pluralistic when it contains a variety of principles that cannot be reduced to or derived from a single master principle (Wenz 1988). Among the principles that constitute our conceptual framework are the following:

1. *The Amazon is a heterogeneous social space*. Amazonia defies classification as a homogeneous region. It cannot be accurately denominated as either a "peasant frontier" or a "big man's frontier." Such conventional social categories, which permeate the structuralist and spatial economy literatures, are too rigid and antiquated to represent accurately the complex patterns of spatial organization in contemporary Amazonia. Rather, the Amazon frontier is an idiosyncratic social space constituted simultaneously by diverse social groups and spatial structures. This primary fact leads to a second principle:

2. *The configuration of settlement systems in Amazonia is irregular and polymorphous, disarticulated from any single master principle of spatial organization*. Diverse settlement configurations abound. Overall, there is considerable asymmetry in the regional organization of settlement systems and little evidence of the orderly nested hierarchies predicted by central place theory. This diversity is compounded over time by the divergent logics of settlement location during different economic epochs. In general terms we observe

that the contemporary frontier, when regarded as a sociospatial continuum along the extensive margin of the nation's space, is socially constructed into zones that vary in their relation to the capital-power-class structure of Brazilian society. At roughly opposite ends of this continuum are "populist" and "corporatist" zones, each of which engenders a distinctive urban form in the area of its domain. But these are not mutually exclusive social spaces, and there are numerous other urban formations in between, such as the increasing urbanization of Amazonian indigenous tribes into archipelago settlements, points of political empowerment for Brazil's most threatened ethnic minorities (Escobar 1995b).

   3. Urban growth is functionally disarticulated from agricultural development in many parts of Amazonia. The irregular and polymorphous patterns of settlement system formation arise in part from the uneven spread of capitalist relations of production in the region. While capital has temporarily penetrated the Amazon, capitalism as a system of wage relations of production has never become very well established in the region's agricultural sector. Indeed, urbanization does not depend on the growth of agriculture in the Amazon. True, towns and cities provide important agricultural support services to their surrounding rural hinterlands, and in many cases the expanding agricultural frontier is the raison d'etre of nearby urban centers (Margolis 1973; Rondinelli 1986). Agricultural expansion, however, is only one of several dynamics inducing urbanization in the region. Donald Sawyer (1987:46) addressed this point when he wrote: "The growth of the rural population is localized, but urban growth in frontier regions is generalized. . . . In other words, the agricultural frontier has become an urban frontier."

   4. The dynamics of Amazonia's rapid urbanization are disarticulated from the process of regional industrialization. Once a rural nation, producing primary commodities such as rubber, sugar, and coffee beans for the world market, postwar Brazil has become a medium-income, newly industrialized country (Schneider 1991). Historic patterns of industrial growth spawned an increasingly extensive, nationally integrated, but unevenly developed urban system in Brazil. The country's historic urban transition, the conversion from a predominantly rural to an urban-based society, occurred during the 1960s (table 1.1). By 1990 fully three-quarters of all Brazilians lived in urban centers, and over a third of the national population clustered in twelve metropolitan areas with more than a million people each. Two sprawling megacities, Rio de Janeiro and São Paulo, each with populations now exceeding ten million, increasingly merge in the nation's southeastern socioeconomic core to form a Brazilian megalopolis of some thirty million people, rivaling in scale the Boston-Washington corridor in North America (Godfrey 1991; Gottmann 1961).

In the general context of Brazil's massive urban-industrial transition, Amazonia's rapid urbanization should not be entirely surprising. However, the dynamics of Amazonia's rapid urbanization diverge in some important ways from those of the Brazilian southeastern core. For example, except for a few enclaves of industrial activity in the major metropolitan areas, industry is virtually nonexistent in Amazonia. Rather, urbanization is inextricably intertwined with the exploitation of natural resources and accompanying migrations to the inland settlement frontiers. Much of the resource extraction is informal in organization, as in small-scale gold-mining (*garimpo de ouro*), but it is closely integrated into contemporary economic cycles. Amazonia functions more as a high-cost extension of the internal consumer market than as an area of production in Brazil.

5. *Urbanization in Brazilian Amazonia is variously linked to economic forces operating at the global level, but is not subordinated to a world economic system.* The rapid growth of the urban population in Amazonia may be unsustainable in the context of the region's economic base. Put differently, the Brazilian Amazon may be "overurbanized," a condition in which the rate of urban growth exceeds the sufficient level of local economic development, industrial, and technological change to make it viable (Herbert and Thomas 1990:51). Alternatively, we posit that the economic rationality and function of frontier urbanization extends beyond the borders of the frontier itself and is variously articulated to forces operating on multiple levels of the global economy. In other words, frontier urbanization in Amazonia is disarticulated from its own regional development and unbounded by the internal limits to growth that the region's pace of development might impose. Nowhere is this more evident than in the emerging role of selected rainforest cities as satellites in a world system of "technopoles" (Castells and Hall 1994). Because of the growing popularity of globalization studies we devote chap-

ter 10 to an examination of the mechanisms of global surplus extraction in the Amazon.

6. *Urban centers in Amazonia are technological crossroads that link specific activities to global circuits of information and exchange.* Modern technologies, now evident in the far corners of the Amazon region, link once isolated areas to an extensive web of external influences and migrant flows: recent additions to the regional landscape include satellite dishes, telecommunications towers, paved roads, mammoth hydroelectric dams, high-speed railroads, and international airports. The often abrupt, even paradoxical, juxtaposition of new and old features in Amazonia indicates the imposition of a global postmodern economy on the Brazilian periphery.

7. *The contemporary Amazonian urban frontier is largely a geopolitical creation but remains politically disarticulated within the centralist state.* As Becker (1990:46) has argued, the frontier now emerges already urbanized as a strategy of regional occupation by the Brazilian central-state apparatus. While the role of the state has been central in both the construction of planned cities and fomenting the conditions for spontaneous urbanization, it has been neither uniform nor definitive regionwide. The role of the state has been heterogeneous with regard to social class interest. The state is a complex, historically evolving creature, which speaks with many voices. The state's ideological position varies internally among government institutions responsible for the regional development of Amazonia. Different state institutions, often vying with each other for bureaucratic hegemony within the state, typically give their institutional imprimatur to the local character of the frontier areas they regulate. Accepting these premises and caveats, the patterns of urbanization we observe in Amazonia today are disarticulated from any coherent or unified ideological position of a centralist state. Frontier urbanization reflects the agglomeration of multiple and often conflicting state policies toward the Amazon region.

8. *Established dichotomous categories of rural and urban become problematic when applied to Amazonia due to complex and regionally heterogeneous local migration patterns.* On the frontier permanent residence is a rarity. Urban pioneers are constantly moving between rural and urban environments. Intense local migratory flux, pervasive occupational mobility, and rapid communication and transportation

links are the benchmarks of the frontier that obscure clear functional lines of division between rural and urban sectors.

9. *Finally, environmental change, including tropical deforestation processes, are increasingly mediated by regional, urban-based interests.* Commonly thought to be related to unclear rural land ownership and misguided government incentives to rural ranching elites, deforestation in Amazonia is increasingly becoming an urban-centered problem. A growing proportion of the urban population owns rural property, and these rural assets are managed in ways that enhance urban accumulation strategies, not necessarily with long-term forest conservation in mind. Moreover, other major developments entailing large-scale regional environmental changes (hydroelectric expansion, mining, and the explosive growth in peri-urban shantytowns) all arise from regional urbanization.

It is axiomatic among scientists that theoretical inquiry involves the search for regularities. We propose a somewhat different approach, necessitated by the sheer complexity of our subject matter. Instead, this book presents the findings of our search for the "structure of irregularity" (Stewart 1989:216) in the multitudinous manifestations of the urbanization phenomenon in Brazilian Amazonia.

One of the streams of methodological practice invigorating our conceptual framework of disarticulated urbanization is drawn from applied systems ecology. Not to be confused with the simple biological metaphors of the classic Chicago school of human ecology, systems ecology emphasizes the complexity of systemic interactions at different levels, ranging from the local to the global scale. As explained further in chapter 4, we adopt an urban systems approach, which begins with the premise that structural changes in any individual settlement cannot be adequately explained without reference to the larger, regional urban network or system of urban places to which it belongs (Meyer 1986).

Like many ecologists, we as social scientists make limited claims about the representativeness of our research. No specific communities, even studied comparatively, can fully represent the urban diversity of Amazonia. Yet we maintain that our empirical case studies, viewed through the theoretical construct of disarticulated urbanization, allow us to analyze the range of regional settlement patterns. In a sense, it is the process of urbanization within distinctive social spaces of the Amazon frontier that is our subject, not the search for common characteristics among the population of urban areas per se. By adopting the regional urban system as the relevant unit of analysis of urbanization, we depart from (and thus hope to improve on) conventional case study approaches, which typically make inferences from the limited experience of individual communities in isolation from their functional family of central places.

AMAZON FRONTIER URBANIZATION
AS A RESEARCH PROJECT

Our understanding of the importance and complexities of the Amazon's rapid urbanization lags behind fast-changing regional realities. Popular images of the region still emphasize idyllic river towns and isolated native villages or, perhaps more recently, new penetration roads and cattle ranches impinging on the rainforest. Even the updated versions of traditional scenes are highly oversimplified. The contemporary differentiation of the Amazon into more complex geographic patterns invalidates facile regional generalizations. This book, in examining the process of frontier urbanization in contemporary Brazilian Amazonia, depicts a region largely ignored by commentators and still relatively unknown to outside observers.

This book focuses on the factors that determine the location and socioeconomic structure of new settlements in the Amazon and their evolving relationships to and functions in both Brazil's national urban system and the global economy. The project, which includes both a comprehensive review of secondary sources and comparative empirical case studies, seeks to shed light on the larger political and economic role of frontier urbanization in the process of Brazilian and global capital accumulation.

In this study we report on field research in dynamic areas of contemporary settlement in eastern and western Amazonia. We are broadly concerned with developments at various levels of the urban hierarchy: metropolitan centers, *município* (county) seats, local service centers, and pioneer settlements. Our study examines whether the apparent contemporary demographic and economic decentralization illustrated by the region is a viable, long-term phenomenon. It is possible that new forms of urban primacy and regional polarization are emerging on the Amazon frontier as a result of the instability of agricultural settlement and resource-extractive cycles. Progressive migrations farther into newly accessible areas shift the dynamics of growth away from established urban centers, which often stagnate or decline. With the depletion of valuable

resources, towns serve reduced local populations engaged largely in cattle ranching. As the frontier economy decomposes, social stratification of
the urban population becomes increasingly evident in the juxtaposition of
underemployed peasant farmers and landed merchants with multiple
holdings. In sum, it appears that the variable patterns of frontier urbanization ultimately produce new forms of regional economic centralization and social class polarization.

The book is organized into ten subsequent chapters. Chapter 2 situates
frontier urbanization in the larger theoretical context of urban geography
and the field's major intellectual traditions. Chapter 3 examines the historical dynamics and contemporary policies of frontier expansion in the
Amazon. Chapter 4 presents our framework of disarticulated urbanization by illustrating two different patterns of settlement formation originating from the distinctive institutional histories of our two study sites
(Rondônia and southern Pará).

The empirical core of the book departs from the most common
approach to research on Amazon towns, the ethnographic case study of a
particular community, which emphasizes idiographic features over regional
comparison. Instead, we present a systematic comparison of urban growth
in various metropolitan and frontier regions. In chapter 5 we review the
metropolitan development in the large cities of Belém, Manaus, and Porto
Velho. Chapters 6 and 7 analyze the history and development of two distant
frontier urban systems: the Rolim de Moura–Alto Alegre corridor in
Rondônia and the Xinguara-Tucumã corridor in southern Pará. Both these
settlement systems began to emerge in 1976–77 and both of us have followed their evolution virtually since their inception.

Our comprehensive study is based on data collected initially in 1977
(Xinguara), 1984 (Rolim de Moura), and then again in 1990 for both
settlements. This last period of intensive field research, carried out
jointly from late June to early August of 1990, used questionnaires on
leading economic sectors—including small businesses, timber firms,
gold-mining activities, and regional freight movements—as well as on
residential households in the Xinguara and Rolim de Moura regions. We
completed over 1,600 household questionnaires, randomly selected by
urban sector in six different towns, representing the most thorough
survey of these communities after the Brazilian census. In addition to
the primary study sites identified above, these data include two secondary urban settlements in each study area. In both Pará and
Rondônia, the various settlements chosen differ in terms of their social

histories, demographics, and resource bases. Most of the towns now serve as municipal political seats, but they clearly perform different functions in the regional urban system. They lie along integrated corridors of settlement and constitute different types of urban centers in rapidly evolving frontiers. In these case studies, then, we hope to capture both the general patterns and local idiosyncrasies of frontier urbanization in the Amazon.

In chapter 8 we consider the complex patterns of local migration and social mobility that characterize the urban population of Amazonia. Chapter 9 addresses the relationship of rainforest cities in the global periphery to the larger national and world economies by examining resource flows. Chapter 10 considers how urbanization has affected (and is affected by) environmental change, especially deforestation, in the Amazon Basin. The final chapter offers a synthesis of our major research findings in relation to enduring theories of frontier urbanization and regional development in the global periphery.

Notes

1. Brazilian governmental agencies define Amazonia in two principal ways. The Brazilian Institute of Geography and Statistics (IBGE) long defined the "North Region" as the states and former territories of Pará, Amazonas, Acre, Rondônia, Amapá, and Roraima. The IBGE now also includes Tocantins, created in 1988 from northern Goiás. The lack of comparative census data from earlier years, however, generally prevents this study from including the new state in historical analysis. In 1991 the six-state North Region officially had a population of 9,337,150 (6.2 percent of the national figure) and encompassed 3,554,933 square kilometers (42 percent of the national area). The Superintendency for Amazon Development (SUDAM) defines the "Legal Amazon" as the North Region and Mato Grosso, Tocantins (formerly northern Goiás), and western Maranhão. This area spans roughly six million square kilometers (58.8 percent of Brazil's national territory) and has a population of roughly 15 million (10 percent of the national population).

2. The principal definition of "urban" in Brazil depends, in practice, on the political status of a locality: a city necessarily serves as a municipality- (*município*) or county-level seat (*sede*) of government. Generally, a population of at least several thousand is necessary to obtain municipal status and subsequent access to state and federal funds. In effect, as the population of a place increases in Brazil, a locality exerts political pressure on state authorities to be granted municipal status. Although this political criterion complicates a precise measurement of the urban population of Amazonia, the contemporary proliferation of new municipal seats in Amazonia is in itself a striking indicator of regional urbanization, as we shall see later in the case studies.

3. A national census normally is carried out every ten years in Brazil, but in 1990 it was delayed for a year because of budget cuts.

4. This and other quotations from Portuguese-language sources are translated by the authors.

5. The municipality, roughly the equivalent of county-level government in the United States, provides the most convenient local enumeration unit for longitudinal analysis of demo-

graphic change in Brazil. Each decennial census counts the population of all Brazilian *municípios* and breaks down these municipal figures by rural and urban proportions along with other factors. Since the most reliable demographic data refers to the municipal seats, emphasis is given to the municipal urban populations in this study.

6. The rapid expansion of small urban centers in Latin America has been widely noted regionwide for some time now (Ingram and Carroll 1981; Klaassen et al. 1981; Hardoy and Satterthwaite 1986). However, the growth dynamics of small urban centers remain largely unknown (Hinderink and Titus 1988; Rondinelli 1983) but certainly call into question the traditional urbanist focus on Third World metropolitan centers (Mohan 1984).

7. Manaus increased its share of the population of the state of Amazonas from 32 percent in 1970 to 48 percent in 1990; Belém's share of Pará's population fell from 30 percent in 1970 to 25 percent in 1990. These divergent patterns reflect the more asymmetrical urban hierarchy of Amazonas, where the cities of Itacoatiara and Parintins, both with populations numbering well under 100,000, dispute claims to be the second largest urban center in the state. Manaus, with a metropolitan population of about one million, overwhelms the state's other cities in size. In Pará larger intermediate-sized metropolitan areas have emerged, such as Santarém, with a population estimated in 1985 at 227,412; and Marabá, estimated at 133,559 (IBGE various years).

8. Like Gurupá, contemporary Amazon frontier towns often press for designation as municipal seats as soon as their demographic growth supports a strong case before the state government. Municipal autonomy helps ensure a settlement's survival and benefits vested local interests.

9. Although the term *caboclo* often takes on negative connotations—much like "hillbilly" or "rural bumpkin" in English—it is nonetheless more widely employed than neutral terms like *ribeirinho* (river dweller) to describe long-standing communities along the Amazon watercourses. For insight into the *caboclo* cultural complex, see Parker (1985).

10. Historian Frederick Jackson Turner (1920) regarded the American western frontier as the "meeting point between savagery and civilization," a moving line of pioneer occupation, tending in time toward more permanent settlement. Revisionist histories, of course, now regard urbanization as essential even in the initial phases of North American frontiers, as pointed out by such authors as Reps (1980) and Barth (1988). The geographer Preston James (1938) argued that the settlement of the Brazilian South did not follow the Turnerian sequence of steadily increasing population densities; rather, it ultimately promoted rural depopulation, a consolidation of landholdings, and environmental degradation, leading to a "hollow frontier."

11. Walter Christaller's classic work *Central Places in Southern Germany* (1933, translation 1966) proposes a deductive model of spatial organization of urban settlements in a stepped hierarchy in which higher-order (larger) settlements offer more specialized services and functions than smaller-order settlements. Applying minimum distance (cost) location criteria, the most efficient arrangement of spatial organization is a regular hexagonal lattice (see chapter 2 for further discussion of central place theory).

2
Theoretical Perspectives
on Frontier Urbanization:
Toward an Urban Systems Approach

❦.❦

Of the few active frontiers left in the world, the vast Amazon region is pre-eminent, both in scale and complexity. Brazilian Amazonia provides an expansive and opportune setting in which to observe firsthand the genesis and evolution of urban systems and to reconsider the accumulated treasure of urban geographical wisdom. The rapid transformation of the Amazon into an urban frontier also invites human geographers, now accustomed to a potluck of intellectual specialties, back to a simpler fare of basic staples: the origins and growth of individual towns, the dynamics of urban network formation, and the growing urbanization of the human population. These interests are treated in diverse literatures that are loosely grouped into three broad perspectives: spatial economics, cultural geography, and polit-ical economy. In this chapter we review a few of the most important con-tributions of each perspective to the discourse on urbanization.

SPATIAL ECONOMIC PERSPECTIVES

Central Place Theory

Few urban geographers believe that human settlements are randomly sit-uated in the landscape. Although the origins of towns at any given time

may be serendipitous, the eventual formation of spatial systems of settlements is widely thought to follow an economic logic. The notion that
human settlements are configured in regular hierarchical relations with
one another is usually associated with central place theory (Christaller
1933; Losch 1954). Walter Christaller, in his classic study *Central Places in
Southern Germany* (1966, orig. 1933), proposed a deductive theory to
explain the sizes and distribution of towns in terms of the types of services they provide to their surrounding hinterlands. Central places are
locations where suppliers concentrate, and the size of their respective
market hinterlands is determined by the minimum economic size of supplier activities and the maximum range of goods in demand. Christaller
observed that for every large or higher-order central place, each providing a wide range of basic and specialized functions (i.e., goods and services) to large hinterlands, there are numerous smaller settlements
located close together that provide only basic (lower-order) functions to
geographically more localized populations.

Christaller deduced three types of nested hexagonal hierarchies of settlements, the most important of which was organized by the "marketing
principle."[1] Acknowledging the fact that settlement systems did not all
necessarily emerge principally as rural market towns on a frictionless
plane, Christaller also considered hierarchies that evolved on the basis of
efficiencies afforded by access (the "transportation" principle) and on the
basis of administrative efficiency (the "administrative" principle).[2]

Christaller was unspecific about the temporal sequencing of settlement system formation; his subject matter, agricultural service centers in
southern Germany, was already an established and densely inhabited artifact of the feudal landscape he observed. From his starting assumption of
the existence of first-order centers, we can surmise that lower-order centers emerge later as a frontier expands to fill the demand for lower-order
functions not efficiently provided by the initial primary center.

The dynamics of urban system formation following central place principles probably resemble the sequence outlined in figure 2.1. In time
period 1, two settlements appear within a common space, each providing a relatively homogeneous range of services to geographically separated populations. Presumably they grow over time as consumers fully
occupy the surrounding hinterland. In time period 2, lower-order settlements emerge to fill in the market area for lower-order goods and functions that are no longer efficiently supplied by the original, now higher-
order centers; a two-tiered nested hierarchy emerges. In time period 3,

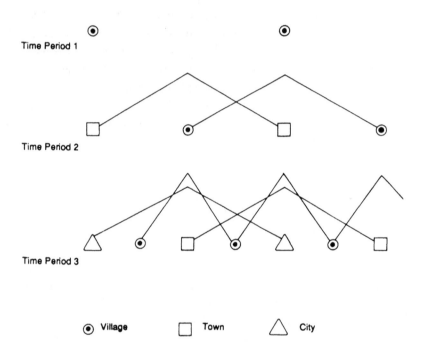

Figure 2.1. Evolution of an urban network based on central place theory
Source: Adapted from Morrill 1974:77

the process repeats itself until the demand for all functions is efficiently supplied. Then, the settlement system presumably stabilizes.

   Christaller's original formulation, however, remains essentially static, and the significant modifications proposed by August Losch (1954, orig. 1940) do not materially affect the limited utility of central place theory in explaining the dynamics of settlement system formation in frontier areas. Empirical observations provide little consistent verification of regular central place tendencies in human settlement patterns, except in special cases. Distortions arise from numerous sources (Haggett 1966). Subregional networks often overlap, thereby truncating neat pyramidal forms of settlement hierarchy. Some tiers develop more rapidly than others. That the original central place models are too restrictive in their assumptions to be easily applicable to developing countries was recognized long ago (Chisholm 1962; Berry 1967).[3] We can only extrapolate the dynamic elements and tendencies in the evolution of frontier settlements from the underlying assumptions adopted by Christaller and Losch in specifying their models. Nevertheless, these fundamental ideas were influential in shaping later conceptualizations of the nature of regions and the dynamics of urban system change (Perroux 1950, 1955; Boudevile 1961, 1966; Dean et al. 1970; Parr 1973; Reiner 1972; Rozman 1978; Rondinelli 1987). Therefore, we cannot dismiss central place theory as irrelevant to our analysis of settlement systems in contemporary Amazonia.

Mercantile Models of Settlement

Classical central place theory emphasized the functional role of towns as centers for the supply of goods and services to a surrounding rural population. The emergence and growth of central places was seen as a response to the spatial expansion and intensification of rural settlement. Under what conditions do settlement systems expand?

James E. Vance, Jr. (1970), in his benchmark study of wholesale trade, proposed a mercantile model of settlement. In contrast to Christaller's endogenous approach, in which growth in a settlement system derives from the increase in internal demand for trade goods, Vance's model is exogenous. Central place networks evolve within the lineaments of the externally based mercantile system. Long-distance trading ties result in the formation of new settlements, initially trading entrepôts, "points of attachment." Over time the successful new settlements become established wholesale centers with their own hinterlands, following central place hierarchies. The dynamic variables determining the growth of urban systems on the frontier include changing forms of commercial information communication, technological improvement in transportation, and supply innovations (i.e., development of new products). Vance (1970, 1990) was concerned mainly with explaining the settlement pattern of North America in terms of colonial mercantile ties to Europe. Viewed more broadly, his emphasis on the operation of commercial forces external to the local urban system in determining the structure and location of the system provides the dynamic missing link in central place theory (fig. 2.2).

A related conceptualization posits that lower-order centers are linked to only one single higher-order center and there are no nesting patterns. The resulting dendritic system has been used to characterize the export-

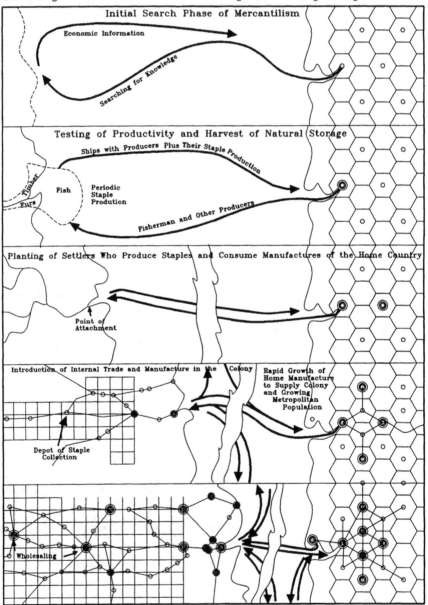

Figure 2.2. The mercantile model of settlement
Source:Vance 1970:151

oriented nature of spatial organization commonly found in developing countries (Johnson 1970). In Latin America, particularly, the centripetal focus of trade and transportation on key gateway ports has heightened asymmetrical urban hierarchies (Roberts 1978). This export-oriented model of the development of urban systems presumes that lower-order centers are fundamentally similar in function and economic structure, serving specifically as points for mobilizing exports and concentrating merchant capital. In the case of Amazonia, as we shall see, the mercantile model of settlement applies best to the rubber boom and other historical cycles of natural resource extraction.

Transportation Improvements and Selective Frontier Urbanization

Neither classical central place theory nor mercantile models of settlement explicitly address the "selective growth" observed among urban settlements following their initial nucleation in actual urban nodal systems on a frontier. Responding to these shortcomings, Muller (1977) proposed a model of selective urban growth drawing from core-periphery conceptualizations of regions and from the export-base theory of regional economic growth based on the nineteenth-century North American experience. Muller's model of historical linear stages distinguishes three periods of frontier development. In the "pioneer periphery" period the frontier is characterized by relative isolation from external markets, low population densities, and weak development of commercial agriculture. Transportation, confined to natural routes, limits communications and commerce, particularly for export. The pioneer economy generates low levels of regional income and, with the high cost of imported goods, low levels of consumption (Muller 1977:24). Most urban centers on the pioneer periphery offer little more than basic goods and commercial services for localized areas. The most dynamic centers arise at the junctions of regional and interregional transportation arteries. Such centers become regional entrepôts, with smaller urban centers appearing as rural settlement spreads into the interior of the region, providing local services and periodic commercial connections with the regional entrepôt.

In the second period, the "specialized periphery," major improvements of interregional connections reduce the frontier's commercial isolation, enhance the competitive position of traditional export products, and accelerate the influx of new settlers. Agricultural prices rise, following improved accessibility to external markets, allowing expansion and specialization of cash crops and associated processing industries. In the specialized periphery, the regional entrepôts concentrate much of the increase in interregional commerce and the related development of manufacturing. For other towns, growth is differentially determined by their location relative to the expanding transportation network and the corresponding access to expanding or intensifying hinterlands. "Thus, . . . , selective urban growth during the second period of staple specialization depended upon location within areas of most rapid rural expansion and the maintenance or enhancement of nodality in the spatial parcelling of service and commercial functions" (Muller 1977:26).

In the third period, the "transitional periphery" becomes fully integrated into the national transportation system (in the North American case by railroads). The ubiquitous access now extended to the frontier enhances the competitive positions of traditional export commodity producers but also encourages the diversification of agricultural activities and secondary manufacturing, usually within or around the regional entrepôt. The reduction of transport time and costs further concentrates many functions at major nodal points at the expense of more localized district trade centers. Meanwhile, improved access throughout the periphery leads to the reduced importance of regional entrepôts as centers of mercantile activity. Such centers turn their attentions to developing varied industries that can capture the new advantages of improved access to national markets.

Muller's model of selective growth of peripheral settlements focuses on the interplay of central place and mercantile forces, activated by transportation improvements over a series of historical stages. In several respects Muller's interpretation of the North American experience diverges from that of Brazil's Amazon. First, the pioneer settlement of the contemporary Amazon frontier is not confined to natural routeways but is afforded more ample latitude by virtue of the region's highway network. As a result, pioneer agriculture in the Amazon was commercial from the start. Second, the development opportunities afforded by the expansion of a highway network (Amazon frontier) are quite different from those made possible by the expansion of a network of railroads (North American frontier). Spatial patterns of settlement, it may be surmised, would be different as well. Third, this difference reduces to a large extent the possibilities for industrial development, which follow more easily from the railroad option than from a highway network. Fourth,

Muller's model, based as it is on an export theory of regional development, largely disregards the important role of the frontier as an expanding consumer market for extraregional manufacturing (i.e., the consumption function of the frontier). Since these manufactured goods are
often more cheaply produced in the industrial center than on the periphery, transportation improvements to the frontier may undermine peripheral industrial development and distort settlement morphology accord-
ingly. Finally, Muller's analysis of selective growth, like central place theory, is entirely endogenous in scope. The possibility that selectivity may
be induced by forces operating at the global level is not clearly specified.
Nevertheless, Muller's conclusion provides a concise, but problematic,
starting point for the analysis of frontier urbanization in Amazonia:

> During the course of regional settlement the bases for urban
> growth shifted from commercial servicing of local hinterlands to
> the development of secondary manufactures. . . . New transport
> modes tended to reinforce the existing circulation network and
> nodal points. The spread of transportation facilities and the integra-
> tion of regional movement with the evolving national network
> resulted in significant changes in the nodality of towns and thus
> affected their growth.   (Muller 1976:199)

Summary

The organizing master principle of the spatial economics tradition, the
"marketing principle" of spatial organization, has enduring value for our
understanding of the contemporary process of frontier settlement
expansion. The powerful draw of metropolitan centers on the frontier
contrasts with the local tugs and pulls of frontier boomtowns, each one
receding in importance as the settlement frontier expands. But, as our
case studies will show, the physical manifestations of the marketing principle in the morphology of Amazonian settlement systems are strewn
with contradictions and irregularities. Market forces may provide an
underlying logic for the location of settlements, but distorting forces
everywhere intrude, as we shall see in the case of southern Pará (chapter 7). It is in the disarticulated nature of frontier urbanization, rather
than in some idealized field of centrifugal market forces, that we must
search for the structure of irregularity in the frontier urbanization experience of Brazilian Amazonia.

CULTURAL GEOGRAPHIC PERSPECTIVES

Frontier Studies

The contributions of cultural geographers to the study of frontiers have been far-reaching and diverse. In this short section, we recall the important influences of Frederick Jackson Turner and Carl Sauer, two North Americans whose ideas continue to frame contemporary analyses of frontier studies.

In 1893 Turner presented a paper before the American Historical Association entitled "The Significance of the Frontier in American History." Though the paper apparently sparked little immediate comment, subsequently it was "recognized as the most influential single piece of historical writing ever done in the United States" (Webb 1964:6). Butland (1966:93) noted the enduring importance of the Turnerian frontier concept in studies of settlement expansion:

> The frontier concept as applied to North America has somewhat dominated studies of the settlement of new lands. The idea of a more or less continuous boundary dividing settled territories from those virgin lands to the west, of an ever-moving line of pioneer occupation, has stamped a classic interpretation not only on the historical process of occupation, but on the characteristics associated with such a movement.

Turner regarded the frontier as the "meeting point between savagery and civilization," basically proceeding through a linear sequence tending toward more permanent occupancy. On the American frontier, hunters, trappers, and fur traders were followed by ranchers and miners; then came pioneer settlers on small farms, followed by more commercial forms of agriculture. Turner omitted numerous important elements in the American frontier expansion, such as soldiers, land speculators, artisans, slaves, mill owners, storekeepers, prostitutes, eastern merchants, religious groups, and hoboes (Gulley 1959). In addition, Turner virtually ignored land conflicts in the settlement of the "Great West": "The existence of an area of *free land* [our emphasis], its continuous recession, and the advance of American settlement westward, explain American development" (Turner 1920:1).

The frontier supposedly necessitated a return to the "simplicity of primitive society," which promoted a series of democratic character traits, such as self-reliance, individualism, self-government, idealism, and inventiveness. In short, "American social development has been continually beginning over again on the frontier" (Turner 1920:2).

The identification of cultural renewal with the frontier landscape has been an abiding American tradition in both hemispheres. That the landscape is a code to the nature of the people that dwell in it and shape it—or as Pierce Lewis (1979:12) puts it, "our human landscape is our unwitting autobiography"—encapsulates the enduring interest and contribution of the North American tradition of cultural geography to frontier studies.

Brazilian scholars have developed a similar tradition of cultural frontier studies that bears some comparison to the North American experience. Cassiano Ricardo's classic *Marcha para Oeste* (1970:17), first published in 1942, describes the movement of the *bandeirantes* or the "flagbearers" as fulfilling a nation-building role similar to that of the pioneers in the United States: "It established the rhythm of Brazilian civilization; it traced the physical silhouette of Brazil; it originated the most appropriate mentality for the realization of our destiny, in an opposite sense to that seen by the littoral."

Viana Moog's (1964) *Bandeirantes and Pioneers,* on the other hand, noted fundamental cultural differences in the people who respectively settled Brazil and the United States. The Brazilian bandeirante, according to Moog, was interested primarily in immediate economic aggrandizement, whereas the American pioneer was motivated by long-term concerns for family prosperity. Higgins (1963:171) strikes a similar note, arguing that in Brazil "log cabin–to–riches stories are exceptions rather than the rule. The frontier in Brazil . . . has been a 'big man's frontier' from the very beginning." For Otávio Velho (1976:146) the frontier image is an ideological construct serving to legitimize political forms of social domination: "While for Turner the frontier is the locus for the development of American democracy, for Ricardo the frontier experience led to the development of Brazilian authoritarianism. In both cases, the frontier is utilized as a prime resource for the creation of myths of origin."

Turner's frontier was essentially rural and did not account well for subsequent urban and industrial development. Turner's failure to consider the forces of industrialism, urbanization, and class conflict as well as the diverse cultural origins of the American frontier experience served a nationalist purpose. His view of the supposedly ennobling influence of the frontier upon both the individual and the nation struck a responsive chord as the United States sought to find historical roots and a cultural identity distinct from Europe. In addition, the romantic vision of the frontier as a scene of social mobility tended to diffuse internal social discontent. Turner described the frontier as, in effect, a safety valve for the release of social pressures in American society:

> Whenever social conditions tended to crystallize in the East, whenever capital tended to press upon labor or political restraints impede the freedom of the mass, there was a gate of escape to the free conditions of the frontier. These lands promoted individualism, economic equality, freedom to rise, democracy. (Turner, quoted in Thompson 1973:7)

The Turner thesis inspired enthusiastic successors to apply the frontier concept on a global scale.[4] Interestingly, the idea of the frontier as the arena for social mobility also has been widely propagated in Amazonian countries engaged in frontier settlement activities. Careful regional applications of the Turnerian frontier concept, however, generally have proved disappointing.[5] The mystification of the frontier as a genuine human experience cast in a pristine rural setting has obscured the fact that frontiers have important urban contexts. Reflecting on Turner, Alistair Hennessy in his book *The Frontier in Latin American History* (1978:27) writes, "An understanding of the dynamics of urban growth in historical perspective has assumed new urgency. Latin America's salvation might seem to require a rural myth: there might well be a role for a Latin American Turner, but he would be hard put to fashion a progressive myth out of past agrarian history."

Despite its teleological importance in shaping an "American ethic" and cultural identity, North American geographers following Turner have tended to avoid broad generalizations about the frontier. Carl Sauer was among the first North American cultural geographers to challenge Turner's thesis at its heart. Sauer disclaimed the Turnerian assumption of universal stages in settlement and argued that no unilinear sequence could apply to the human occupation of all areas, although he did admit to the existence of specific "cultural successions" that were reflected in distinct regional landscapes. Contrary to Turner's stress upon the common changes in character and country wrought by the frontier experience, Sauer emphasized the "eternal pluralism of history," arguing that the frontier must be defined as an episode of "cultural succession," the nature of which "was determined by the physical character of the country, the civilization that brought it in, and by the moment of history that was involved" (in Leighly 1963:48–49).[6]

The urban dimension of the frontier experience in North America was never explicitly treated by either Turner or Sauer to a degree that could structure the geographic analysis of urbanization in the Amazon frontier. Yet, both of these voices resonate through an important tradition of frontier studies that originated in North American cultural geography. Perhaps the time has come to commemorate the pastoral ideas of Turner's frontier vision, recognize the abiding idealism it embraces, given its historical context, and move on to a new appreciation of the meaning of "frontier" in its contemporary, largely tropical, and largely urban, contexts.

Spatial Diffusion Approaches

A second approach to the study of settlement systems owes its parentage to the marriage of sociology and cultural geography and focuses on the diffusion of (and resistance to) innovations through society. Efforts to explain the process of diffusion derive in large part from the important works of Hagerstrand (1952), who proposed a four-stage inductive model for the passage of "innovation waves" and Yuill (1965), who studied barriers to the spatial diffusion of information. As in central place theory, the geographical interest in diffusion has aspired to find or explain regularity in the patterns and processes of innovation diffusion and its impacts on development.

In general terms, the application of spatial diffusion concepts to urbanization in the hinterland focuses on the growth and spatial movement of firms and entrepreneurs as agents of innovation. In established urban systems, diffusion is expected to proceed from larger to smaller centers, a regularity termed the "hierarchy effect." Within the hinterland of a single urban center, diffusion is expected to proceed in a wavelike fashion outward from the urban center, first affecting nearby rather than more distant locations, a regularity known as "neighborhood" or "contagion effect" (Brown 1981:21). The emergence of new urban areas in the national hinterland is also partially the consequence of the diffusion of growth impulses transmitted from the primate city and leading to industrial diffusion, referred to as long-distance mercantile ties (Vance 1970, 1982), industrial decentralization, and urban deconcentration or "counter-urbanization" (Vining and Kontuly 1978; Hall and Hay 1980) or "polarization reversal" (Townroe and Keen 1983; Richardson 1980; Storper 1984; Vining 1984). Certain types of (low-wage) firms relocate from metropolitan centers to smaller towns and cities to obtain production efficiencies, such as lower wage rates, reduced transportation costs, etc. (Hoover 1948; Richardson 1972), but their immediate impacts on economic growth and "entrepreneurial innovation" are considered to be limited (Berry 1972; Pedersen 1970). A variant of this approach characterizes the diffusion process as occurring in waves expanding from the highest growth center in the national space economy and dissipating as distance from the center increases (Morrill 1965).

Much of the diffusion research has adopted central place frameworks to examine the spatial spread of innovations. Hagerstrand's (1952) analysis of the diffusion process explicitly adopted a central place context. This process was characterized by three phases: a "primary stage" during which the initial diffusion centers are established; a "diffusion stage" during which neighborhood-type diffusion occurs in areas near initial dif-

Figure 2.3. Schematic portrayal of diffusion viewed at different spatial scales *Source: Brown 1981:45*

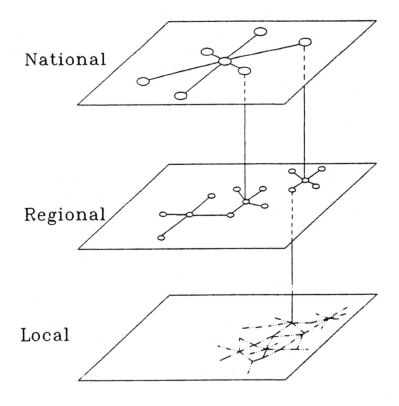

fusion centers and secondary diffusion centers are established in lower-order urban places; and a "condensing" or "saturation" stage during which filling-in occurs and diffusion eventually ceases (Brown 1981:21).

The process of diffusion postulated by Hagerstrand occurs both within and between tiers of the settlement hierarchy. The national, regional, and local levels of spatial aggregation each contain a network of central places within which neighborhood effects of spread operate. Furthermore, a node on one level may establish a linkage with nodes on another level in the hierarchy, thus providing a conduit for the filtering of innovations and growth impulses between levels (Brown 1981:44) (fig. 2.3).

Spatial diffusion frameworks address the dynamic aspects of settlement system formation, thereby filling an important gap in the classical central place models. However, they encounter various conceptual problems when applied to frontier regions in which a stable locational configuration of economic activity is lacking, where market structure is poorly articulated, and where public policy and other state actions intrude extensively upon the normal market mechanisms determining the spatial and sectoral allocation of investment, the direction of population movement, or the accumulation of capital in the hinterland. All of these qualifications apply, in varying degrees, to the Brazilian Amazon. Nevertheless, these concepts of central place and spatial diffusion have clearly influenced regional development programs in the Brazilian hinterland, which have emphasized locational and sectoral growth-center strategies (IBGE 1987). The Transamazon settlement scheme, for instance, was based on a conceptual framework called "rural urbanism" that represented the merger of central place and diffusion ideas (see chapter 3).

POLITICAL ECONOMY PERSPECTIVES: CAPITALISM, URBANIZATION, AND THE STATE

Interpretations of urbanization based on critical political economy have been informed by various structuralist influences stemming from historical materialism and world systems theory. The historical materialist analysis of urbanization, represented in an extensive and diverse literature, generally places urban growth in the political economic context of capitalist development (Castells 1977; Gordon 1978; Harvey 1976, 1978, 1985; Lefebvre 1970; Roberts 1978; Smith 1980). Contradictions inherent in the process of capitalist development in the Third World are reflected in the dualistic physical organization, social composition, and economic structure of Third World cities. Urban areas grow in tandem with the expansion of capitalist relations of production in the country-side, the consolidation of small peasant holdings into agro-industrial estates, and the transformation of the peasantry into an urban-based pro-letariat (the "agrarian transition"). Cities are the artifacts of the appro-priation of surplus labor value through capitalist relations of production; they function to promote the accumulation of capital in a dual class soci-ety (de Janvry 1981; Goodman et al., 1984; Sorj 1980).

Other structuralist writers have attempted to situate the process of urbanization in the broader context of the "world political economy" (Amin 1974; Armstrong and McGee 1985; Chase-Dunn 1983; Frank 1969, 1978; Portes and Walton 1981; Wallerstein 1974a, 1974b, 1991; Walton 1976a, 1976b). The world system operates in three interacting spheres of capital-ist development: production, circulation, and consumption. Third World cities are centers of accumulation in which surplus capital is mobilized and siphoned from the periphery to the center in the global economy. Cities are also vehicles through which Western consumer values are diffused in Third World societies. The expansion of national urban systems serves to extend the control of transnational capital to the global economic hinterland and expand domestic markets for Western consumer products and materialist cultural values (Armstrong and McGee 1985). The ascendant argument of the 1980s that the expansion of the global economy was fostering the for-mation of a homogeneous world culture was abruptly challenged in two benchmark essays on world cities (Friedmann and Wolff 1982; Friedmann 1986) that launched a new direction in the discourse on world systems (Knox 1994; Knox and Taylor 1995).

The discourse initiated by Friedmann's hypotheses about world city formation has led to considerable debate about global market fragmenta-tion, global metropolitan primacy, and global culture formation, a dis-course we highlight as well in this section. In much of this recent world cities discourse the global periphery has virtually vanished. The First World is depicted as pulling away from the Third World, even as the for-mer provides a cultural role model for the latter. By implication, the growth of frontier cities on the "periphery of the periphery" is disarticu-lated from the logic of accumulation at the global level; such growth can-not be explained by linkages to world cities or by global processes of sur-plus value extraction, except in special cases.

Although the historical materialist and world systems perspectives dif-fer in the relative emphasis they give to distinctive aspects of the structure of the accumulation process by which urbanization is explained, neither has explicitly and systematically treated contemporary frontier urbanization as a particular aspect of capitalist expansion. Both general perspectives are grounded in the concept of surplus value appropriation (excepting some recent world cities literature), an issue we explore empirically in chapter 9. Both frameworks have been creatively adapted to Amazonia by a diverse group of Brazilian scholars. Because these recent Brazilian contributions represent a distinctive analysis of frontier expansion, we selectively review this literature in a separate section at the end of this chapter.

To an important extent our understanding of the process of frontier urbanization has been shaped by broader conceptualizations of the process of frontier expansion in Amazonia. Three distinct perspectives constitute the contribution of critical political economy to the analysis of frontier expansion: intersectoral articulation, capitalist penetration, and world systems (i.e., extractive export and world cities) perspectives. We do not propose a thorough review here but rather a selective synthesis of those works that have contributed most directly to the discourse on frontier urbanization.

Intersectoral Articulation Models

Among the enduring themes spanning nearly a century of Marxist debate on the "agrarian question" is the persistence of the peasantry in various social disguises under capitalist development.[7] Oliveira (1972) and Sorj (1980) have emphasized that the reproduction of noncapitalist forms of rural production is functional, not inimical, to the accumulation of urban industrial capital in Brazil (for a review, see Goodman et al. 1984:190). The peasant and urban informal sectors are articulated to the urban industrial sector in that "primitive" production directly subsidizes urban capital accumulation by reducing the long-term reproduction cost of labor employed in the urban capitalist sector. Food crops can be produced with the use of unpaid family labor for less than the subsistence wage rate, i.e., the cost of reproducing the peasant labor force. Hence, the "use-values produced by noncapitalist forms of production subsidize urban capital accumulation by depressing rural real wages, and the real prices of foodstuffs, the main urban wage goods" (Goodman et al. 1984:190; de Janvry 1981:85–93).[8]

Intersectoral (i.e., urban-rural) articulation assumes various forms. Moreira (1985), for instance, argues that the "moving frontier" is created by capital to control the peasantry by continually keeping it on the move ("reterritorialization" of the peasantry) and thereby preventing it from making alliances with the urban industrial working class. Frontier cities, by extension, become peasant control points, or centers in which peasant land dispossession is effectuated either through force of law or violence.

Roberts (1991) demonstrates how the fragmentation of the urban labor force by a state-owned mining enterprise in an Amazon boomtown (Parauapebas, Pará) reinforces this functional dualism between informal and formal modes of production and strengthens the process of capital accumulation on the frontier. The expansion of the urban system on the frontier, then, replicates the pattern of functional dualism found throughout Brazil's dependent industrial economy.

The morphology and economic composition of urban settlements on the frontier develop so as to support this functional articulation of the informal and peasant sector to the dominant capitalist mode of production. According to this intersectoral articulation approach, we would expect to find frontier settlements populated in the main by underemployed peasants who work in the agricultural sector, partly to produce their own subsistence.

"Capitalist Penetration" Perspectives

Articulationist approaches, which centered on the persistence and exploitation of the peasantry, contrast with another branch of neo-Marxian scholarship focusing on the process of capitalist penetration of the countryside under the aegis of state modernization policies (Velho 1972; Singer 1979, among others). Frontier urbanization is explained by the changes in the agrarian sector induced by the penetration of capitalism. Armstrong and McGee (1985:219) put it succinctly as follows: "The penetration of capitalism into the countryside through the expansion of cash cropping, new agricultural production technologies, and the growing proletarianization of the peasantry dislocated rural populations who migrated to the cities or became full- or part-time wage labor for the export-oriented production of agribusiness and local producers."

The intersectoral articulation and capitalist penetration approaches differ in three points. First, the unit of analysis shifts from the peasantry as a discrete social category to modern agro-industrial and traditional *latifundio* factions of capital and the state. Second, while the articulation perspective emphasizes the logic by which precapitalist producers are preserved under capitalism, the latter perspective stresses the logic by which we countenance the dissolution of the peasantry under capitalism (Bartra 1974). Third, the shift from articulation to capitalist penetration also entails a shift of theoretical emphasis from the appropriation of labor value congealed in the product of the peasantry (i.e., foodstuffs) to one focused on the appropriation of the means of production (i.e., land) possessed by the peasantry. In a phrase, *"o colono prepara a cama para outro se deitar"* ("the colonist makes the bed for another to lay in"—Menezes and Gonçalves [1986:42]). The underlying logic of the capitalist penetration perspective implicitly builds on Marx's prediction of declining profit rates and the "underconsumption crisis" and on Lenin's analysis of the imperialistic nature of advanced capitalism. There are several variants of this perspective that require some amplification, especially as they might apply to frontier urbanization.

Explanations of the process of capitalist penetration tend to adopt historical frameworks comprised of distinct stages or "frontier cycles." Sawyer (1984) characterized this process with more explicit reference to its consequences for frontier urbanization. The initial phase of settlement of the frontier can be characterized in terms of peasant fronts, generally followed by capitalist forms of occupation driven by speculative interests. The resulting conflict over land (and other resources) results in the expulsion of the peasantry and the closing of the frontier. In turn, "the closing off of access to land also leads to accumulation of new migrants in small and medium-sized urban centers" (Sawyer 1984:195).

Foweraker (1981) situates the salient moments of agrarian transition in a generally linear framework comprised of "noncapitalist, precapitalist, capitalist" stages. However, Foweraker (1981) and Graziano da Silva (1982) acknowledge that the spread of capitalism over the frontier is uneven and selective; different frontier areas undergo full or partial integration into the dominant capitalist mode of production at different moments relative to their respective social histories. Moreover, different frontiers are likely to differ in the degree of their integration. But while diverse specific geographic manifestations may emerge, the historical process of integration and the "cycle of accumulation" are essentially the same on every frontier. For Foweraker, however, the displacement of peasants from their smallholdings on the frontier and the emergence of a "free" labor market do not necessarily lead to frontier urbanization, or rural proletarianization, but rather to the accumulation of temporary rural workers on large cattle ranch estates and the functional retention of the peasantry as proposed by articulationists (Foweraker 1981:27–57).[9] Foweraker contends that the surpluses produced by small farmers on the pioneer frontier and appropriated by merchant capitals are still of decisive importance to the urban food supply (Goodman et al. 1984:208).

Although efforts to explain the process of frontier expansion in terms of stages of capitalist penetration are varied, there is considerable unanimity among Marxist theoreticians concerning the consequences of this process. D'Incao (1975) argues that the expansion of capitalism on the Brazilian frontier entails certain characteristic transformations: the concentration of land ownership, land speculation, and increased capital intensity of agricultural production with the overall effect of generalizing wage relations in agriculture (Goodman et al. 1984:192). A sizable portion of the rural population, once engaged as resident workers in a variety of forms (tenants, sharecroppers, permanent workers), becomes a more homogeneous class of casual wage-laborers (*bóias frias*) in the process.

A second major variant of the capitalist penetration perspective is presented by Pompermayer (1979, 1982) who has suggested that frontier settlement increasingly was determined by the dynamic of agro-industrial accumulation, and not by merchant capitals, imparting new characteristics to this process (Goodman et al. 1984:208). With assistance from the state, in the form of fiscal incentives and public infrastructure investment, a mixture of industrial capitals bypassed the pioneer stage of expansion presumed in the earlier variants, and immediately took possession of the frontier. Capitalist expansion under this formulation was immediate and direct as opposed to sequential. Colonists, squatters, *bóia frias*, (i.e., migrant farmworkers), and their direct exploiters (merchant capitals) became an "unnecessary complication rather than an asset for such [industrial] capitals" (Goodman et al. 1984:208). We revisit this position in our discussion of corporatist frontiers (chapter 4).

The urban implications of this distinct process of frontier expansion are stunning. State-planned rainforest cities are almost instantly erected to comfortably house the permanent workers of some big development project, sometimes cofinanced with transnational capital. Before long, a motley host of makeshift shantytowns has appeared beyond the security gate to precariously shelter the hopeful contract workers who flock to the area in search of work (Roberts 1991). In Brazil, Jarí, Tucuruí, and Carajás are examples of this "corporatist frontier" phenomenon (see chapter 4). But in virtually every rainforest city of the Amazon there are also protected bedroom enclaves of corporatist capital (e.g., Vila

Eletronorte outside of Porto Velho, Rondônia). Stated differently, capital penetration of the frontier is not always the result of a gradual transition from peasant to proletarian relations of production. Urbanization often begins as the artifact of capital.

A third major variant distantly draws upon Lenin's "farmer road" to capitalism. In the process of agricultural modernization the peasantry differentiates. Some family farms evolve into small capitalist enterprises alongside agro-industrial estates (Muller 1982). The coexistence of agro-industrial capitals with this "residual" form of rural production observed in southern Brazil during the 1960s and 1970s is explained by the specialization of the latter in labor-intensive production that is required, or at least tolerated, by the former. Nevertheless, the small family farm enterprise must participate in a process of continual technical upgrading (i.e., "capitalization") of production to remain competitive. Farm families unable to accommodate the continuous redefinition of their relations to agro-industry and to incorporate the continuous advances of production technology become progressively marginalized. "In the South, it is this layer which has provided the *colonos* to settle the new agricultural frontiers in Mato Grosso and Rondônia in the 1970s and early 1980s" (Goodman et al. 1984:202).

Despite their differences, alternate capitalist penetration perspectives provide a common basis for differentiating the frontier into distinct social spaces (e.g., agrarian frontiers, agro-industrial frontiers, etc.) determined by the specific historical stage of their progressive integration into the dominant capitalist mode of production. As will become evident in chapters 5, 6, and 7, however, a frontier typology based on such social categories is empirically problematic. In reality the frontier, as a spatial construct, is simultaneously constituted by different social classes with varied functional relationships to the different factions of the dominant capitalist mode of production in Brazil.

World Systems: Extractive Export Economies and World Cities

A third political economy perspective begins from a world systems framework but emphasizes two distinct aspects of the global economy: the extractive export function of the local/regional level of the periphery and the overall organization and orchestration of the global economy by transnational capital headquartered in a few northern metropolitan centers, called "world cities" (fig. 2.4).

Researchers studying extractive export economies emphasize the importance of the global exchange economy in regimenting the process of frontier expansion around the extraction of valuable natural resources. Drawing upon diverse traditions, ranging from Canadian staple theory (Innis 1956; Watkins 1967) and "comparative advantage" concepts (North 1955, 1956; Perloff 1963) to paradigms of "dependency," "surplus extraction," and "unequal exchange" (Frank 1969, 1978; Wallerstein 1974a, 1974b, 1991; Emmanuel 1972; Bunker 1985), world systems writers tend to adopt a dual spatial (i.e., center-periphery) framework and a vision of the periphery as a resource frontier. Economic development of the frontier is cyclical, characterized by boom and bust phases associated with the extraction of valuable local resources. Unlike capitalist penetrationist perspectives, which vary in the degree of importance they give to global capital, world systems interpretations typically stress external drivers to capitalist expansion, building on Rosa Luxemburg and Lenin's treatise on imperialism.

Several authors have applied various aspects of world systems theory to the issue of the modernization of Brazilian agriculture (Sampaio 1980; Moreira 1985; Brum 1988). The basic lineaments of the world systems interpretation of frontier expansion in Amazonia extend from this literature and link such expansion to transnational capital investment and international commodity markets. Patterns of development are cyclical. During boom phases population is drawn to extractive activities on the frontier, especially to boomtowns at the lower tiers of the urban hierarchy. Because labor is always scarce and its opportunity cost high during this growth phase, workers are (often forcibly) prevented from engaging in subsistence production. The concentration of labor in the extractive sector and the consequent breakdown of the subsistence sector make extractive frontiers dependent upon interregional food imports (Bunker 1985; Weinstein 1983).

Because the surplus from extraction is appropriated at the level of exchange through various coercive labor arrangements (e.g., aviamento, advance purchase financing, sub-contracting) and is remitted upward through the urban hierarchy and finally abroad to global centers that control the channels of international trade, an investible surplus is never retained by the frontier for its own development (Weinstein 1983; Frank

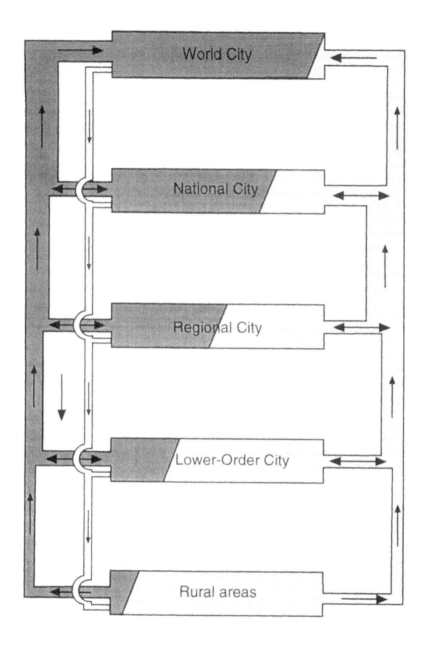

Figure 2.4. World systems surplus/extraction model
Source: Adapted from Armstrong and McGee, *Theatres of Accumulation* (1985)

1978; Levin 1960). With the depletion or substitution of extractive products, or the saturation of their markets due to overproduction in the resource frontier, the extractive economy eventually collapses, stranding population and infrastructure. During the bust phase of the extractive cycle, boomtowns decay, and idled workers in the extractive sector drift into established cities on the resource frontier. Katzman (1976:447–448) alluded to this problem in reference to the decline of the Amazon's rubber economy:

> With the decline of rubber, the descendants of the Northeastern gatherers [lured to the Amazon to work in the rubber trade] comprised a basically stranded population which, since the 1940s have been moving to the cities, making Amazonia the most urbanized region outside of Brazil's industrial Southeast.

The urban implications of the extractive export economy perspective are clear. The expansion of the urban network at the lower tiers is ephemeral. Decentralization of the population is articulated to the expansive period of the extractive cycle. With the collapse of the regional extractive economy, the urban network contracts and population descends once again upon larger metropolitan centers to be held in reserve for the next extractive cycle.

World systems theory is predicated on the assumption that global capitalist expansion not only reproduces capitalist modes of production in the periphery, but engenders in diverse Third World peoples a unifying materialist consumer ethic at the expense of local beliefs and cultural values. For our purposes, the important implication of this analysis is that urban centers throughout the world become staging arenas for producer surplus appropriation and holding areas for workers temporarily displaced by ebbs and flows in extractive cycles. Meanwhile, peripheral urban systems become the networks through which that surplus is channeled to the global level (Armstrong and McGee 1985). In the postmodern world, based on information technologies and "telematics," the physical manifestation of surplus extraction need not express itself as a dentritic urban system integrated by roadways. Rather, we would expect to find a regionally disarticulated system of physical flows in which selective urban centers are directly integrated into the global economy by air travel (Manaus) and others indirectly by land transport and satellite (Xinguara, Rolim de Moura). On the surface, the resulting spatial morphology appears irregular and multicephalic.

The notion that the postmodern world is converging toward such an integrated global economy and homogeneous world culture steeped in materialism has produced some measured criticism from certain world systems scholars. Although it should be clear that the regimentation of the extractive export economy in the periphery is linked to international markets and transnational corporate forces, these writers have distanced their analysis from any claims about the universality or predictability of globalization processes in the Third World. In the post-Fordist era of flexible production and accumulation the First World cum "Fast World" (Europe, North America, and Japan) is decoupling from the "Slow World" (Latin America, Africa, Southeast Asia). Global economic and financial power is increasingly concentrated in an archipelago of three world cities (London, New York, and Tokyo) from which transnational corporations compete for commercial supremacy over the Fast World market of some seven to eight hundred million consumers (in Europe, North America, and Japan) able to sustain materialist lifestyles (see Knox [1994] for overview). The chief beneficiaries of the global localization strategy of transnational corporations have been these three world cities of the core economies, "the 'mass-consumer technopoles' that have become pre-eminent centres of commercial innovation and corporate control, undisputed centres of taste-making, crucibles of consumer sensibility and seedbeds of material culture" (Knox 1994:6–7). It is difficult to believe that urban centers of the Third World could be unaffected by these global transformations.

Anthony Giddens' recent influential work would suggest the impossibility that localities, however remote, can altogether escape the powerful "distantiated" forces shaping their nature:

> In conditions of modernity, locales are thoroughly penetrated and shaped in terms of social influences quite distant from them. What structures the locale is not simply that which is present on the scene; the "visible form" of the locale conceals the distantiated relations which determine its nature   (Giddens 1990:18–19 in Dickens 1992:148)

Other world systems scholars have questioned the concept of a "world culture" (Wallerstein 1991), or have argued against perspectives that dismember national and subregional urban networks from the larger global city system. "All cities are world cities" writes Anthony King (1991:82). Other writers have reemphasized the importance of the nation-state as the central organizing principle of spatial economics and cultural formation. Immanuel Wallerstein (1991:96) writes:

The history of the world has been the very opposite of a trend towards cultural homogenization; it has rather been a trend towards cultural differentiation, or cultural elaboration, or cultural complexity. Yet we know that this centrifugal process has not at all tended toward . . . pure cultural anarchy. There seem to have been gravitational forces restraining the centrifugal tendencies and organizing them. In our modern world-system, the single, most powerful such gravitational force has been the nation-state.

Janet Abu-Lughod, in her paper "Going Beyond the Global Babble" (1991:135), underscores the conditionality of globalization processes in shaping local and regional development:

Globalization is an ideal type of instantaneous, indiscriminate and complete diffusion of all cultural products with no need for intermediate interpretation. We are still very far from that. Rather, what we are experiencing is rapid, incomplete and highly differentiated flows in global transmission.

David Simon (1995) is one of a few world cities scholars to explicitly consider Friedmann's world city hypotheses from the optic of the global periphery. Based on his work in sub-Saharan Africa, Simon identifies three characteristics of peripheral cities that reflect their changing, but subordinate, role in the international division of labor. First, peripheral cities are articulated to the world system through their functions as base points for the extraction of raw materials and as captive markets for imported manufactured goods. Simon acknowledges significant regional variations in this dependent relation, but the process by which different peripheral cities become differentially articulated to world cities remains unclear.[10] Second, peripheral cities have experienced rapid population growth but only limited economic transformation since independence. Again, Simon refers to sub-Saharan Africa, but with few exceptions of economic dynamism (e.g., São Paulo, Singapore), much the same can be said for the majority of Third World capital cities. Third, while most peripheral metropolitan centers have witnessed an unprecedented expansion of the personal informal services sector since World War II, more recently Third World services have been penetrated by technologically sophisticated developments, especially in the communications subsector. The process of global accumulation that engenders in world cities the strongest dynamism in a primary cluster of high-level business services (management, finance, legal services, technical consulting, research and development, and "telematics") produces the same economic restructuring in selected peripheral cities. Yet, the technological deepening of the Third World's service sector has further served to widen the gulf between wealth and poverty separating the system of core cities from those of the semiperiphery and periphery. "So, paradoxically, just as the world is becoming ever more tightly integrated and interconnected by virtue of advances in aviation and telecommunications technology, it is also becoming increasingly differentiated in relative terms" (Simon 1995:137).

Beyond general inferences, the implications of recent world city formation for Third World urbanization go largely uncontemplated. We might infer from this recent literature that the Third World is not entirely jettisoned from the global economic matrix headed by the triad of northern world cities. Rather, international metropolitan centers are selectively integrated into the matrix depending upon their ability to mobilize capital through commercial or production activities or on some comparative advantage in information technology. One such city in Brazilian Amazonia is Manaus. Favored with a free trade zone and a biophysical location in the heart of the world's largest remaining sanctuary of wild germ plasm (the Amazon rainforest), Manaus, an enclave within its own national system of cities, is directly articulated to the circuit of world cities, an issue we take up in chapter 5. But other lower-order rainforest cities indisputably have played a provisional role in regimenting production in specific corners of the Amazon periphery (e.g., mahogany trade, see chapter 6; gold rush, see chapter 7). Clearly, one unfinished task for world cities scholars is to examine more closely the logic of selective and flexible enclosure of the Amazon frontier into the global economy and the mechanisms by which peripheral cities worldwide are selectively incorporated into the extended web of world city influence, an issue to which we return in chapter 9.

Summary

Throughout the structuralist literature the function of the frontier is defined by its subordinate relation to capital. First, the frontier serves as a labor safety valve, a conduit through which excess labor can be channeled and maintained in reserve (i.e., controlled). Second, the frontier serves an important role in the production of wage goods (i.e., food-stuffs), providing capital with the opportunity for surplus appropriation on the frontier while keeping the urban wage rate in the industrial centers depressed. Third, the frontier, insofar as it is capable of retaining disposable income from local production, potentially provides an important consumption function in the national economy. The consumer market expands with the frontier, providing new opportunities for commercial accumulation. Finally, although not explicitly treated in the recent world cities literature, we may infer that the Amazon frontier is marginally relevant to the emergent neo-Fordist regime of accumulation based on "Fast World" markets and centered at the level of an elite group of world cities. The Amazon remains a distant resource frontier, and for the time being of little direct interest to transnational industrial and financial capital. However, insofar as selective cities (e.g., Manaus) may serve as useful duty-free shopping centers catering to a Brazilian national market of 150 million, they serve as urban stepchildren to the world cities.

RECENT BRAZILIAN CONTRIBUTIONS TO FRONTIER URBANIZATION THEORY

Several Brazilian writers have added significantly to the structuralist discourse, infusing it with a distinctive range of analyses of frontier urbanization in the Amazon that warrant special review. Much of the recent Brazilian literature on frontier urbanization in Amazonia has been informed by structuralist frameworks that have held the abiding interest of Latin American social scientists since the 1950s (for a review see Browder and Borello 1992). Core-periphery frameworks figure prominently in the Brazilian structuralist analyses of frontier urbanization. Urban areas on the frontier facilitate the process of the core's extraction of surplus produced on the periphery. For many writers in this camp, the emergence and functions of towns and cities on the frontier cannot be adequately understood apart from the structural changes occurring in the countryside as a consequence of the spatial expansion of capitalist relations of production. Urbanization is a strategy of capital to rapidly occupy the frontier and mobilize the surplus social product (Gentil 1988), and thus it occurs as a function of the agrarian transition and the transformation of the peasantry into a (semi-)proletarian class. However, urban-rural sectors as discrete analytic categories quickly dissolve in this literature. Urban transition and agrarian transition become two expressions of the same process. Different aspects of this general approach are emphasized by different writers.

For Torres (1988), frontier urbanization has a provisional character and is sustained only in relation to rural land markets in a "closed frontier." With all available land on the frontier occupied by either agro-industrial capital, small capitalist farmers, or marginalized peasants, frontier urban areas serve as temporary holding stations for migrants while they wait for rural land to appear on the marketplace. In essence, Torres adds another step to the intersectoral articulation model, weaving together the agrarian and urban transition strands into a single process. Frontier urban centers, therefore, are transitory; when rural land no longer becomes available (i.e., it becomes fully consolidated into permanent commercial estates), the urban population declines to a level necessary to service the consolidated countryside. Fundamentally, these frontier urban centers support the larger function of real estate markets in transferring peasant holdings (and the labor value congealed on them) to more capitalized social groups.

Monte-Mor (1980a, b) suggests that the constant flow of peasants and landless migrants creates a labor surplus on the frontier that demands urban nuclei as organized labor markets. Additionally, rapid labor market expansion exerts pressures on the land market, thereby multiplying opportunities for land speculation and different levels of capitalization on the basis of access to land. This perspective draws from de Janvry's "merchant road" approach to capitalist penetration, with a dominant role for the state (de Janvry 1981). Urban spaces enhance state control of the frontier, exerted in the service of landed elites. The state induces urban concentration to more easily produce the necessary means of collective consumption for the labor force and to create the general conditions of production required for capitalist accumulation (Monte-Mor 1980b). Land speculation becomes a major source of profits in rural and urban areas involving capital in several forms: small farmers, the urban petite bourgeoisie, corporate capital, and peasants and landless workers who gain access to land through state colonization. In Rondônia, Monte-Mor observes, even as urban real estate prices rose more than twice as fast as those for rural land in the mid-1980s, urban dwellers (especially public servants and small merchants) were quick to purchase rural lots as the best long-term investment. The agrarian frontier is eventually subordinated to urban-based interests comprised of a mixed constellation of merchant capital, state bureaucrats, the petite bourgeoisie, and professionals.

Becker (1985) argues that urban nuclei serve three important functions in the organization of frontier space. First, they serve as the base of the organization of the frontier labor market, as "points of agglutination" of new migrants seeking land, as residential spaces for small producers (rural landowners), and as enclaves for semiproletarians or former small producers displaced by capitalist land consolidation. In this regard, Becker emphasizes the labor market function of frontier urban nodes. For both Torres and Becker, such nodes are ephemeral; they decline with the dislocation of labor fronts and with the expropriation of peasant land.

Second, Becker views urban nuclei on the frontier as serving a "political-ideological" function by which the state creates the tension-alleviating illusion of an alternative urban social space for the displaced peasantry, an option that is ultimately foreclosed by urban land speculation and conflict (paralleling the same process in rural areas). Moreover, these urban places are points of diffusion of modern consumer values and ideologies. Finally, Becker recognizes the function of the urban nucleus as a center for the social differentiation of the urban population, a place where some migrants achieve higher socioeconomic status. For example, small urban merchants apply their petty surpluses to acquire small rural parcels on which they produce food stocks for their urban shops or which they use as financial investments (see also Ozorio de Almeida 1987), a tendency that complicates any clear functional distinctions between urban and rural places. The frontier town is a heterogeneous social space that absorbs various social groups, all of which are related in different ways to the capitalist transformation of the frontier. For Becker, frontier expansion and urbanization are two complementary phases of the same process.

The merchant exploitation of peasants through unfavorable terms of exchange and eventual expropriation from their holdings continues with the latter's arrival in urban centers. The rapid growth of urban centers on the frontier, due to rural property consolidation and peasant expulsion, creates additional opportunities for capital appropriation not only through the possibility of wage relations but more immediately in the regulation of urban real estate. Peluso de Oliveira (1987) refers to the mercantilization of the urban labor force in her case study of Ceres, Mato Grosso. Displaced peasants move to local towns, and finding few opportunities for home ownership, they occupy vacant areas on the urban fringe until they are expelled by existing property owners or by local government and forced into private rental housing or government-built housing. In the latter case the intention of local politicians is "to regiment the migrant labor force, making it participate in local politics as a captive client" (Peluso de Oliveira 1987:43).

The view that frontier urbanization is directly linked to the capitalist transformation in social relations of production in the countryside, turning small rural producers into landless urban semiproletarians *(bóias frias)*, recurs throughout the Brazilian research literature (see also Ferreira 1987; Jardim 1987). This argument tends to evoke the image of frontier towns and cities teeming with displaced peasants and a small cadre of petty merchants, government bureaucrats, and professionals.

A different view is offered by Sawyer (1987), who observed that nearly two-thirds of the Amazon's total agrarian area has witnessed net out-migration, leaving a vast "hollow frontier" in its wake. Meanwhile comparatively high urban population growth rates are uniformly found throughout the frontier in areas undergoing both net in-migration and out-migration alike. However, Sawyer's analysis suggests that agrarian transition and urbanization on the frontier are distinct processes; or at least their functional relationship is more complicated than most structuralist interpretations allow. Indeed, Sawyer suggests that these divergent demographic patterns require a fundamental reconceptualization of the Amazon frontier. "The 'agricultural' frontier has become an urban frontier. . . . One cannot just speak of urbanization in the frontier, but rather a veritable urbanization of the frontier" (Sawyer 1987:46).

Sawyer suggests that frontier urbanization should be understood in the larger context of national urbanization in Brazil. Among the reasons given to explain urbanization generally are: (1) the partial mechanization of agriculture that has reduced permanent agricultural employment; (2) the growth in demand for modern agricultural services; (3) reduced access to land associated with land consolidation; and (4) reduced employment opportunities in the large metropolitan areas. These trends, which appear throughout Brazil, have led to the proliferation of "pioneer settlements" in the frontier. Sawyer identifies three types of pioneer settlements: private "company towns," "official" settlements established by government agencies as part of official land settlement programs (e.g., agrovilas along the Transamazon Highway, the rural service centers *[núcleos urbanos de apoio rural]* of polonoroeste in Rondônia), and "spontaneous" settlements that emerge in the agricultural frontier with the extension of the road network or as the result of resource extraction (gold and timber).

However, Sawyer downplays the importance given by Becker to the state in directing the process of capitalist expansion and frontier urbanization (Sawyer 1984). In their study of the swamp *favelas (baixadas)* of Belém, Mitschein et al. (1989) support Sawyer's contention: "The rapid growth of the cities in the Eastern Amazon, having long since surpassed the managerial ability of the State to structure them by means of urban planning, deserves the title 'wild' *[selvagem]*, and is not the expression of a 'deliberate strategy' of the state," as Pompermayer (1979) and Becker (1987) claim (Mitschein et al. 1989:22–23 in Roberts 1991:180–181).

Sawyer's prognosis for the future of the "pioneer" settlements in the Amazon frontier is dim, especially for smaller settlements heavily dependent on resource extraction. Boomtowns have become ghost towns in much of the region owing to the transitory nature of their bases in mining and timber extraction.

The Brazilian school in the main emphasizes the function of the frontier settlements as centers for the accumulation of producer surplus within the superstructure of the global economy. Key differences within the Brazilian school appear in the focus on the relative importance of land markets versus labor markets, the orientation of the dominant social faction in the frontier (merchant versus agro-industrial capital), and the emphasis given to the role of the state in shaping the process of urbanization in the frontier. The structure of the processes of surplus appropriation that occur simultaneously within the local economy of the frontier and in its relationship to national urban capital remains somewhat vague. Neither spatial nor temporal dimensions of the urbanization process are elaborated in any detail in this Brazilian structuralist discourse.

In broader reference to the political economy literature overall, efforts to systematically classify and examine the different functions in the accumulation process presumably provided at different levels of the urban hierarchy have been minimal, as have efforts to periodize this process in relation to urbanization. Moreover, political economy interpretations of postmodern Amazonia have come under recent criticism for the superficiality of their customary analytic categories (e.g., peasantry, proletariat) and for the typical confusion between the penetration of capital and the spread of capitalism (Cleary 1993). Yet, the structuralist school provides a powerful overall framework for examining the general patterns of economic expansion on the frontier within which to formulate more specific hypotheses concerning frontier urbanization.

Summary

The evolution of ideas concerning the spatial organization of human settlements spans a broad spectrum of intellectual interests and influences. Initially framed by the organizing principle of central places, urban geography has reached out to cultural geography, sociology, and critical political economy of various stripes (table 2.1).

Table 2.1 Characteristics of leading theories of frontier urbanization

CHARACTERISTICS	CENTRAL PLACE THEORY	MERCANTILE THEORY
Conception of frontier Market development process:	Homogeneous surface/plane. Endogenous, progressive, segmented.	Natural resource frontier. Exogenous, selective, based on trade routes.
Rural sector impacts of frontier development:	Increased spatial integration increases prices of agricultural land and commodities, inducing innovation, specialization and urban growth.	Increased integration increases competition among stockholders, inducing specialization or relocation to frontier.
Urban sector impacts of frontier development:	Specialization varies with urban size and location (spacing) within emergent hierarchy; autonomous regional growth.	Specialization and industrialization concentrated in trade "gateway cities."
Urban settlement location criteria:	Minimum transportation and spatial interaction costs.	Natural resource location, minimum transportation costs.
Urban system growth process:	Evolutionary, sequential.	Linear, sequential.
Urban systems morphology:	Symmetrical, hierarchical, hexagonal lattice.	Dendritic, axial.
Principal urban functions:	Scale graduated: rural producer service functions.	Trade service functions.

CHARACTERISTICS	DIFFUSION THEORY	INTERSECTORAL ARTICULATION THEORY
Conception of frontier: Market development process:	Abstract space, hinterland. In waves, cyclical, stepwise from higher to lower order settlements or "propulsive" growth poles.	Periphery, labor safety valve. Uneven spread, spatial/functional dualism: disarticulated (luxury v. basic goods) markets.
Rural sector impacts of frontier development:	Unspecified.	Rural poverty. Peasantry pushed to extensive margin to produce food goods with unpaid household labor to subsidize urban expansion.
Urban sector impacts of frontier development:	Unspecified.	Industrialization based on luxury goods behind protective tariffs. Small formal wage sector, large informal sector.
Urban settlement location criteria:	Reduce interaction costs or maximize production efficiency.	Maximize capital class control over labor.
Urban system growth process:	Selective or general spread, based on productive innovation sequential.	Truncated growth, uniform tier of small towns with minimal social overhead.
Urban systems morphology:	Symmetrical, hierarchical, hexagonal lattice.	Unspecified, presumably irregular.
Principal urban functions:	Scale-graduated: rural producer service functions.	Centers of labor agglutination, control points.

CHARACTERISTICS	CAPITALIST PENETRATION THEORY	WORLD SYSTEMS THEORY
Conception of frontier:	Contest zones (capital v. labor).	Global periphery, resource frontier.
Market development process:	Linear progression of stages: peasant fronts, capitalist and speculative expansion, eventual frontier closure.	Exogenous, selective, cyclical (boom/bust), dependent on transnational capital investment.
Rural sector impacts of frontier development:	Rural poverty. Peasantry alienated from land becomes day workers, land consolidation by national elites.	Peasantry survives. Rural poverty. Peasantry displaced from land to make way transnational modes of production. Peasantry dissolves.
Urban sector impacts of frontier development:	Homogenization of urban network as temporary "labor landings."	Enclave export production, disarticulated to domestic markets, contingent on tnc investment.
Urban settlement location criteria:	Minimum distance to active peasant frontier.	Maximize TNC surplus extraction from frontier.
Urban system growth process:	Contingent upon movement of labor.	Conditional upon global market forces.
Urban systems morphology:	Unspecified. Presumably truncated, single asymmetrical tier system of small towns.	Hierarchical, multitiered center-satellite structure.
Principal urban functions:	Penetration staging areas, centers of class conflict resolution.	Theaters of TNC accumulation and demand diffusion.

In this chapter our attempt to integrate these multiple influences on the urban geography of settlement systems leads us to the following conclusions. First, as noted at the beginning of this chapter, recent work in urban geography reflects a growing interest in the explanation of process and less so in the traditional description and classification of pattern. This is by no means a new development, but it is one that is well represented in the more recent city systems literature. Mercantile models, the diffusion literature, and selected writers in the political economy tradition present rather graphic interpretations of the dynamics of urban system formation, albeit with vastly different results.

Second, with a few exceptions, efforts to characterize the process of urban system formation have relied extensively upon various "transition models," in which this process is viewed as a linear sequence of stages. Spatial economy and capitalist penetrationist perspectives are particularly inclined toward an evolutionist approach. From such deductive frameworks, distinctive but predictable patterns emerge at different periods of time. The process of system change is determined by an underlying logic derived from the application of a single master principle (e.g., capitalist expansion, market area segmentation, diffusion, etc.). In general, these are highly constrictive frameworks that dismiss or gloss over a wide range of urban "misfits" and outliers. Nevertheless, such models have been instrumental in focusing debate on the empirical difficulties of formulating reliable settlement typologies.

Third, there is a decided tendency away from endogenous and toward exogenous explanations of urbanization. This is partially manifest in the rapidly growing interest in global frameworks for interpreting local phenomena and diverges from the traditional inductive penchant among geographers to construct explanations from the ground up, that is, beginning at the level of the "operational taxonomic unit," the smallest items that are combined to form classes.

Fourth, there is increasing emphasis on more complex units of analysis. We seem to have evolved beyond an interest in the merchant, the entrepreneur, the pioneer yeoman to consider in their place the multinational corporation, social classes, and the state as the relevant points of departure for an analysis of the spatial organization of settlements on new land. Even so, many of these frameworks are burdened with antiquated or empirically reductionist analytical categories, such as the "market principle," or the "proletariat" that grossly oversimplify a more complex human reality. Correspondingly, our understanding of the urbanization process cannot be completed by individual case studies of representative settlements but requires more rigorous comparative approaches that focus on the dynamics of change emerging from nascent urban systems.

Notes

1. Given an array of "first-order" centers evenly distributed over an idealized landscape, an array of "second-order" centers will emerge at the six corner points of the hexagon circumscribing the efficient market area of the first-order center. Within its market area, the first-order center will supply itself and one-third of each the six outlying centers with the range of functions it offers. Christaller called this a k = 3 settlement system (referring to the number of subsidiary-order centers circumscribed by the center of the next highest order, combining fractions). Lower-order functions (e.g, highly perishable food products) not offered efficiently by either first- or second-order centers to distant consumers will provide a market space for the emergence of third-order centers, and so forth.

2. Although somewhat dated, there are several excellent reviews of central place studies: Haggett (1966:114–152), Herbert and Thomas (1990:65–84), Johnson (1970), Miller (1979), and Morrill (1974:74–80).

3. Central place theory typically assumes a closed economic system, a conceptualization of space as a homogeneous surface or plane, and a homogeneous range of services available at each step in the hierarchy, assumptions that have very limited empirical validity, especially in emerging market economies.

4. Webb (1964) interpreted European overseas expansion in light of the Turner thesis in *The Great Frontier*. Bowman (1931) advocated land settlement as "a new form of nation building" in *The Pioneer Fringe*. Wells (1973:12) contends: "The history of the world . . . has been in large measure the history of ever-widening expansions followed by the ebb and succession of ethnographic frontiers." Meyer (1956) interpreted the nineteenth-century development of the Calumet region of northwest Indiana as a series of transitional stages between indigenous/extractive social forms and subsistence farming.

5. Loy (1973) tested the Turner thesis in eastern Colombia and found a "static frontier." Zavala (1957) applied Turner's concept of the mobile frontier to northern Mexico, finding that the idea of a safety valve did not hold. Moran (1975) concluded that the behavior of colonists on Brazil's Transamazon Highway did not show a transformation due to the influence of the frontier environment, as the Turner thesis suggested.

6. In his critical essay on world culture formation, Immanuel Wallerstein (1991) rejects the two classic explanations of cultural change—the secular tendency toward one world among distinct cultures and the stage theory of human development. Instead, he argues for a "model of successive historical systems" as a useful starting point.

7. For instance, see the special issue of *Latin American Perspectives* entitled "Peasants: Capital Penetration and Class Structure in Rural Latin America" (vol. 27, parts 1–3, 1980) and D. E. Goodman and M. R. Redclift, *From Peasant to Proletarian: Capitalist Development and Agrarian Transitions* (Oxford: Basil Blackwell, 1981). Among the prominent Brazilian scholars contributing to the recent articulationist genre are Silva (1979), Sorj (1980), Graziano da Silva (1982), and Muller (1982).

8. Brazilian articulationists were undoubtedly influenced by published observations of the effects of intersectoral capital transfers (via overvalued Brazilian currency, export taxes, and labor migration) from agriculture to industry in Brazil during the 1950s and 1960s that suggested a type of functional articulation between urban capital accumulation and rural poverty.

9. Based on the evidence available to researchers at the time (ca. 1960s to early 1970s), Foweraker concluded that the economic destiny of the peasants formerly occupying the smallholdings subsumed by capitalist enterprises (mainly cattle ranches supported by government subsidies) was on those same ranches, not in urban areas: "Certain small townships had developed, it is true, but the majority of the population worked in the countryside. The only place to look for this population is in the cattle estates themselves" (Foweraker 1981:53–54). This clearly proved to be incorrect.

10. Simon notes that this primitive form of dependency is characteristic mainly of sub-Saharan Africa and would not apply to several Latin American countries that pursued import substitution policies or to the dynamic centers of export production in Southeast Asia and the Pacific Rim region, in which very different forms of urbanization emerged, e.g., the "kotadesatie" process of specialized industrial development giving rise to "extended metropolises" (cf. Ginsburg et al. 1991).

3

Geopolitics, Regional Development, and Urbanization: Historical Dynamics of Amazon Frontier Expansion

The Amazon is not a green hell nor a lost paradise! It is a vast area where a whole
generation anxiously and confidently awaits the splendid dawn of a prosperous,
different, and promising tomorrow. It is time, truly, for man to command life
in the Amazon, ceasing to be a slave to the river. . . .What we want is an integrated
Amazonia, but forever Brazilian.
—General Rodrigo O. J. Ramos (quoted in Mattos 1980:29–30)

Before the middle of the twentieth century, urbanization of Amazonia
entailed a long-term process of expansion from favored locations on the
Amazon River and major tributaries. An externally oriented regional
economy linked river towns to distant markets through the agency of
political and commercial institutions in the major gateway cities, Belém
and Manaus. The regional settlement system evolved according to the
geopolitical designs of different state regimes and the locational logics of
various commercial cycles in natural resource extraction. Three principal
historical periods of urban expansion can be discerned before the emergence of contemporary frontier urbanization: first, the colonial period
after the founding of Belém and missionary villages in the seventeenth century; second, the mercantile period of the Companhia Geral Grão Pará
and Maranhão in the late eighteenth century; and third, the rubber boom
from approximately 1850 to 1920 (Correa 1987, 1991).

Contemporary urbanization of the frontier selectively integrates Amazonia into the national city-system by opening the *terra firme* uplands to regional in-migration and settlement along new road corridors and around airstrips in the interior. Since 1960 new interregional transportation connections have alternately either depopulated or reinvigorated the older river towns of Amazonia. In both cases, the expansive space economy of Brazil's southeastern core region links both preexisting river settlements and new frontier towns to national and even global urban hierarchies. Despite the changing emphases of regional settlement, contemporary migrants to Amazonia thus continue a long-term historical project of geopolitical expansion into the interior of South America, which began with the Spanish and Portuguese colonial conquests.

This chapter explores the historical geopolitics of frontier expansion and urbanization in Amazonia. While the legacy of the rubber boom and other previous historical epochs remains evident in the region's spatial structure, contemporary patterns of frontier urbanization largely arise from a series of recent Brazilian government programs to control and develop the region. The major changes in public policy toward the Amazon and the influence of competing governmental institutions have created particular favored forms of occupation in different areas of the Amazon. The aggregate result of urbanization is an array of settlement systems that are selectively integrated into the national economy and functionally disarticulated from the process of autonomous regional growth in the Brazilian Amazon.

COLONIAL URBANISM IN AMAZONIA

Luso-Brazilian expansion into the South American interior began in a global context of European imperialistic competition. Before the late seventeenth century virtually all Portuguese settlements except São Paulo were coastal cities with convenient ports and defensive fortifications to guard against attacks by Spain, France, England, Holland, and other rival powers. Beginning in the late sixteenth century, Brazil's national heroes, the *bandeirantes* or flag-bearers, pushed from an initial base in inland São Paulo into the backlands *sertão* of the Northeast, the Amazon, and the South. In their search for valuable resources and Indian slaves, these Brazilian pioneers also expanded the cattle-ranching frontier into the interior and founded new settlements along important transportation routes. The *bandeirantes* discovered gold in inland Minas Gerais (General Mines) in 1695, as Caribbean competition devastated the Brazilian sugar industry and sent coastal areas into a prolonged economic depression. The subsequent gold rush, along with the discovery of diamonds in 1729, provoked a mass in-migration in Minas Gerais, where the population exploded from virtually nothing to thirty thousand in 1709 and to half a million by 1800 (Boxer 1962). Further mineral discoveries in Mato Grosso, Goiás, Bahia, and Maranhão promoted distant frontier settlements. Although the mineral boom created great wealth, it did not provide a self-sustaining economic development; smuggling to avoid taxes was rampant, and the regional economic dependence on Europe continued (Dickenson 1982; Moog 1964).

In Amazonia the Portuguese founded a settlement at Belém in 1616 as both a defensive outpost against foreign incursions and a base for further expansion into the interior. Belém's original military fort, cathedral, and residential quarter sit atop a peninsular promontory now known as the Old City (Cidade Velha), overlooking a sheltered harbor in Guajará Bay, located in an estuary at the entrance to the vast Amazon River network (plate 3.1). From the original Amazon nucleus at Belém, the Portuguese fanned out to establish such towns as Brangança in 1635 and Cametá in 1635. Operating out of monasteries and convents in Belém, the Jesuits, Carmelites,

Plate 3.1. View of ramparts of the Forte do Castelo and the spires of the Nossa Senhora da Graça Cathedral, Belém, 1988 *Photo: Brian J. Godfrey*

Franciscans, and other Roman Catholic orders established missionary settlements, either in preexisting Indian villages or next to military fortifications, to evangelize the native peoples of Amazonia. Such cities as Alenquer (then Surubiú), Monte Alegre (Gurupatiba), and Manaus (São José do Rio Negro) originated as military-religious settlements during the seventeenth century. Due to the Amazon's isolation and vast extent, the Portuguese crown created in 1621 the separate northern state of Grão Pará and Maranhão, which encompassed the Lower Amazon region, with its capital at Belém (Correa 1991; Godfrey 1991).

Besides serving as Portuguese strongholds in the region, Amazonia's early colonial settlements also served as collection points for the *drogas do sertão*, resources gathered from the interior, such as woods, spices, guaraná, dyes, animal skins, and other primary commodities. The religious orders, especially the Jesuits, benefited more than even the Portuguese crown and merchants from the regional trade. Exempt from taxes, the religious orders controlled the forest resources and native labor in the missionary villages from where the commodities were transported to Belém and then to Lisbon in return for imported merchandise (Correa 1991; Hecht and Cockburn 1989).

In the late eighteenth century, influenced by the Bourbon economic reforms in Spanish America, Portugal attempted to promote economic activity and to generate greater government revenues in the Amazon. In 1755 Portuguese Prime Minister Pombal, a virtual dictator, created a state-sponsored mercantilist company, the Companhía Grão Pará e Maranhão. Pompal curtailed the power of the Catholic religious orders by secularizing the missions and freeing the Indians to live and work elsewhere. Between 1755 and 1760 forty-six missionary villages were elevated to the status of towns. Growing conflicts between the commercial interests of the Jesuits and Portuguese merchants led the crown to expel the Society of Jesus and to expropriate all Jesuit property in 1759. Land grants (*sesmarias*) were awarded to Portuguese soldiers and colonists, who increasingly intermarried with native women, promoting miscegenation in the region. In addition, African slaves began to be imported to work the plantations of cacao, coffee, tobacco, and other products. Cattle ranches also expanded in the Amazon. Incipient *aviamento* networks of debt peonage emerged in the eighteenth century before a period of relative economic stagnation overcame Amazonia after 1780 (Burns 1993; Hecht and Cockburn 1989).

In 1755 Pombal created the Captaincy of São José do Rio Negro in the western Amazon, roughly equivalent to the current state of Amazonas, and the Captaincy of Grão Pará in the Lower Amazon Basin. Both captaincies were part of the State of Grão Pará and Maranhão, of which Belém remained the capital. In 1772 this state was dismembered with the creation of the separate states of Maranhão and Grão Pará (Correa 1991). These changes in regional administration recognized the growing Luso-Brazilian presence in Amazonia by the late colonial era. The borders of Portuguese America extended far beyond the restricted limits of the Treaty of Tordesillas, which in 1494 had divided the New World between Spain and Portugal. Although vast areas of the Amazon remained to be effectively occupied, the era of Portuguese colonial rule shaped the general contours of the Brazilian national territory, provided an enduring economic tendency to rely on boom-and-bust cycles in natural resources, and established settlements according to mercantile principles of location (Dickenson 1982; Moog 1964).

URBAN IMPACTS OF THE AMAZON RUBBER BOOM

Amazonia became part of a united independent Brazil after 1822, but the region remained marginal to national life. Prevailing ocean currents long made it easier to sail from Belém to Lisbon than to Rio de Janeiro. Amazonia's incorporation into Brazilian and international affairs accelerated in the late nineteenth century during the rubber boom. After about 1850 Brazilian migrants to the Amazon region, largely from the Northeast, clustered initially along the major watercourses at entrepôts formed for the extraction and transport of rubber (plate 3.2). Through repeated boom-bust cycles of resource extraction, native Indians often were displaced or acculturated into Brazilian society, forming a miscegenated *caboclo* population along the rivers (Burns 1993; Parker 1985).

An integrated extractive-mercantilist economy connected the rural *caboclo* formation of the interior to the regional metropolitan centers. Relying on a regime of debt peonage known in Portuguese as *aviamento*, import-export companies (*aviadores*) in the major cities, backed by foreign investment capital, advanced credit to primary producers (*aviados*) in the Amazon rainforest through a chain of subsidiary agents. Basically, the metropolitan *aviadores* encouraged the owners or lessees of extraction areas (*seringalistas*) to set up trading posts (*barracões*) in the interior, operated by small merchants (*comerciantes*). From the trading posts along the Amazon

Plate 3.2. Balls of rubber latex piled up at a modern-day *barraca* near São Félix do Xingu, 1978   *Photo: Brian J. Godfrey*

waterways, *aviadores* advanced credit to the rubber tappers (*seringueiros*) to purchase imported goods in return for forest products (Despres 1991). From the local perspective, this socioeconomic organization of the rubber trade tied domestic *caboclo* households in the interior to an extensive chain of local *comerciantes*, mercantile interests in the capital cities, and ultimately national and foreign markets. In spatial terms, the *aviamento* system relied on regional river transport and local forest trails to link a series of hierarchically ordered settlements (Correa 1991). In terms of the regional urban system, six basic elements stand out in the *aviamento* extractive-mercantilist regime. These six levels of the urban system were linked hierarchically by political-economic power, capital flows, and internal trade and labor migration patterns, as suggested in figure 3.1.

1. At the top of the world system, metropolitan centers, such as London, Paris, and New York, directed the global forces of mercantile capitalism. These global metropolises regulated commodities markets, furnished international capital for overseas investments, and exerted political pressure on the areas of primary production. For example, the capital invested in the establishment of import-export *aviador* firms in Belém and Manaus during the rubber boom originated largely in London. In addition, English companies took leading roles in building the port facilities and transportation infrastructure of turn-of-the-century Amazonia.

2. At the national level Brazilian commercial elites and political power brokers operated mainly from the national capital of Rio de Janeiro and the emerging commercial center of São Paulo. The national bourgeoisie was often marginal to Amazonia in the laissez-faire economies of the late nineteenth and early twentieth centuries. National authorities extracted tax revenues from the rubber trade, but Brazilian metropolitan elites essentially competed with foreign interests for control over the Amazonian commerce.

3. Within Amazonia the metropolitan centers of Manaus and Belém featured large commercial houses and financial interests specializing in the export of primary materials and natural resources to external markets. To facilitate international trade, these commercial-financial *aviadores* of the regional metropolises also provided consumer goods and instruments of work to smaller merchants in the interior. These regional elites of the metropolis, linked both to foreign interests and to areas of primary production, predominated in local political affairs and enjoyed considerable autonomy from national authorities during the rubber boom.

4. In the interior, strategically located river ports, such as Santarém, Porto Velho, and Marabá, served as intermediary cities linking the regional metropolitan centers and local producers in the extractive-mercantile economy. These medium-sized cities often were the home of agrarian interests, the *seringalistas*, who controlled the areas of forest extraction, and of commercial emporia for the surrounding lower-order settlements. The river ports served as points of transshipment in the extractive-mercantilist economy, collecting forest products and accumulating petty commodities for distribution in the interior.

5. Farther up the watercourses of Amazonia, small-scale entrepreneurs plied the rivers from the local central places of the regional economy. The small commercial *aviador* operated a local collecting post, known in Portuguese as the *barracão* or "big hut," where primary producers bought imported commodities and sold forest products. The trading posts also served as the starting point for travel on small river tributaries and forest trails leading to the areas of production in the Amazonian extractive-mercantile economy of the late nineteenth and early twentieth centuries.

6. Finally, *caboclo* families in scattered *barracas*, or small huts, of the interior engaged in subsistence agriculture and collected rubber latex, Brazil nuts, and other natural resources along forest trails. The *caboclos* traded their forest products for commodities they could not produce themselves, receiving financial credits for their inevitable financial shortfall in a system of unequal exchange.

The collapse of the rubber trade in the early twentieth century precipitated a prolonged period of economic stagnation and demographic stabilization in Amazonia. Yet the basic urban organization of the *aviamento*

Figure 3.1. Amazon urban hierarchy under the *aviamento* extractive- mercantilist regime

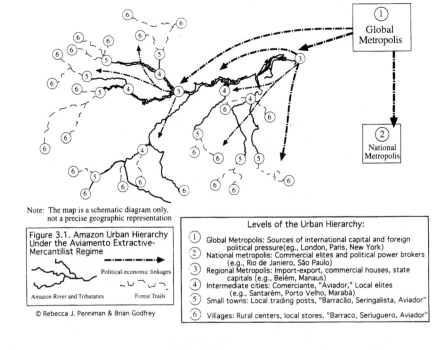

Note: The map is a schematic diagram only, not a precise geographic representation

Figure 3.1. Amazon Urban Hierarchy Under the Aviamento Extractive-Mercantilist Regime

Political-economic linkages

Amazon River and Tributaries   Forest Trails

© Rebecca J. Penniman & Brian Godfrey

Levels of the Urban Hierarchy:

1. Global Metropolis: Sources of international capital and foreign political pressure(eg., London, Paris, New York)
2. National metropolis: Commercial elites and political power brokers (e.g., Rio de Janiero, São Paulo)
3. Regional Metropolis: Import-export, commercial houses, state capitals (e.g., Belém, Manaus)
4. Intermediate cities: Comerciante, "Aviador," Local elites (e.g., Santarém, Porto Velho, Marabá)
5. Small towns: Local trading posts, "Barracão, Seringalista, Aviador"
6. Villages: Rural centers, local stores, "Barraco, Seriuguero, Aviador"

extractive-mercantilist regime survived intact. Trade in such natural resources as Brazil nuts, minerals, and animal skins became the mainstay of the local extractive economies though generally to lesser degree than rubber had been at the height of its extraction. The expansion and retraction of the economic cycles in primary products determined the demographic fluctuations of the Amazon river towns until after World War II, when other productive sectors began to develop to a greater degree (Velho 1972; Wagley 1962).

Despite deepening regional poverty, rural out-migration, and slow urban growth rates in the early twentieth century, the basic socioeconomic dependency relationships persisted in the region. The *caboclos* of the interior continued to depend for their livelihood on local merchants, landlords, and metropolitan entrepreneurs. For their part, regional elites could no longer be so extravagant in consumption as they had been in the boom years; moreover, they saw their position weakened in external dealings. Nevertheless, many wealthy families continued to prosper on the basis of their urban commercial establishments, their vast cattle ranches, and claims to rubber tapping, Brazil nuts, and other forest products in the interior. Indeed, although the national government attempted to monitor Amazonian affairs and influence events in the region, the interests of local elites generally prevailed over those of the distant regimes in Rio de Janeiro before World War II. Prewar Amazonia was dominated by the landed gentry and urban merchants, often interrelated by marriage and socioeconomic ties, whose interests generally prevailed in state and local politics.

POST–WORLD WAR II ANTECEDENTS TO AMAZON INTEGRATION

"National integration" became a primary policy objective of successive developmentalist regimes after World War II. Previously Amazonia's boom-and-bust cycles in natural resources had been driven primarily by the profit motives of merchants and private landowners: extractive cycles of rubber latex, Brazil nuts, diamonds and minerals, and other primary commodities ran their course along the region's natural system of waterways with little direct governmental assistance. Although national sovereignty and effective occupation of the Amazon were important geopolitical concerns of the Brazilian governments before World War II, postwar development programs succeeded in more profoundly altering the region's human geography. With the spate of development programs in recent years, the traditional local elites often have been supplanted by new elites of army officers, civil servants, and entrepreneurs—most originating outside the region—who have benefited from new financial incentives and communication ties to the national system (Miller 1985).

Recent resettlement efforts have encouraged out-migration from rural problem areas plagued by poverty, modernization of agriculture, or simply an inequitable distribution of landholdings. The prevalent forms of frontier urbanization in contemporary Amazonia are largely the product of specific strategies by successive Brazilian national governments and various state agencies as well as organized private interests to establish political authority and more lucrative economic activities along the country's vast northern periphery. Because various regional development strategies have been directed to different microregions within Amazonia, highly irregular patterns of urbanization have resulted.

An early battle cry from the national core to integrate the vast Amazon region more effectively into the mainstream of Brazilian life came from populist leader Getúlio Vargas, who in 1940 proclaimed a "March to the West." In an effort to promote national unity, Vargas proposed "to conquer and dominate the valleys of the great equatorial torrents, transforming their blind force and their extraordinary fertility into disciplined energy" (Hecht and Cockburn 1989:105). In 1953, during a later presidential term, Vargas created Amazonia's first major regional development agency—known as the Superintendency for the Economic Valorization of the Amazon (SPVEA), reconstituted in 1966 as the Superintendency for Amazon Development (SUDAM). This agency was charged with developing the transportation and communications infrastructure, health facilities, and other projects in the region. Before his dramatic suicide in 1954, Vargas succeeded in doubling the country's network of roads and vastly increasing the number of airports from 31 to 512 (Revkin 1990:87). But most of these improvements to the national transportation network, along with the rest of his program of industrialization and modernization, occurred outside the Amazon. Even if Vargas did not succeed in breaking the isolation of insular Amazonia, he articulated a nationalistic ideology that subsequently propelled the forces of popular and corporate expansion into the northern frontier.

The northern region remained on the margins of national life until the presidency of Juscelino Kubitschek from 1956 to 1961. President Kubitschek continued efforts to industrialize the country, to extend the internal market, and to expand into Brazil's interior. In his most dramatic gesture of national integration, Kubitschek moved the national capital from coastal Rio de Janeiro to the planned administrative center of Brasília, located near the geographic center of the country. In addition, he presided over the construction of the first major road into the Amazon, the Belém-Brasília Highway, inaugurated in 1960 (paved by the early 1970s), providing for the first time access by land transportation to a major northern city.

Occupation of areas near the Belém-Brasília Highway established a precedent for the disorderly process of land enclosure that would accompany the construction of other new penetration roads in Amazonia. Despite a large-scale in-migration of settlers, mainly from the Northeast, the rampant speculation in unsurveyed federal lands (*terras devolutas*) resulted in the enclosure of vast areas along the road by private owners of immense landholdings, encouraged by governmental regional development agencies. Between 1959 and 1963 about 5.4 million hectares were transferred from public to private hands in Pará alone. A legacy of competing land claims, fraudulent titles, and rural violence still plagues this region (Schmink and Wood 1992; Hecht and Cockburn 1989). Small farmers, who received little governmental assistance in obtaining land titles or rural credit, were steadily displaced from agricultural lots or employed as cheap farm labor (Hebette and Acevedo 1979; Ianni 1979). The eventual consolidation of small land claims into large landholdings along the Belém-Brasília Highway has promoted a specific form of urbanization: tenement towns. Lacking ready access to land, many migrants had little choice but to congregate in a series of burgeoning cities along the route of the Belém-Brasília Highway in the states of Goiás, Maranhão, and Pará.

Although SPVEA began the momentous process of opening up the Amazon to more direct control by central government forces, the agency itself was fatally tainted by claims of insider corruption and favoritism to entrenched local elites, which ultimately led the Brazilian military regime to favor stronger central control of regional development efforts. In 1966 SUDAM was created to centralize further development planning in Amazonia (Schmink and Wood 1992; Hecht and Cockburn 1989).

Brazilian national authorities have long been obsessed with achieving better control over the vast northern periphery. The Brazilian military, in particular, long saw Amazonia as the basis for a prosperous national future through the development of an inland resource empire integrated with existing economic centers along the coast. National concern over ambiguous borders and the reputedly covetous designs of neighboring countries and northern powers, coupled with the lack of a firm Brazilian presence in much of Amazonia, created a persistent anxiety regarding the region and a popular basis for establishing a national presence there. Thus, Brazilian campaigns to occupy Amazonia more effectively increased in frequency and intensity along with other contemporary efforts at national development (Reis 1982; Sternberg 1987).

After World War II, the Brazilian military became increasingly concerned with issues of national integration and development as well as with issues of national security. The creation of the National War College (Escola Superior de Guerra), founded in Rio de Janeiro in 1949 to train upper-level officers, brought new geopolitical currents to the fore in Brazil. More and more the military saw its role not only as the guardian of national security but also as the promoter of economic development. By the 1950s the Brazilian military firmly subscribed to a philosophy of security and development (*segurança e desenvolvimento*), which extended the traditional concern with military strategy into areas of economic growth, resource management, and national integration (Foresta 1992).

EMERGENCE OF A MILITARY-AUTHORITARIAN REGIME (1964–1985)

The early 1960s witnessed a steady radicalization and polarization of Brazilian politics. After the resignation of Janio Quadros, who served for less than eight months in 1961, his successor João Goulart (1961–1964) steadily leaned leftward in his policies, proposing various kinds of structural change, redistribution of agricultural lands, literacy campaigns, and workers' rights (Leff 1968; Pereira 1984). Meanwhile, the country was beset by a rash of strikes, growing inflation, economic stagnation, and political deadlock. With firm U.S. support, the Brazilian military carried out a coup d'état, deposing Goulart in 1964 (Burns 1993). Although many observers expected the military to remain in power only long enough to reestablish order and guarantee private property, the generals ultimately ruled the country for twenty-one years with the aim of fundamentally restructuring Brazil, eliminating the threat of communism, and promoting the free-enterprise system (Schneider 1991).

To stifle internal opposition, the military regime promptly eliminated thirteen existing political parties and created two encompassing political organizations: the Alliance for National Renovation (ARENA), and the opposition Brazilian Democratic Movement (MDB). A massive infusion of official foreign investment encouraged the proliferation of the activities of multinational corporations (MNCS) in Brazil. The authoritarian regime created a hospitable environment for investment through the provision of tax incentives, liberal profit remittances, and forced political stability. Coupled with massive borrowing from abroad, the military led the country into a period of rapid economic growth in the late 1960s and early 1970s popularly known as the "Brazilian miracle."

During this period national income, as distributed by sector, continued the postwar shift from a concentration in the primary activities of agriculture, cattle, and resource extraction to secondary industrial activities. The benefits of industrialization, however, accrued primarily to the upper-income groups and became regionally concentrated in the Southeast, particularly around the São Paulo core area, creating one of the world's most unequal societies (Wood and Carvalho 1988). Brazilian geographer Milton Santos (1979) has called this uneven growth the product of an "industrialized underdeveloped country."

The Brazilian model of "dependent development," according to Evans (1979:51–52) featured "capital accumulation at the local level accompanied by increasing differentiation of the economy, which is to say by some degree of industrialization." In terms of class alliances, the military-authoritarian regime forged a "triple alliance" of state, national private, and foreign capital, known in Portuguese as the *tri-pé*. This arrangement served several useful functions: it integrated the economy under central government stewardship; it gave national businesses the opportunity to gain technological and managerial skills, and it gave the MNCS political legitimacy and local protection from nationalist critics. Internationally, the *tri-pé* legitimated the MNCS in the host country, and the state depended on this legitimation to raise development funds on the international market. Although the interests of members of the triple alliance sometimes diverged on particular issues, as in the case of the Brazilian computer and arms industries, generally the different state and private factions agreed on a political-economic strategy of increased income concentration, which effectively excluded the masses from the benefits of development.

During two decades of military rule, a series of development programs transformed the face of Amazonia. The military regime, regarding regional integration and economic development as geopolitical imperatives, saw the Amazon as an important arena in which to justify its rule. Rather than bowing to the traditional autonomy of regional elites as it had done in the past, the military attempted to centralize authority and promote the interests of more broadly based national elites and multinational corporations. Geiger and Davidovich (1986:290) sum up the military's general centralizing approach to regional problems in the following terms:

> The regime of 1964 took political autonomy away from the states and counties (*municípios*), and thus directly weakened the class fractions that might have dominated decision making at those levels. But the regime simultaneously penetrated those economies at all those levels much more deeply, and delegated parts of the administration of state enterprises directly to the various levels. This, in turn, reconstituted power relations at each spatial level, but did so in line with the will of the central state.

Since World War II, and especially during the military-authoritarian period (1964–1985), the Brazilian government created the conditions for Amazonian occupation and regional development, and its control over the exploitation of resources became more extensive and precise. The accelerating pace of change in Amazonia during the late twentieth century stems directly from the emergence of the authoritarian Brazilian state, intent on ensuring national sovereignty through regional investments in infrastructure, interregional migration, and economic development. Despite different policy emphases over time, the contemporary programs of national integration have selectively linked the Amazon hinterland via modern modes of transportation and communications to national and foreign metropolitan centers. Although the federal role has diminished since civilian authority was restored in 1985, the general contemporary spatial patterns of settlement and urbanization have arisen largely from the central government's post–World War II expansion into Amazonia.

In political-economic terms, after World War II the national state interposed itself between foreign and regional interests in Amazonia through the imposition of centralized taxes, regulatory agencies, policies of import substitution and export promotion, free trade zones, and favored development poles. Private corporations, some receiving lucrative subsidies from the central government, took the lead in developing export-oriented industrial complexes, most notably at Manaus, where the international airport began to receive direct flights from foreign and national centers. State-supported national firms, such as the Companhia Vale do Rio Doce mining company, mounted mechanized resource-extractive operations at Carajás and elsewhere. Large-scale cattle ranches, encouraged by fiscal incentives, dominated vast swaths of the Amazon region's interior. Yet, small-scale agricultural, mining, and other activities held sway over large areas as well. The overall result of the national integration of the Amazon has been a disarticulated urban system, as indicated in general schematic terms in figure 3.2.

Despite its central role in guiding development processes during the military-authoritarian period, the Brazilian government has not acted as a homogeneous force in Amazonian affairs. The governmental presence has often proved internally fragmented by different agencies operating at the national, state, and local levels. It would be incorrect to see the central government's drive for national integration as an uninterrupted, unitary process of political centralization. Rather, the central government has expanded its control by fits and starts, reflected in varying and sometimes competing policies, which have left behind different spatial patterns of regional settlement. It is important, we shall argue here, to understand

Figure 3.2. Disarticulated urban systems of contemporary Amazonia

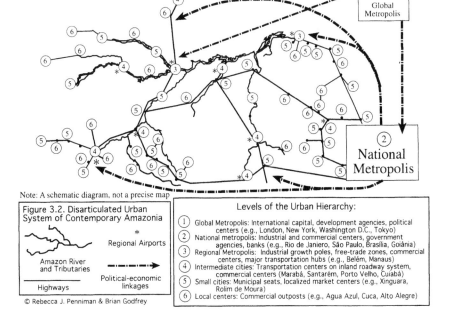

Note: A schematic diagram, not a precise map

Figure 3.2. Disarticulated Urban System of Contemporary Amazonia

*   Regional Airports

Amazon River and Tributaries

▬▬▬ Political-economic linkages

Highways

© Rebecca J. Penniman & Brian Godfrey

Levels of the Urban Hierarchy:

(1) Global Metropolis: International capital, development agencies, political centers (e.g., London, New York, Washington D.C., Tokyo)
(2) National metropolis: Industrial and commercial centers, government agencies, banks (e.g., Rio de Janiero, São Paulo, Brasília, Goiânia)
(3) Regional Metropolis: Industrial growth poles, free-trade zones, commercial centers, major transportation hubs (e.g., Belém, Manaus)
(4) Intermediate cities: Transportation centers on inland roadway system, commercial centers (Marabá, Santarém, Porto Velho, Cuiabá)
(5) Small cities: Municipal seats, localized market centers (e.g., Xinguara, Rolim de Moura)
(6) Local centers: Commercial outposts (e.g., Agua Azul, Cuca, Alto Alegre)

different governmental efforts at integration of the Amazon in terms of their own particular internal rationales and distinctive regional impacts.

In addition, the restoration of civilian power in 1985 reversed the process of political concentration orchestrated by the military-authoritarian regime. Under civilian control, in response to economic pressures from abroad, the centrist state in Brazil became selectively dismembered. These recent developments have led to the fragmentation of regional interests into a cacophony of specific state and local voices. The unifying, regimenting role of the central state has largely fallen away. The region, such as it is, has politically disintegrated.

A brief glance at the spatial distribution of urban centers suggests that the region's variegated settlement structure is not easy to inventory or classify. Urban settlements on the frontier tend to emerge in groups rather than as isolated developments. The internal spatial configuration and socioeconomic orientation of settlement systems have evolved differently in relation to changing programmatic priorities of different government institutions at the federal and state levels.

Broadly, settlement systems can be identified along a sociopolitical continuum. At any given point along the continuum, a distinctive typology of individual settlements may be proposed. At one such generalized point is the populist frontier, areas characterized by colonization by small farmers, independent miners, petty merchants, and others engaged in various forms of labor-intensive activity. The Brazilian colonization and agrarian reform agency (INCRA, then MINTER) has been the leading institution fashioning the populist frontier in diverse regional areas, such as Rondônia and the Transamazon Highway region. The ideological *force motris* of populist expansion has been the national security imperative.

The polar opposite of the populist expansion front in our typology is the corporatist frontier. In the corporatist-driven fronts, spatial organization is dominated by capitalized enterprises, both in public and private hands, pursuing such activities as corporate cattle ranching, agribusiness, large-scale resource extraction and mining, and hydroelectric projects. The Brazilian Amazon development agency (SUDAM) and the state-owned Companhia Vale de Rio Doce (CVRD) have been two of the principal lead institutions promoting the expansion of the corporatist frontier. The ideology supporting corporatist expansion has been characteristically developmentalist. To a large degree the history of contemporary Amazonia is the story of competing populist and corporatist expansion fronts put into motion at specific locations by different factions of the Brazilian state and society. Yet it is important to stress that the Amazon is not neatly subdivided into mutually exclusive social spaces, each oriented to one or the other frontier type. Rather, it is the concentration and intermixing of populist and corporatist social modalities that gives the frontier its unique heterogeneity, its fluidity, and its mosaic like nature.

The distinctive and irregular forms of urbanization found in the Brazilian Amazon today are, in large part, recent historical artifacts of competing bureaucratic, political, and economic interests. To better understand how divergent state actions of the military-authoritarian regime promoted distinctive forms of frontier settlement and urbanization, we briefly review the most important programs of the period, summarized in table 3.1.

OPERATION AMAZONIA, 1966–1970: THE CORPORATIST FRONTIER

During the military-authoritarian period ambitious governmental plans for regional development were motivated by a broad set of geopolitical concerns. First, preoccupation with national sovereignty in a remote, sparsely populated region led to a desire to "plant the flag" through a larger Brazilian presence in Amazonia. Second, it was thought that a better regional demographic balance would be achieved by encouraging permanent migration to and settlement of the interior, diverting growth away from the crowded coast. This resettlement also was intended to help avoid politically charged programs of agrarian reform in heavily settled rural areas, like the Northeast and South. Finally, the state planning apparatus reasoned that the construction of regional infrastructure and the implementation of cattle ranches, agriculture, resource extraction, and other activities would promote economic development in both the Amazon and Brazil generally. The military's ambitious regional development programs sparked massive waves of spontaneous migration to the region, thereby affecting far more people than just the participants in the specific projects themselves (Becker 1990; Mattos 1980; Reis 1982).

Table 3.1 Amazon regional development: Key programmatic periods of military-authoritarian period

Time Period	Operation Amazonia 1966–1970	National Integration Program (PIN) 1970–1974	POLOAMAZONIA 1975–1979	POLONOROESTE 1981–1985
Program Goals	*Secure national sovereignty	*Promote national integration, *Social reform	*Promote regional economic growth	*Regulate spontaneous migration, *Promote regional agricultural development.
Program Areas	Pará, northern Mato Grosso	Transamazon (Pará, Amazonas)	Regionwide (15 growth pole areas)	Rondônia, western Mato Grosso
Lead Institutions	SUDAM, BASA	INCRA	SUDAM, BASA, CVRD, SUDECO, SUFRAMA	INCRA, SUDECO, World Bank
Key Program Instrument	Corporate fiscal incentives	Land settlement, rural credits	Corporate fiscal incentives	Land settlement, rural credits
Dominant Frontier Type	Corporatist	Populist	Corporatist	Populist
Emergent Urban	*Nonlinear array of crossroad settlements	*Rural urbanism, *Irregular network of tenement towns.	Polymorphic array of: *Dendritic export central places *Truncated bimodal subsystems	*Polygonal system of irregular central places

The military's strategic and centralizing designs for the Amazon Basin became apparent in a series of ambitious regional development programs. The very name of the first development plan, Operation Amazonia (1966–1970), connoted the strategic importance of the region for the Brazilian military. Operation Amazonia put into action a post-ISI philosophy of national security and developmentalism: no longer would Brazil be content with merely substituting for foreign imports; rather, the internal market would be actively enlarged through national integration. To replace SPVEA, regarded as corrupt and tied to vested local interests, the military regime created a new regional agency, the Superintendency for Amazon Development (SUDAM), more amenable to central control by Brasília. In addition, a new regional development bank, the Bank of Amazonia (BASA) was to channel development funds and fiscal incentives to the region. A free trade zone was established in the city of Manaus.

Thus, SUDAM and BASA jointly worked to create a corporatist frontier in Amazonia based on investments by the state and private capital.

Operation Amazonia basically sought to lure private investment to the region through the construction of new infrastructures and the availability of various fiscal incentives for corporations. The financial benefits of investing in Amazonia included a reprieve from corporate income taxes for a period of ten to fifteen years, lower import duties on materials needed for land development, and tax credits allowing up to fifty percent deductions on total liability, even for enterprises outside the region, as long as the savings were invested in approved Amazon development projects (Browder 1988a; Hecht 1985; Mahar 1979; Revkin 1990). In practice, SUDAM's first five-year plan, covering the period of 1967–1971, emphasized the construction of transportation infrastructure to improve access to the region. Transport projects received 40.5 percent of the total SUDAM incentives budget during this period; on the other hand, only 1.4 percent of the agency authorizations were allocated to small-farmer colonization (Millikan 1988:32). Federal policy took a generally laissez-faire approach to small-farmer settlement throughout the 1960s. As in the case in the Belém-Brasília Highway, peasant farmers in areas such as southern Pará and Rondônia received little public assistance and generally were displaced by speculative cattle ranchers (Hebette and Acevedo 1979; Ianni 1979). The new transportation infrastructures of Operation Amazonia served to intensify the migratory influx into the region and to encourage the further growth of cities. Not coincidentally, Amazonia's urban transition of the 1970s followed the military-authoritarian regime's initial efforts at regional integration.

THE PROGRAM OF NATIONAL INTEGRATION, 1970–1974: THE POPULIST FRONTIER

The next military plan for Amazon development was launched by General Emilio Médici, president of Brazil from 1969 to 1974. In 1970 Médici proposed a National Integration Program (PIN), featuring a utopian colonization scheme along an extensive new road corridor cut linearly through the central Amazon, the Transamazon Highway. Ultimately, PIN was to be the most ambitious colonization project in the region up to that time. As opposed to Operation Amazonia, PIN was motivated by social concerns (landlessness, famine, poverty in the Northeast) and actively encouraged small-farmer settlement as an important factor in national security. The National Institute of Colonization and Agrarian Reform (INCRA) was created in 1970 to oversee an orderly agricultural settlement of Amazonia. For a brief time INCRA assumed federal control of the planning and settlement of the Transamazon Highway scheme, PIN's centerpiece. Eventually, however, political tides turned and INCRA was subordinated to SUDAM and increasingly powerful state land agencies in Pará (Browder 1988a; Fearnside 1978, 1984; Moran 1976, 1981, 1984a, b; Smith 1976, 1981, 1982).

In the first five years of the program, one hundred thousand families (more than five hundred thousand persons) were to be settled along the three-thousand-kilometer road, stretching from the Brazilian Northeast to the western edge of Amazonia at the border with Peru. After a decade, a million families (more than five million persons) were to be resettled on the Transamazônica INCRA 1972). In short, the Transamazon Highway represented an attempt, as President Médici put it, "to settle a people without land in a land without people." Yet by the mid-1970s, when new resettlement was virtually discontinued, only about 7,500 families—roughly 7.5 percent of the projected number— had officially been settled INCRA 1975) there.

Plate 3.3. View of a colonist's house on the Transamazon Highway, between Marabá and Altamira, Pará, 1977. Burned Brazil-nut trees stand like skeletons against the sky in the practically cleared colonist lot.
Photo: Brian J. Godfrey

The Médici administration expropriated a hundred-kilometer-wide strip on each side of the Transamazon Highway, of which a ten-kilometer-wide section beside the road was divided into lots of 100 hectares each for distribution to incoming colonists (plate 3.3). The remaining public lands were reserved for sale to private companies in lots ranging from three thousand to sixty-six thousand hectares. The largest concentrations of population on the Transamazon Highway were in the eastern state of Pará, particularly between the cities of Marabá, Altamira, and Itaituba; in addition to the lands set aside for small farms, about 3.7 million hectares along the stretch of the road from Marabá to Itaituba were cleared for cattle ranches. In the state of Amazonas, large stretches of the road remained virtually uninhabited (Smith 1976 and 1982).

The colonist families could buy lots of one hundred hectares each from INCRA for about U.S.$700, repayable over a twenty—year period. Many colonists also were provided with a four-room, wood-frame house for about U.S.$100 (Smith 1976). According to INCRA's estimates the government investment per settler family was the equivalent of about U.S.$4,333 (at the official exchange rate for 1973 dollars) and included the "value of lands, settler's house, topographic survey, part of the road along the frontage, machinery, transportation of settler and his family, hand tools, six month's support credit, medicines, seeds" INCRA 1973).

In addition, INCRA designed an ambitious settlement system of planned service centers along the road. Known as rural urbanism (*urbanismo rural*), this plan consisted of a tier of lower-order settlements, called *agrovilas*, of forty-eight to sixty-four families to be built every ten kilometers along the highway (plate 3.4). A middle-tier of *agrôpoles* were to be located one hundred kilometers apart as intermediate administrative centers. At the top of the Transamazon's urban hierarchy were *rurôpoles* of up to twenty thousand inhabitants every two hundred kilometers to serve as commercial and administrative centers for the region (Cunha Camargo 1973). In this fashion a hierarchy of central places was to be superimposed onto the settlement frontier along the Transamazon Highway. By 1976 two *agrovilas* had been constructed in the Marabá project, while twenty-two *agrovilas* were in operation along the segment between Altamira and Itaituba. One *agrôpolis*, Brasil Novo, had been constructed at kilometer forty-six near the eastern entrance to the highway, and the *rurôpolis* Presidente Médici had been built at the junction of the Transamazon and the Cuiabá-Santarém Highways.

Theoretically, the spatial organization of the rural-urbanism scheme was supposed to be flexible, but in practice the division of colonist lots along the highway followed preconceived geometric forms, taking no account of microgeographic variations in topography, soil types, and access to potable water. Agricultural lots along the highway had five hundred meters of frontage and were two thousand meters deep. The agricultural lots of farmers settling in *agrovilas* were the same hundred-hectare size, but they were located up to ten kilometers away from the official family residence in the planned village center. Since the farm lots located far away from the family house were inefficient for labor inputs and exposed untended crops to hazards of pests and poaching, many *agrovila* colonists therefore constructed improvised shelters on their lots and returned to the service centers only occasionally. Most of the *agrovilas* fell into a state of abandonment and disrepair, and one journalist referred to them as "rural slums" (Rodriguez Pereira 1975).

In late 1976 six of the original eight agrovilas in the Altamira project had an occupancy rate roughly equal to or less than 50 percent, and only two showed most of the residences occupied.[1] The two more fully occupied

Plate 3.4. The agrovila Alto Turi along the eastern Transamazon Highway in Maranhão. It illustrates the radial agrarian settlement model of the PIN philosophy of "rural urbanism." *Photo: Robert Skillings*

agrovilas were located conveniently close to the highway, where colonists were able to rent their houses to either migrant workers or government officials. Production indices in the *agrovilas* away from the highway proved to be quite low compared to that of road-front lots. A survey of two *agrovilas* located twenty-three kilometers from the highway in the Altamira project showed a dismal production of 66 sacks of rice per colonist family, as compared to a mean of 256 sacks on the road-front farm lots.

Clearly spatial organization and the location of agricultural plots were important factors in the disparities of income among colonists along the Transamazon Highway, penalizing those located away from the favored roadside lots. During the heavy rains of 1975 the feeder roads off the Transamazon Highway were impassable, leaving about two thousand settler families away from the main highway without means of selling or storing their produce. According to INCRA estimates, 20 percent of the 900,000 sacks of rice and corn produced in the Altamira project were lost (Rodriguez Pereira 1975).

In terms of interregional migration patterns, original government plans called for a 75 percent component of settlers from the states of the impoverished, drought-plagued Brazilian Northeast. Regarded as hardworking but ignorant of modern production and marketing techniques, the Northeasterner was expected to benefit from the example of what were considered more prosperous, commercialized settlers from the South. Yet the number of Northeasterners along the Transamazon Highway fell steadily from 67 percent in 1972 (Smith 1976) to 41 percent in 1975 (INCRA 1975). The displaced Northeastern colonists did not necessarily leave the Transamazon Highway: many *nordestinos* sold their lots to incoming farmers from the Brazilian South, especially to *gaúcho* migrants from the state of Rio Grande do Sul, and remained on the land as wage-laborers or *peões*. Interviews among 101 wage-laborers on 78 colonist lots indicated that most of the migrant workers came from the Northeast, mainly from the states of Maranhão (33 percent) and Ceará (28 percent).[2] These laborers were generally young males, more than half of them between the ages of twenty and thirty, who served as a mobile rural workforce on the expanding Amazonian agrarian frontier.

The estimated remigration rate among Transamazon Highway colonists within the first five years of settlement was about 25 percent. Given this high turnover rate, the growing disparities of income along the road, and the relatively small number of colonist families established on the road, the Transamazon Highway indicated that such official colonization schemes would be incapable of providing a mass transfer of Brazil's rural population. The relatively few colonists who managed to create profitable operations along the Transamazon Highway constituted a minute class of small farmers, contrasting with the hoards of landless agriculturalists who entered the region during this period to join the swelling ranks of day workers.

Such basic factors as farm location, soil fertility, and type of subsistence strategy were important in the settlers' adaptation to the Transamazon Highway (Moran 1981; Smith 1982). The failure of PIN to reach its settlement goals stemmed in large part from an utter ignorance of local realities on the part of the geopolitical strategists and regional planners who had so optimistically laid out the program. In addition to the technical flaws in the rigid spatial organization of the Transamazon Highway, the competition of other socioeconomic forces for land and power also worked against the implementation of a truly massive program of official small-farmer resettlement in the Amazon.

The record of the Transamazon Highway indicated that the "frontier solution" of demographic displacement served more as a diversion than as an effective policy response to problems of underdevelopment in Brazil's older rural communities. The utopian notion that entire settlement systems, inspired by central place theory, could be planned and implanted wholesale in the rainforest landscape proved to be ridiculously naive. Nevertheless, PIN served strategic purposes: by justifying the construction of new transportation and communications infrastructure in the region, displacing local native groups, and providing a reservoir of migrant laborers, the program prepared the way for the corporatist frontier of the late 1970s. This frontier witnessed a concentration of the population in frontier tenement towns and a consolidation of lands into large holdings.

POLAMAZONIA, 1975–1979: RETURN OF THE CORPORATIST FRONTIER

The administration of General Ernesto Geisel (1974–1979) deemphasized the resettlement of small farmers in the Amazon and reinstituted the program of fiscal incentives to encourage corporate investment in the region. The Second Amazon Development Plan (1975–1979), or POLAMAZONIA program, identified fifteen areas in the region considered suitable for large-scale development projects. POLAMAZONIA witnessed the rise of pseudo-pragmatic growth center models and the return of monumentalist regional developmentalism, the ideology promulgated by the region's various development institutions (SUDAM, SUDECO, SUFRAMA, the Companhia Vale do Rio Doce). Lands formerly intended for "men without land" reverted to large-scale enterprise, and corporate cattle ranching became the region's privileged investment sector (Browder 1988a; Hecht 1985). By 1983 the real value of all government subsidies and tax breaks to Amazon cattle ranches exceeded U.S.$2.5 billion (Browder 1988b). Additional billions of dollars in (mostly) public and private capital found their way into various megadevelopment ventures in the region.

POLAMAZONIA emphasized a corporatist mode of frontier hegemony, administered largely by the self-interested, bureaucratic agendas of SUDAM, SUFRAMA, and SUDECO. These institutions, representing large national and multinational investors, encouraged the formation of truncated, polarized urban systems. A distinctive form of urban settlement system emerged from the corporatist frontier—planned corporate rainforest cities and the squalid squatter settlements that spontaneously appeared as satellite slums surrounding them. Initially, the corporatist frontier landscape imposed clear physical and socioeconomic divisions between the planned corporate center and the spontaneous popular periphery. Over time these divisions tended to collapse and sometimes dissolve in tandem with the spatial vacillations of capital movement and government action in Brazil, creating an amalgamation of previously separated social places. Cities clearly articulated to social class and position blend into a socially and locally disarticulated jambalaya (chapter 4). The case of one corporate city, Tucumã, and its sibling slum, Ourilândia d'Oeste in the Xinguara–São Félix do Xingu corridor of southern Pará illustrates the dynamics of urban change in the corporate frontier (chapter 7).

While inviting global capital into the Amazon during the corporatist period of the Second Amazon Development Plan, the Brazilian government could not entirely ignore less privileged, populist sectors. In response to populist pressures for agrarian reform in the South and Southeast, a very different frontier experience unfolded in Rondônia during the 1970s. The Northwest Region Development Program (POLONOROESTE) was groomed from the template of growth pole ideology, POLAMAZONIA writ large. Yet, this episode of frontier expansion and its implications for urbanization were markedly different; populist forces in agriculture were conceded a significant portion of the frontier. Under the aegis of Brazil's agency for land reform and colonization (INCRA) and with financial participation of transnational capital from the World Bank, Rondônia became a populist agrarian frontier.

POLONOROESTE, 1981–1985: EXTENSION OF THE POPULIST-AGRARIAN FRONTIER

Although arising from POLAMAZONIA, the policy framework of the corporatist frontier, the institutional and ideological content of expansion into Rondônia was decidedly populist in nature. Spontaneous migration to Rondônia of landless farmers, land speculators, and gold and tin miners from the South and Southeast had been occurring since the early 1960s. The mass consolidation of family farmlands into corporate soybean, citrus, and sugar plantations in these regions during the early years of the military-authoritarian period added momentum to the settlement process in Rondônia. In the closing years of the 1970s more than two thousand families per month were crossing the border from Mato Grosso to seek forestland in Rondônia. In 1981 the World Bank financed the Northwest Region Development Program (POLONOROESTE) to help bring order into the chaotic process of spontaneous settlement that had overwhelmed INCRA planners.

Unlike the disjointed, typically bimodal urban subsystems that appeared on the corporatist frontier of southern Pará during the late 1970s, the urban system emerging in Rondônia during this period was geometric. Indeed, the marketing principle of central place theory provides the initial underlying logic for spatial organization in most of the settlement areas of Rondônia during the 1970s and 1980. Disarticulated pockets of self-contained urbanization could be found in mining sectors of the state and between the contemporary settlement frontier and the older riverine settlements established during the rubber boom. We return to a more detailed discussion of Rondônia and the populist frontier in chapter 6.

Recent years have witnessed a further fragmentation of regional development policy in Amazonia. Various new sectors emerged in the 1980s and 1990s, selectively reshaping growth in particular areas. The small-scale *garimpeiro* mining sector became a leading economic sector in several regions, as discussed in chapters 7 and 10. Hydroelectric dams became the new megaprojects of choice for development planners, creating sizable growth poles and equally significant environmental controversies, as noted in chapters 4 and 10. Throughout the region, informal settlement on the urban periphery took the form of massive shantytowns. This contemporary mix of growth sectors has reinforced the over-all pattern of disarticulated urbanization in Amazonia.

SUMMARY OF CONTEMPORARY TRENDS IN AMAZON FRONTIER URBANIZATION

Several trends emerge from this review of contemporary urban location and development in Amazonia. First and foremost is the role of the public sector in regional planning and social overhead investment. The vast majority of new towns emerged within the sphere of influence of the three major highways built by the Brazilian government through the *terra firme* of Amazonia over the last twenty-five years—the Belém-Brasília, the Transamazon, and the Cuiabá-Porto Velho Highways. Yet the Brazilian government's effect on urbanization in Amazonia has been neither uniform nor predictable. In some instances, urban systems have been imposed upon a natural landscape. For example, the preposterous "rural urbanism" scheme for the Transamazon during the early 1970s reflected the Brazilian generals' misunderstanding of basic regional economics. Within twenty years, the grand rural urbanism stratagem for the conquest of the Amazon under PIN had deteriorated into a series of local urban-rural disasters. In Rondônia, government planners also worked feverishly to plan entire urban systems. But unlike the heavily directed Transamazon scheme, Rondônia began as a populist melee. While many town sites in Rondônia were selected (often arbitrarily) by INCRA planners, the larger market forces regimented Rondônia's settlement system into formations pat-terned, albeit imperfectly, after central place principles.

Second, many new towns emerged spontaneously, especially at the intersections of national and state highways and local feeder roads. Many such nodes provide a limited range of functions associated with their strategic locations, such as auto repair, restaurants, hotels, and other ser-vices, and they often support only a small resident population with, pre-sumably, limited potential for growth or structural development. However, other nodes at network junctions have "taken off" and become important locations for extractive forest industries, crop processing, regional commerce, banking, and government services. The selectivity of urban settlement survival remains an on-going research concern. Yet the comparative history of contemporary Amazonian urbanization should indicate that extraregional linkages to strategic locations (i.e., seemingly obscure crossroads) are not accidental, even if the emergence of some rainforest settlements seems spontaneous.

Third, the vicissitudes of resource extraction by independent operators—facilitated by the state's investment in infrastructures—also greatly affect the location, rate, and rhythm of frontier urbanization, further perplexing any taxonomic device based upon dichotomous social categories. Indeed, the highest rates of urbanization in contemporary Amazonia are found in the inland settlement frontiers at early stages of timber extraction, mining, and other resource sectors. Although many expeditionary resource settlements are ephemeral, some do evolve into local service centers. In addition, a significant proportion of the labor force in the larger municipal or county-level service centers now engages mainly in resource extraction and related commercial activities. Public policy would be well directed at ameliorating severe environmental problems, such as mercury contamination in the small-scale gold mines, and at providing basic social and health services for the migrant populations.

Finally, the most obvious and fundamental conclusion must be stated: frontier expansion in Amazonia has not resulted in a single or unified settlement system. As different frontiers have evolved over time, each articulated differently to the national economy through the mediation of distinct governmental institutions, so have regional urban systems diverged. The next chapter further develops a conceptual framework to distinguish the salient dimensions of contemporary Amazonia's two prevailing frontier types, populist and corporatist expansion fronts.

Notes

1. Data on Transamazon Highway colonists were collected in collaboration with the disease surveillance program of the Pan American Health Organization (PAHO). Brian Godfrey accompanied the PAHO field teams during their bimonthly visits to colonists along the road from June to November of 1976, administering questionnaires to approximately two colonist families in each five-kilometer *gleba* (ten roadside lots) along the road, which PAHO had previously selected as a representative sample of the total settler population. Only the author, of course, is responsible for these data. This material originally was presented by Brian J. Godfrey in a paper at the Conference of Latin Americanist Geographers, "Brazil's Transamazon Highway: Migration and Agricultural Production on the Amazon Frontier," Sonoma State College, California, June 8–9, 1978.

2. These interviews were carried out from November 16–20, 1976, by Brian Godfrey.

4

A Pluralistic Theory of Disarticulated Urbanization

The lack of comprehensive theories of urbanization is perhaps not surprising.
Less excusable is our failure to articulate a common framework that
would enable us to relate individual city studies to the larger vision of
process and structure in the formation of human settlements.
John Friedmann (1988:80)

Place has a kind of fluidity. . . . It is a mosaic within larger mosaics. . . . The subtle
and many-layered cosms of the universe have found their own way into symbolic
structure and have given us thousands of tawny human-language grammars.
Gary Snyder (1990:77)

Having reviewed both the major theoretical perspectives on urbanization (chapter 2) and the historical background of Brazil's Amazon expansionism (chapter 3), we now present our own theory to explain the diverse urban forms found in Amazonia. After revisiting the concepts "urbanization" and "frontier," we outline a conceptual framework inspired by recent innovations in ecological rationality, evolutionary theory, and critical realism. Because no single master principle fully explains Amazonia's urban diversity, we propose a pluralistic theory of "disarticulated urbanization." This framework allows us to view the frontier as a confluence of diverse social spaces, a sociospatial continuum spanning two ideal types: populist (agrarian) and corporatist systems. We do not suggest that these are the only types of frontier systems found in Amazonia or that their spatial incidence is mutually exclusive. Indeed, the character of the frontier at any given time will likely entail some mix of both types leading, at the regional level, to a pattern of disarticulated urbanization.

URBANIZATION

Traditionally, urban geography has been concerned with the study of three main phenomena: the emergence of new urban settlements, the formation of city-systems and urban networks, and the growth of the urban population in existing centers.[1] We have observed that the contemporary trend toward growth of the urban population has been constant throughout the Amazon region, varying only in terms of growth rates (chapter 1). The processes by which the frontier population has become urbanized, however, are not regionally uniform, but differ according to the distinctive types of social spaces that have emerged in Amazonia. Similarly, the location of individual settlements and the evolution of settlement systems reflect distinctive patterns from place to place that cannot be understood without considering the institutional history of each place in the frontier. Different government institutions, corporate interests, and local social groups continuously redefine their interrelationships as they alternately compete and collaborate for resources on the frontier. Urban centers in the frontier go beyond being simple theaters for mobilizing capital in the historical project of global capitalist accumulation (Armstrong and McGee 1985). Such centers become arenas in which these diverse social groups strive for economic and political power. Ultimately, the social orientation of the Amazon frontier is influenced by forces emanating from local, regional, and global levels.

THE "FRONTIER"

The challenge of formulating a coherent conceptual framework to explain Amazon urbanization is heightened by the inherent ambiguity of the term *frontier* as an analytic category. Historically and culturally based definitions are often locally specific and typically ignore the larger commonalities of frontier experiences. Other problems arise from considerations of scale. By shifting from local to regional levels of analysis, the very character of the frontier appears to change. Similarly, ambiguity arises from the different economic activities that characterize different areas of the frontier. Is an "extractive frontier" based on industrial bauxite mining fundamentally distinct from an "agricultural frontier" based on low-input family farms? At the closest level of resolution these are vastly different social spaces. At a more abstract level, however, both are based on the extraction of energy and differ only in form and intensity. What, then, is a *frontier*?

In an effort to enhance the analytical utility of the term *frontier*, it is useful to distinguish five criteria that define the term in practice: the demographic, political, economic, social, and cultural. The first criterion is *demographic*. All frontiers are characterized by relatively low population densities and high population growth rates in both urban and rural zones, due mainly to in-migration. However, populations at specific locations within the frontier tend to be unstable and boomtowns and ghost towns are common settlement conditions.

Second, frontiers are creations of the *political* life of nations and may be distinguished by their institutional histories, particularly in those cases where state action is central in opening new regions to national occupation. An area that is opened by the government's colonization agency is likely to be very different in composition, character, and function than one opened by the state-owned mining enterprise. The frontier bears the imprimatur of the institution(s) that open and organize it.

Third, frontiers are typically characterized by the extractive *economic* base supporting their occupation. The generic term *resource frontier* or *extractive frontier* dissolves into more specific commodity fronts—such as coffee fronts, mahogany fronts, and gold-mining fronts—that typically exist for relatively short cycles. However, in each cycle the social character and economic organization of the frontier space may be quite different.

Fourth, a frontier is defined by reference to the predominant *social* group or mode of production found there. Areas dominated by small farmers might be considered agrarian frontiers. Areas in which corporate forms of occupation prevail might be considered corporatist frontiers: for example, large-scale cattle ranches, mining projects, and plantation agriculture. Of course, the frontier is rarely carved up into neat and mutually exclusive social spaces. In Amazonia the intensity of resource competition and social conflict gives rise to what Marianne Schmink and Charles Wood (1992:13) call contested frontiers:

> Contested claims to land, gold and timber existed simultaneously across the landscape, as did competing forms of labor control and political authority. In addition, the actions of regional elites and the grassroots mobilization of peasants, Indians, rubber tappers, and

independent miners repeatedly subverted the military's agenda and the institutions that large capital attempted to impose on the region. The result was not a single process of linear change but instead a diversity of contested frontiers with highly varied outcomes.

Finally, some frontiers possess a metamorphic quality as a meeting place of distinctive *cultures*. For the North American poet Gary Snyder (1990:14), "[a] frontier is a burning edge, a frazzle, a strange market zone between two utterly different worlds." The spatial edges of a homogeneous ethnolinguistic or religious community represent cultural frontiers, the geographic boundary beyond which cultural changes in the landscape quickly appear. The frontier as a momentary experience of conjuncture between distinctive cultures at a given place is also a moment of cultural ambiguity.

In addition to the above criteria, the concept of the frontier as an extension of the national space economy has a finite temporal connotation. It is usually assumed that frontiers only exist for relatively short periods of time. Once fully incorporated into the national space economy, there is little to remind one of the frontier past. This conceptualization is Turnerian in essence. Eventually the frontier "closes." Based on this concept of a discrete opening and closure, frontier studies have sought to describe the process of transition from frontier to something else. This process is usually depicted as a general sequence of stages. Simple conceptualizations of the frontier change as a linear progression toward some "closed" or urbanized state, grossly oversimplify a more complex reality. We resist adopting strictly evolutionist approaches. Yet we are also unsatisfied with the frontier conceptualized as a riot of unpatterned forces.

We suggest that the frontier be viewed as a sociospatial continuum along the extensive geographic margins of national space in which different socioeconomic groups coexist. The interactions between these groups, often mediated by state institutions, give different settlements on the frontier continuum their own specific social characters, which may vary from place to place in several respects: in degree of intensity of capitalization and energy consumption, in nature of social organization and political participation, in impact on the natural environment, and in urban form. The urban systems that coevolve with these dominant groups will also vary in their morphology, functions, and linkages to the outside world. Hence, the frontier is a heterogeneous and fluid array of social spaces. We posit that no single theory adequately explains the diverse patterns of urbanization found in contemporary Amazonia, and so we reach toward an ecological pluralism of regional analyses, gathered together in a framework of disarticulated urbanization.

TOWARD AN ECOLOGICAL PLURALISM IN REGIONAL ANALYSIS

"Pluralism" is a loaded word, and our use of it requires some explanation. First, the term is not used here in the traditional historical anthropological sense to describe the spatial conjunction of enclosed, self-regulating social groups under a dominant rule (Kuper and Smith 1969). Nor does our use correspond to widely held political applications as indicating, for instance, a "polyarchic" form of governance (Dahl 1956) or a critique of the basis of state authority (Nicholls 1975). Rather, we seek a postmodernist ecological meaning of pluralism. The philosopher Frederick Ferré (1988:94) observes: "Postmodern images of reality that come from ecology portray the world as an endlessly complex network of organic and inorganic systems locked in constant interaction." The outcomes of this interaction are not easily predicted.

For at least the last forty years, the discourse on development and underdevelopment, framed by an oligopoly of intellectual orthodoxies, has failed to produce a conceptual synthesis that preserves an awareness of the nuances and details of social phenomena in dynamic interaction. This failure of hegemonic theory leads many social scientists, ourselves included, to adopt what Peter Wenz (1988) calls a "pluralistic theory," containing "a variety of principles that cannot be reduced to or derived from a single master principle." The preeminent physicist John Archibald Wheeler (quoted in Swimme 1988:51) demands that the details of nature be given the same attention we give to the unifying theories: "There is no law [in nature] except the law that there is no law." These suppositions do not imply the absence of structure or patterns in the phenomena we seek to explain. Yet, the theoretically guided search for regularity in experience has become a target of the postmodernist critique of objective science and its fundamental reliance on what Weber called "ideal types" (Bendix and Roth 1971). "Every significant theoretical concept reflects an idealization . . . not directly met in the unkempt world of experience," writes Ferré (1988:88).

Our efforts to elaborate a pluralistic conceptual approach to frontier urbanization and regional analysis take note of analogous developments in three other fields: ecological rationality, evolutionary biology, and human geography (structuration theory and critical realism).[2] Because of the influence of postmodernist ecologists on our thinking, both in terms of framing the subject matter of urbanization into a systems perspective and in providing a transdisciplinary argument for pluralism, we denominate our approach *ecological pluralism*.[3] We do not, however, suggest a return to the simple biological metaphors of the Chicago school of human ecology or the mechanistic Spencerian notion in sociology that societies undergo processes of change that replicate those observed in the natural world.[4]

Ecological Rationality

The movement toward conceptual pluralism in regional analysis adopts three principles of ecological rationality.[5] First, like ecosystems, urban systems exhibit a high degree of *interpenetrability*. Every urban system exchanges materials and energy with other urban systems. Urban systems are necessarily open systems, and their economic and social boundaries shift around over time. Second, the extent of interpenetration within and among urban systems produces *emergent properties*. An emergent property is any characteristic of a system that is not predictable from a knowledge of the elements of that system (Dryzek 1987:26). The introduction of exotic social groups into an ecological (or regional) system may induce systemic changes in the form of emergent properties transforming the system and giving it a new or hybrid character. Social speciation (the formation of new or hybridized indigenous social groups) may also occur internally through processes that cannot be reliably predicted on the basis of theories about the nature of the system itself. The region as a homogeneous field of regularly placed settlements is subject to continuous flux, especially over the long run, as the result of emergent properties. The third principle concerns the evolutionary tendency toward organized *complexity*. "Organized complexity involves systematic and variable interactions among the elements of a system. The behavior of such systems cannot readily be captured on a statistical basis, let alone subsumed under general laws" (Dryzek 1987:28–29). As urban centers grow in size, they become more diverse in social composition and more complex in function.[6]

Our analytical use of ecological rationality has clear limits. The principle of self-regulation, the ability of ecosystems to maintain their essential structures and functions (i.e., homeostasis) in the face of exogenous shocks (Dryzek 1987:27), is an unlikely principle in any framework of urbanization. Immanuel Wallerstein (1991:104) suggests we recognize "the inherent lack of long-term equilibria in any phenomena—physical, biological, or social." Nonetheless, ecological rationality provides a useful corollary for understanding some key aspects of urbanization through a pluralistic lens. Urban systems as interpenetrated, dynamic systems exhibiting emergent properties cannot be neatly explained by concave theories through which circular rationalities lead to conceptual closure upon a single point, a master principle.

The New Biology

A second discourse supporting conceptual pluralism may be found in recent innovative advances in evolutionary biology.[7] In their provocative essay on pluralism and convergence in evolutionary theory, Ho and Saunders (1984:5) write:

> Pluralism is a predominant feature of the emerging paradigm of evolution. . . . Evolution is a complex phenomenon and it is to be expected that different kinds of explanations will be appropriate to different aspects. In particular, higher level explanations cannot always be collapsed or reduced into lower level ones. Thus, pluralism ought in principle to be a permanent feature of evolutionary studies.

Much the same must apply to the social inquiry into regional change. Ho and Saunders (1984:261) conclude that "there will be different theories and, indeed, types of theory, each dealing with some different situation or problem." In a similar vein, evolutionary biologist and natural historian Stephen Jay Gould (1995:9) disputes the view of evolution as predetermined, pursuing only sensible directions: "Evolution is only the summation of its fortuitous contingencies, not a pathway with predictable directions."

Speaking to a broader audience Samuel Hays (1987:247, in Oelschlaeger 1991:281) echoes much the same idea: "Environmentally oriented inquiry has not led to a single system of thought such as social theorists might prefer, and it would be difficult to reduce its varied strands to a single pattern." These are more than convenient biological metaphors for the present study. They reflect an interdisciplinary undercurrent of discontent with the epistemic status quo, dominated by hegemonic master theories.

Human Geography: Structuration Theory and Critical Realism

A third set of influences upon our conceptualization of ecological pluralism originates in the social sciences with "structuration theory" (Pred 1983, 1984, 1985; Thrift 1983, 1985; Giddens 1979, 1984, 1990; Gregory 1986) and "critical realism" (Bhaskar 1975, 1986; Harré 1986; Sayer 1984, 1989).[8] Two aspects of structuration theory draw our attention. First, Anthony Giddens, the principal progenitor of structuration theory, argues that social structures and human agency are mutually enabling and dialectically related dimensions of social life, changing each other over time (Giddens 1979:69; Dickens 1992). As we shall see especially in the case study of the contested corporate frontier (chapter 7), one social space transforms into another over time as a result of the interaction of social structure and human agency. This position provides yet another basis for conceptual pluralism. Second, we are drawn to the writings of geographer Allan Pred on the nature and constitution of "place." Pred (1985:338) describes place as a "historically contingent process": "Any place or region expresses a process whereby the reproduction of social and cultural forms, the formation of biographies, and the transformation of nature and space ceaselessly become one another . . . in ways that are not subject to universal laws but vary with historical circumstances" (Pred 1985:344, quoted in Cloke et al. 1991:117). Structuration theory, in seeking to define the middle ground between universal laws and individual agency, emphasizes the importance of emergent tendencies, novel generative structures, and, in general, contingency in causality. Peter Dickens (1992:xv) writes, "these [generative] structures and tendencies are not observable in an unmediated form. They emerge and combine in complex ways with contingent relations and tendencies." These positions, although by no means universally agreed upon by social scientists, are among the ones that provide an epistemological basis for the conceptual pluralism underlying our theoretical framework of disarticulated urbanization.[9]

Critical realism arises as a critique of positivism's search for, and prediction of, empirical regularities and its claims to universal statements about the causes of those regularities.[10] Cloke, Philo, and Sadler (1991:146), argue that "realism presents an alternative [to positivism] by assuming a stratified and differentiated world made up of events, mechanisms and structures in an open system where there are complex,

reproducing and sometimes transforming interactions between structure and agency."

These ideas argue for a plurality of theoretical insights: "Concrete
research invariably has to mobilize and integrate concepts from several
theories, in order to capture the many-sided nature of its object" (Sayer
1989:223 in Cloke et al. 1991:167). Not surprisingly, such an "elastic
epistemology" has attracted some formidable criticism from both struc-
turalists and humanists who view it as being easily assimilated into other
broader epistemologies (including positivism) for its contradictory
reliance on empiricism and for its dangerous tendency to reduce causal
explanation to a mass of contingencies in a "cacophony of voices" (Harvey
1987; Cloke et al. 1991). While we recognize these shortcomings of realism as a coherent epistemological coda, or as a substitute for positivism,
we acknowledge its contribution as critique. We tend to concur with
Cloke, Philo, and Sadler's (1991:168) conclusion that realism is "an
accommodating and synthesizing epistemology that asks new questions,
prompts a more sensitive analysis and participates in the waymarking of
new directions from which to approach human geography."

Our theoretical framework, disarticulated urbanization, is founded on
what we see as an emergent interdisciplinary movement toward conceptual pluralism. Our study acknowledges the contributions and limitations
of the leading theoretical positions on urbanization and frontier expansion (reviewed in chapter 2). We recognize the importance of beginning
this study with a conceptual framework that enables readers to identify
the patterns of pluralism on the social landscape of Amazonia. Our framework consists of two sets of constructs: analytical and socio-spatial.

ANALYTICAL CONSTRUCTS AND MULTIPLE
LEVELS OF ANALYSIS

Our analytical construct recognizes the multidimensionality of social
phenomena and the inadequacy of master principles and ideal typologies,
especially those that tend to focus on a single level in social analysis. This
view also takes some further inspiration from ecology. As the prominent
ecologist Eugene P. Odum states:

It is important to emphasize that findings at any one level aid in the
study of another level, but never completely explain the phenomena occurring at that level. When someone is taking too narrow a

view, we may remark that "he cannot see the forest for the trees." Perhaps a better way to illustrate the point is to say that to understand a tree, it is necessary to study both the forest of which it is a part as well as the cells and tissues that are part of the tree. (Odum 1983:4, quoted in Ferré 1988:94)

As outlined in chapter 2, some theoretical approaches (e.g., central place theory, capitalist penetration theory) focus on the local levels or micro-units of analysis (the relationships between the cells and tissues in Odum's analogy). Other approaches dwell on the macro-level range of forces, the big picture of the "forest" (e.g., world systems theory, mercantile theory). And others, to varying degrees, emphasize what might be called meso-levels of analysis, the ambiguous mid-range of the region and sectors of activity (e.g., diffusion theory, intersectoral articulation theory). None alone, however, quite captures the integrated essence of the phenomenon of urbanization. We do not suggest that our analytical construct easily overcomes these limitations. But we urge upon our readers a perspective that allows awareness of the importance of those nuances and details at various levels of analysis, those "events beyond law" (Wheeler, quoted in Swimme 1988:51) that slip between the cracks of ideal types and defy understanding through hegemonic master theories.

Spatial Levels of Analysis

As a simple heuristic device, we suggest that the phenomena of urbanization in different places be seen as differentially articulated to forces emanating at three highly general spatial-functional levels of analysis: local, regional/national, and global. These are obviously relative and permeable concepts, and consistent with the ecological theory of organic systems we would expect events arising at one level to have repercussions for events at other levels. We would not expect, following Odum, a theory that is adequate in explaining the emergence of a phenomenon at one level to be also adequate in explaining its emergence at another level, hence our plea for conceptual pluralism.

Local levels include households, land-based communities (e.g., municipalities) and environmental resources that have fixed spatial distributions specific (or endemic) to certain comparatively small areas. At the local level, the courageous or visionary actions of individuals may be more important in shaping the spatial distribution, growth, and functions of urban settlements than the grand forces arising from the level of the global economy. Local growth coalitions include kinship groups, labor unions, cooperatives, church congregations, political organizations, and charismatic leaders. Such coalitions are important personae shaping spatial organization in the frontier as well as in established urban settings (Rudel and Horowitz 1993; Molotch 1976; Logan and Molotch 1987). While such local actors are often acknowledged in social analysis, they are typically reduced to consumers maximizing utility (central place theory) or subordinated to a master principle, such as class conflict (Marxist capitalist penetration theory). In practice, such actors can behave with considerable autonomy and in ways that are often unpredictable and unpatterned; that is, their actions seem disarticulated from the logic of the hegemonic paradigms.

The global level of analysis needs little clarification to most readers. This is the domain of the transnational enterprise, often in confederation with national corporate interests and specific governmental institutions. Here we also find multinational lending institutions, international networks of nongovernmental organizations, international religious organizations, and occasionally bilateral governmental institutions.

The meso-level or mid-range is a more problematic domain to define. Generally, we associate regional phenomena with the midlevel. State and regional government institutions, regional political and social movements, and trans-local/subnational organizations representing specific business interests constitute this level of analysis. Here, too, we situate national economic and political groups and institutions of governance, national nongovernmental organizations, and various parastatal entities operating at the national level. These analytical categories are purposely imprecise. Others have elaborated similar frameworks to integrate various scales of analysis of urbanization.[11]

Sociospatial Constructs

Our sociospatial constructs of populist and corporatist frontiers are not intended to be deterministic models. Rather, they serve as probabilistic constructs to illustrate how different social spaces pattern events by restricting the possible outcomes of socially indeterminate processes (Sheldrake 1988). Implicit here is the notion that these distinct frontier types operate on the basis of different internal logics. For example, it might be argued that the corporatist frontier operates on the logic of capital accumulation and the populist frontier on the logic of household welfare. Distinctive forms of spatial organization are patterned by these divergent logics.[12] This notion, however, is also somewhat arbitrary as these logics are not necessarily absolute and mutually exclusive. Political institutions crystallize the divergent logics into spatial patterns. The social character of frontier spaces (and all spaces have a social character) is mediated by the dynamics of interinstitutional relationships.

By *institution*, we mean "the sets of working rules that are used to determine who is eligible to make decisions in some arena, what actions are allowed or constrained, what aggregation rules will be used, what procedures must be followed, what information must or must not be provided, and what payoffs will be assigned to individuals dependent on their actions" (Ostrom 1986:51). But beyond this definition we acknowledge that institutions are typically nested in larger fabrics of socioeconomic and political interest. Institutions, given (partial or integral) responsibility for transforming a frontier space, cannot avoid impressing upon that space those sets of "working rules" that most favorably project the particular interests of client groups. Since most frontier spaces combine diverse institutional interests, the dynamics of institutional relationships lead to numerous variations in their socioeconomic character.

Scale is another important determinant of the social composition of space. Distinct social and cultural forces construct space to be functional at specific scales within which different institutions are intended to operate. Neil Smith (1993:101, in Knox 1994:9) puts it this way:

> The construction of scale is not simply a spatial solidification or materialization of contested social forces and processes; the corollary also holds. Scale is an active progenitor of specific social processes. In a literal as well as metaphorical way, space *contains* social activity and at the same time provides an already partitioned geography within which social activity *takes place*.

The frontier writ large is not neatly subdivided into corporatist and populist frontiers, a point we repeatedly emphasize. But, we hypothesize, the dynamics of social change are fragmented in relation to the social composition and institutional orientation of different frontier areas and at different scales within the same area. Local institutions of corporatist and populist social spaces are differentially articulated to forces operating at the mid-range and global levels, creating a three-dimensional matrix of possible outcomes. The emergence of contemporary urban networks is the physical manifestation of these multileveled interacting processes.

TOWARD A THEORY OF
DISARTICULATED URBANIZATION

Urban Amazonia has evolved historically from a hierarchical form of spatial organization, based on an extractive-mercantile system dominated by the primate cities of Belém and Manaus, to a more complex and fractured array of spatial configurations. Along with its expanded spatial extension and intensified functional complexity, the contemporary regional urban system has selectively strengthened connections to the national urban hierarchy. Many of the Amazon's contemporary urban centers are linked directly to the national core regions of southeastern Brazil, as explained in chapter 3. Other centers have attracted multinational investment, sanctioned by the state, and are directly linked to global markets, as in the case of the Manaus Free Trade Zone (*Zona Franca de Manaus*). Still other settlements are functionally disarticulated from the local or regional system of settlements in which they are located, as in the case of Porto Velho, Rondônia.

The uneven quality of contemporary urbanization erodes older patterns of regional metropolitan primacy driven by the extractive-mercantile economy. Manaus in particular, favored by government policies to attract transnational capital to the free trade zone, has outstripped Belém in modern industrial development and links to global capitalism, as discussed in chapter 5. In addition, as the new urban centers of Amazonia became selectively incorporated into the national space economy, they also became locally disarticulated from the traditional regional metropolises.[13] Amazonia's differential articulation challenges the hegemonic conception of urban system formation based on either central place theory or world cities hypotheses. The main function of the Amazonian hinterland historically has been to provide raw materials and new markets for global, and sometimes national, economic expansion. The fundamental orientation of inland settlement, however, changed from regional waterways to more flexible modes of movement, especially national roadways, air-traffic corridors, and satellite communications systems. These more advanced interaction technologies in turn have permitted greater selectivity of regional spatial integration into the matrix of national and global economic forces, creating a mosaic of fragmented social spaces. A schematic representation of the resulting overall pattern of regional disarticulated urbanization is presented in figure 4.1.

Principles of Frontier Urbanization

Our theory of disarticulated frontier urbanization rests on the following basic principles, first identified in chapter 1 and elaborated on here. First, the Amazon is a heterogeneous social space. Different social groups came to the region at different times for different reasons and engaged each other in different ways.[14] Conventional social taxonomies found in much of the structuralist and spatial economy literatures, which seek to homogenize this diversity, now seem too rigid and antiquated to succeed in the task of accurately describing contemporary patterns of regional urbanization.[15]

Second, the region's urban system is irregular and polymorphous. Urban systems form in which individual settlements do not necessarily

Figure 4.1. Functional relationships of frontier settlements

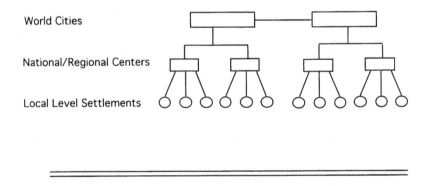

A. Articulated Systems

World Cities

National/Regional Centers

Local Level Settlements

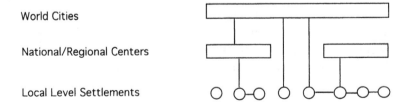

B. Disarticulated Systems

World Cities

National/Regional Centers

Local Level Settlements

emerge in hierarchical relation to one another over time, as explicitly predicted by both central place and world systems theories. Rather, we find evidence of multiple and overlapping networks. Some urban subsystems are nested incongruously in others. This does not mean that central place tendencies are nowhere evident in the region's overall urban structure. Indeed, we find considerable localized evidence of symmetrical subsystems. But these are often spatial fragments nested in larger dendritic networks. Nor is the region's settlement structure accurately portrayed as bifurcated between two or three regional metropolitan entrepôts reigning over a myriad of extractive outpost settlements, as during the last major economic epoch, the rubber boom. A layer of secondary cities has emerged since 1970 that gives the regional urbanization added scale and dimensionality.[16] Yet some of these secondary cities function independently from the region's metropolitan centers and have even supplanted some of the latter's economic influence.[17] They are, in short, disarticulated from the larger settlements in their proximity.

Third, the fragmented patterns of spatial organization in Amazonia partially reflect the uneven and incomplete expansion of the market economy in the region. While capital has temporarily penetrated the Amazon to varying degrees, capitalism as a system of wage relations of production has never become very well established in the region in either agricultural or industrial sectors.[18] Urbanization is disarticulated from the process of local economic development. In a thoughtful critique of political economic interpretations of Amazon development, David Cleary (1993:344–345) argues:

> Within Amazonia, it was increasingly obvious that the trend [during the 1980s] was towards greater heterogeneity in both economic structure and social relations, rather than a convergence towards capitalism. . . . The Amazonian economy has proved to be much more autarchic, and much less capitalist, than frontier theory predicted.

Fourth, to a considerable extent this urban disarticulation reflects the region's sectoral disarticulation (de Janvry 1981). In other words, urban growth occurs independently of both agricultural development and regional industrialization. This disarticulation occurs in the realms of both consumption and production and at two spatial levels. At the local (subregional) level, productive sectors emerge in some urban centers that do not serve local consumer markets. Examples of this form of export-enclave disarticulation include urban centers where local food crops or wood products are exported, compelling local markets to depend upon more expensive interregional imports of these same goods to meet local consumption needs. At the larger regional and national levels, productive sectors emerge that are not vertically integrated within the national market economy but are based on short-term contractual arrangements between transnational capital and national merchant capital (e.g., trading companies), a flexible investment pattern typical of post-Fordist modes of production.

Frontier urbanization in Amazonia is frequently mediated by global economic forces and transnational capital. Regional development projects financed by the World Bank have given additional momentum to urbanization. Merchant capital infusions into episodic extractive cycles have boosted urban growth associated with selected natural resources, such as mahogany and gold. Certainly, selected urban centers in the Amazon play important roles as technological crossroads linking specific local activities to global circuits of information and exchange. Yet it is implausible to argue that the region as a whole has become integrated into the global economy. The region's convoluted urban morphology, composed of distinctive subregional city-systems, not only embodies the accumulated artifacts of distinctive historical modes of production, but it also reflects various ways in which different contemporary socioeconomic interests in Amazonia are linked to local, national, and global economic forces.

Fifth, while various processes of organized expansion were initiated by the centralist state, Amazonia remains largely aloof from government control. Indeed, many frontier areas effectively integrated into the national economy through state and private investment during the 1960s and 1970s have rapidly fragmented after the recent contraction of the centralist state. While the government was the driving force of frontier expansion during the military-authoritarian period, the process of urbanization has progressed apart from the actions of the government since the return to civilian rule in 1985. Urbanization in the Amazon has become largely disarticulated from the centralist state. Cleary (1993:344) correctly observes that

> wherever one looks in the Amazonian economy, the state is in retreat: unable to finance tax breaks or build highways without the aid of multilateral banks, unable to include more than one per cent of the rural population in official colonisation schemes, unable to control land titling or land conflicts, unable to register or tax the

greater part of the Amazonian economy, unable to enforce federal law on more than a sporadic basis.

Since 1985 no one development model has given unity or functional cohesion to economic life in the frontier. Rather, a plurality of local models are arising from the ruins of the centralist state: extractive reserves, farmer cooperatives, and rural workers unions and the political fragmentation of both states and *municípios*. Urban centers are important staging areas for these locally based changes. Although the centralist state is currently in disarray, governmental institutions still have profoundly influenced the social orientation and spatial organization of the frontier. Here, too, it is important to disavow a conception of the Brazilian state as a unified or politically coherent assemblage of government institutions acting in unison. Rather, different agencies pursued different development agendas for the region and followed different, often conflicting, legitimization strategies with their respective client groups.[19] The multidimensional social composition of the Amazon and the subregional urban networks that appear in the landscape are, in large part, the legacy of this multifaceted institutional sponsorship embedded in the local history of specific places.

Sixth, we observe that the processes that lead people to settle in urban areas on the frontier are diverse (see chapter 8). Urban residency in the frontier is a fluid phenomenon; intramunicipal migration is the most important pattern of demographic movement in the frontier. Urban population growth is due mainly to urban-urban, not rural-urban, migration. But migration is often cyclical, involving both rural and urban-based activities. Separating rural and urban sectors as functionally discrete categories creates a problematic dichotomy.

Finally, as we discuss in chapter 10, regional urbanization has led to significant environmental changes in Amazonia. Tropical deforestation, water resource contamination, and hydroelectric development are all linked to urbanization.

We find that various processes lead to urbanization, and each urban center tends to reflect those processes differently. For example, different institutions have been instrumental in shaping the conditions leading different social groups to assume urban residency. It would thus not be accurate to assert that frontier urbanization is the result of the monolithic driving force of capitalist expansion dispossessing the peasantry of the land and causing rural workers to pile up in "cities of peasants." The process of urbanization at each settlement has a distinctive history: some recent settlements are creations of farmers; others were built by private corporations; yet others owe their origins to gold miners; some were planned by the state; and others arise as enclaves of indigenous Amerindian communities, the last sanctuaries of cultural resistance.

In the regional aggregate, frontier urbanization is a fragmented process, reflecting the interplay of multiple forces at different points in time and leading to a disarticulated pattern. As indicated in our definition of *frontier*, the Amazon as a regional space economy cannot be conceptualized accurately as *a single* frontier; rather, it is a mosaic of frontiers within larger mosaics, each linked differently to often competing political institutions and social groups.

Amazonian rainforest cities, however rustic and minimalist, are flexible and "restless" human landscapes, whose social histories, economic orientations, and physical designs adapt to their changing functions in local, national, and global spheres. Taken as a whole, the frontier surface produces various "bubbles" of activity, some purely extractive, others transformative, others speculative, and each linked to urban centers. Such bubbles are localized social formations displaying contingent relationships between various socioeconomic groups and classes (capital and labor) and engendering distinctive forms of spatial organization in the landscape. Basically, our challenge lies in understanding the unique internal dynamics of change within each of these social constructs and their relationships to national and global networks of production and exchange. Despite the high degree of local variability, frontier urbanization in Amazonia displays some underlying regional patterns, two of which we review below.

PATTERNS OF FRONTIER URBANIZATION

We observe two underlying patterns of urban network formation in Brazilian Amazonia, "populist" and "corporatist" in nature. Each pattern reflects a distinctive economic rationality, and each displays different forms of internal organization and dynamics of change. Spatially, these processes overlap and intrude upon each other at different times, distorting the original pattern of regularity that each, by its own logic, would otherwise display. Frontier urbanization reflects both local and global forces, and yet, over the long run, the aggregate effect of this medley is a regional urban system that is disarticulated from both internal forces of local agency and external forces of global structuration. Again, we hasten to emphasize that the frontier cannot be neatly dichotomized into these two ideal types. Rather, the frontier might be more accurately viewed as a spatial continuum on the extensive margin of the nation space where populist and corporatist forces periodically intermingle to seek control of resources in an open-access environment.

By *populist* we mean a social space dominated, for a time, by autonomous, self-employed labor. Access to natural resources and land is relatively equitable for all newcomers arriving at the same time; that is, early arrivers typically get the best land. Participatory local decision making is often the norm in populist settlements, at least initially, or until formal representative forms of governance take hold. Sometimes, however, traditional patrimonial forms of political domination emerge. The precise process of spatial change is rarely predictable. Over time, social class distinctions (new emergent properties) typically develop from within the population. New extractive resource fronts, such as timber or gold mines, may be discovered and contort emergent regularities in the spatial configuration of settlements on the populist frontier. In this parsimonious definition we intentionally avoid confusing our usage with more conventional meanings of the term *populism*, especially in the Latin American political context.[20]

By contrast, the *corporatist* frontier is typically a centrally planned and capital-intensive development project, entailing corporate forms of organizational management and strong bureaucratic linkages to state institutions with regional development mandates.[21] Labor participation is carefully regulated and flexible wage and short-term contractual relations prevail. Access to resources in the corporatist frontier is controlled exclusively by the corporation, although populist incursions are frequent, creating what Schmink and Wood (1992) call "contested frontiers." Local political decision making is centralized to ensure the maximum control of the resource base by the corporation and its institutional ally. As in the populist frontier, the corporatist frontier is dynamic, and frontier change is not altogether predictable over the long term. Yet one universal pattern stands out: the intensity of corporate investment attracts hordes of underemployed migrant workers. A bimodal urban network typically emerges, in which the lead corporation attempts to segregate its workers and production space from the informal shantytowns that quickly emerge around it. Instability in this system arises from several sources: the incapability of the lead institutions to contain the popular pressures on its perimeter; the depletion of the resource the corporate complex was designed to extract; the withdrawal of transnational financing; or the environmental collapse of the entire complex due to overcrowding, pollution, or disease.

Two fundamental differences apply to populist and corporatist forms of spatial organization. First, corporate management seeks to regulate socially indeterminate processes by centralizing control of its environment. Conversely, populist political behavior seeks the same outcome by diffusing power and limiting central authority. Second, commodity and labor markets are more fluid and open in populist spaces, moving in tandem with the expansion of the general population. In corporatist-dominated regions, market structures are bifurcated; the corporation directly provides most of the goods and services required in its space, and a closed market economy emerges. The residual of this autarchic supply is "dumped" into unregulated parallel local markets serving the local population outside the controlled corporatist space. We consider the dynamics of urban growth associated with each frontier type in greater detail below.

Urban Change on the Populist (Agrarian) Frontier

The Amazon region's urban system is comprised of various types of settlement subsystems, each differentially functional to local, regional, and global processes of production and exchange. The populist frontier is an idealized type that includes an array of distinctive urban morphologies associated with diverse labor activities ranging from popular gold mining to family farming, and the agrarian frontier is a subtype of it. These settlements on the populist frontier are unified in their origins in autonomous labor, equal access to land, reciprocal relations of labor exchange, and production for both subsistence and surplus. Over time, formal market mechanisms directly linked to national capital in various forms may selectively appear on the populist frontier. These markets serve multiple functions (see chapter 9). First, financial capital establishes markets to channel the surplus of the frontier to more profitable investments in the national economy. Second, industrial capital establishes markets for consumer and intermediate goods and floods the frontier with goods manufactured outside the region, including, ironically, food products. Third, merchant capital develops its own niche in the local exchange of goods and services and in the arbitrage of real estate, providing limited opportunities for capital accumulation in the frontier itself. The institutional mechanisms through which these various forms of capital flow to and from the frontier define the extent and depth of the frontier's integration into the larger national economy. This section briefly updates and summarizes our generalized model of urban transition on the agrarian frontier first introduced to readers in 1990 (Browder and Godfrey 1990).

To avoid confusion, the "agrarian" frontier is a subset of the populist frontier connoting only that independent agriculture, typically comprised of small to medium-sized family farms (as opposed to corporate farming), is the defining social form on this frontier. This does not suggest that other economic activities (e.g., mining, timber extraction, urban commerce) are excluded or unimportant. Rather, populist agriculture is the leading sector, both in terms of economic output and social orientation, in the agrarian frontier.

We hypothesize that the initial underlying force driving spatial organization in the agrarian frontier is the marketing principle of central place theory. After its opening, the spatial expansion of the frontier reinforces the central place hierarchy of the urban system serving it. Over time, however, older dominant settlements lose economic importance to their hinterland. Unable to specialize in the production of higher-order goods, the older cities become relics as younger towns closer to the perimeter of the frontier usurp their role in providing more efficiently the limited range of basic goods and services demanded by pioneer farmers. This pattern of change is not usually uniform, smooth, or linear. Rather, the process is interrupted in fits and starts associated with the discovery and feverish exploitation of a new "hot resource" (e.g., gold, mahogany, cassiterite) or with the emergence of a dominant elite family in local politics able to marshal the resources of the state more effectively than elites in other, neighboring communities, or with the sudden appearance of a government development project, or with the disappearance of a locally important credit program. Distortions in regular settlement network patterns are varied and ubiquitous. Given these qualifications, the general pattern of urban change associated with agrarian frontiers involves four benchmark periods, each of which may vary in duration from a few months to years (fig. 4.2).

1. *Resource-extractive frontier:* The discovery of valuable natural resources in the unoccupied region—such as commercial timber, minerals, or rubber—stimulates the competitive penetration of extractive enterprises, both populist and corporatist. Typically, these firms import their own workers to the frontier for a primitive, labor-intensive process of resource extraction. If the resources are valuable enough, or if credit policies provide sufficient incentive, corporate firms may step in from the outside and eventually

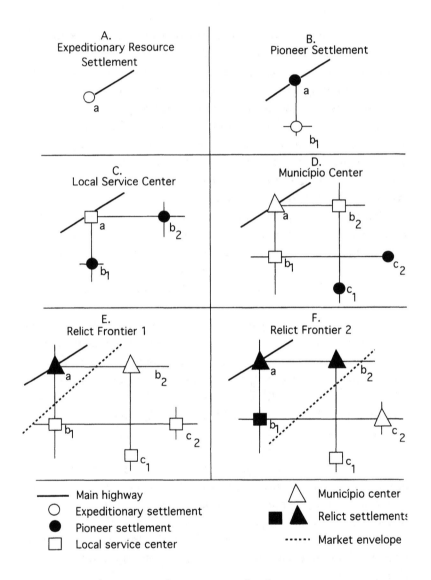

Figure 4.2. Urban system change on populist frontiers

dominate the local extractive process, for a time shifting the basic character of the frontier to a corporatist mode. This resource-extractive frontier can last a long time, even generations, as a series of boom-and-bust commodities is exploited in succession. Alternatively, the extractive frenzy can end abruptly after a few months. During the bust periods the remnant population declines and drifts to towns created as points of transportation and trade during the boom periods. The infrastructure created for transportation and communication—river transport facilities, penetration roads, sometimes isolated airstrips for especially valuable commodities—tends to attract speculators, agriculturalists, and ranchers to the area.

Dominant urban form: the expeditionary resource settlement. Extractivists, sometimes accompanied by land speculators and pioneers, establish an outpost off the main highway or interregional transport link (settlement *a*, fig. 4.2A). The outpost serves a small, transitory, predominantly male labor force, usually extracting a single resource, such as timber or minerals, but sometimes including prospective farmers. Urban functions are limited to services of extraction—the fuel depot, mechanical equipment repair shop, canteen, a dispensary of foodstuffs, usually distributed on credit (*barracão*), rudimentary lodgings—and there are virtually no amenities. Such places are highly mobile, almost proto-urban communities serving a temporary population. Most do not usually evolve into permanent settlements although they may facilitate subsequent occupation by pioneer farmers (plate 4.1).

2. *Pioneer agricultural frontier:* The in-migration of small peasant farmers often follows in the wake of extractive activity and represents the next generalized phase in the expansion of the agrarian frontier. Private and public road investments, often following Indian trails or logging roads, usually determine the general area into which settlers expand although in some recorded instances (see chapter 6) intrepid pioneers walk for days from transitable roads to claim homesteads on the frontier. These settlers primarily employ family labor to cut down the forest and to plant food crops for household consumption. Most of these pioneers retain a residence in established towns nearby until transport improvements justify a full-time presence on the farm and the expansion into cash

Plate 4.1. Expeditionary resource settlement—a tin mining camp in Rondônia *Photo:Willem Groenveld*

cropping. In anticipation of such improvements, land speculation and conflict run rampant. Road and communications improvements that allow access to regional markets are prerequisites to the formation of inland (nonriverine) Amazon settlements. Once access is established, however precarious, the settlement swells with in-migrants prospecting unclaimed land. In the melee of this initial stage of pioneer front expansion, pioneer settlements (settlement *b1*, fig. 4.2B) become focal points of activity (vehicle repair, rustic hostelries, and diners) serving the influx of new settlers while older settlements acquire a more diverse array of functions associated with local service centers (settlement *a*, fig. 4.2C). Farming and resource extraction are typically combined into household survival strategies. For example, most farm households will deploy one or more family members in off-farm extraction work (e.g., gold mining, logging, etc.). Conversely, professional loggers and lumber entrepreneurs will often invest some of their profits in ownership of local land and cattle ranching. Similarly, strong rural-urban linkages form as farmers hold on to urban dwellings while penetrating the frontier for farmland. Retailers from larger towns, attracted by the brisk commerce of the expand-

ing frontier, relocate from the metropolitan core to pioneer settlements and then often acquire their own urban and rural properties
as speculative investments.

   Dominant urban forms: the pioneer settlement and local service centers.
The opening of the frontier to permanent settlement is a group
endeavor, often initiated by an extended family, and begins with the
clearing of forest and planting of annual food crops. A congregation
of small homesteads locates near a river or a stream, at the terminus
of a spur road, along a logging trail, or next to a lumber mill. Most
of these pioneers previously resided elsewhere on the frontier, often
in urban areas with which they maintain contact, and they often
encourage the arrival of other family members, friends, and
acquaintances. Meanwhile, word of unoccupied land travels fast, and
the number of unrelated settlers also increases. The settlement
becomes a "landing," serving recent migrants on a short-term basis
as they prepare to occupy the surrounding forest.

   Urban functions in the pioneer settlement are generally limited
to service activities: rudimentary hostels; canteens with petty gro-
ceries and a billiard table, often operated by middlemen merchants
extending informal credit; auto repair shops; perhaps a small
schoolhouse and an evangelical church. As pioneers fill in the surrounding countryside, the range of goods and services demanded
increases, and selected pioneer settlements grow into local service
centers. In these towns, especially where resource extraction is
lucrative, commercial banks may establish branch offices as do
other services specializing in particular resource sectors (e.g.,
band-saw repair). However, as the urban population grows, so too
does the consumer market, and the marketing principle becomes
the underlying driving force of economic location on the frontier.
Basic processing activities (e.g., lumber, ceramics, grains, etc.) also
appear in local service centers during this phase of frontier expansion (plates 4.2. and 4.3).

   3. *Independent municipal center:* Political leaders emerge in local
service centers and begin to lobby the state for municipality designation, a status that accords important political and financial privi-
leges to the *municipio.* Typically, towns and surrounding hinterlands
with combined populations of more than ten thousand residents are
separated from existing larger municipalities in a progressive

Plate 4.2. Pioneer settlement—Alto Alegre, Rondônia, 1989
Photo: John O. Browder

Plate 4.3. Local service center—Santa Luzia, Rondônia, 1990
Photo: John O. Browder

Plate 4.4. Private company service town—Cassol, near Santa Luzia, Rondônia, 1990. It illustrates the emergence of a petty corporatist settlement within a predominantly agrarian frontier. *Photo: John O. Browder*

process of jurisdictional fragmentation that accompanies the expansion of the populist frontier (settlement *a*, fig. 4.2D).

Dominant urban form: The município *center.* With municipality status comes a host of public services and patronage jobs: selected state agencies establish operations, and the conventional political process is extended. A post office is opened, the military police establishes a base, regular bus transport commences, and additional shops and services emerge, as do migrants in search of wage labor opportunities. Commercial services increase in number and scale: several hotels and restaurants, a supermarket, an air field, furniture manufacturers, movie theaters, bakeries, schools, and churches of diverse denominations. Several commercial banks and various government agencies establish branch offices in the settlement. Land prices in both rural and urban sectors stabilize after an upward spiral. The town expands outward, decrepit public housing appears on the peri-urban fringes to accommodate newcomers from other urban and rural areas (plates 4.5. and 4.6).

4. *Relict frontier:* Although the municipality status provides an important economic injection of public sector patronage jobs, in the absence of productive specialization (especially in manufacturing), it does little to enhance or protect the economic influence of the *município* center over its market catchment area. If the populist frontier continues to expand, then gradually the *município* center may lose its competitive edge as a supplier of basic goods and services to newer service centers located closer to the settlement fringe. The market envelope of the frontier shifts outward beyond the catchment area of the *município* center (fig. 4.2E and 4.2F).

For those *município* centers located at the rear of an expanding populist frontier, the next phase of frontier expansion is frequently characterized by consolidation and closure. The frontier may witness a concentration of landholdings by large ranchers—often

Plate 4.5. Município center—Rolim de Moura, Rondônia, 1990
Photo: John O. Browder

based outside of the region—the liquidation of productive assets in extractive enterprises, and the dispersal of private providers of agricultural services. In the absence of a new extractive resource cycle or the suspension of institutional rents to subsidize agricultural expansion, the agrarian frontier collapses into a socioeconomic mosaic as successful small farmers commingle with larger agricultural and ranching enterprises. Out-migration from the local countryside continues or accelerates; peasant farmers and landless workers move further out to new active frontier areas, or they migrate to regional secondary cities and metropolitan centers in search of work. Some, but not many, may stay for a time in interior towns.

Not all farmers, however, are doomed to failure. Perhaps as many as one half of the original agrarian population overcomes the

Plate 4.6. Decrepit public housing project on the perimeter of Rolim de Moura, Rondônia, 1990 *Photo: John O. Browder*

various obstacles to survival (Browder 1995). The successful ones are often those farmers who have also established themselves as permanent urban residents while pursuing one or more urban-based activities such as the sale of produce from their farms.

Dominant urban form: the relict município center. Once the center of vibrant commerce and political activity, the município center *a* diminishes in economic importance as it yields market area to the younger, more efficiently scaled and located service centers *b1* and *b2*, the latter of which becomes in due course a município center (fig. 4.2E).[22] As the frontier continues to expand outward, center *b2* gradually relinquishes its economic dominance over the frontier to younger, more remote pioneer settlements (settlements *c1* and *c2*, fig. 4.2F). For as long as the market envelope of the populist agrarian frontier shifts outward, the general dynamic of spatial transition roughly follows this pattern. However, as we shall see in chapter 7, this predictable pattern is frequently interrupted and distorted by various forces operating at the local level (in individual settlements). An asymmetrical, polymorphic urban network develops, in which the traces of original central place tendencies are only vaguely evident.

Finally, we mention two other tiers of the frontier's urban hierarchy: the regional secondary city and the regional metropolitan center.

The regional secondary city. The agrarian frontier, unlike extractive zones on the populist frontier, develops regional secondary cities. The population of such cities typically grows to more than one hundred thousand, and they become centers of political, commercial, and social life for substantial portions of the frontier population. Examples of regional secondary centers in Rondônia include Jí-Paraná and in Pará, Santarém, Marabá, and Conceição de Araguaia. However, not all regional secondary cities are functionally articulated to the lower-order centers emerging in the settlement frontier. In Rondônia, for instance, the secondary city, Jí-Paraná, is virtually the economic capital of Rondônia's agrarian frontier. Porto Velho, the state's capital city, became a relict of the rubber boom, and its political importance is accentuated only by global capital infusion through its state bureaucracy (chapter 5).

The regional metropolitan center. At the top of the regional city system are two metropolitan centers with populations exceeding one million: Belém and Manaus. Historically important commercial centers, these cities are fully integrated into Brazil's national city system. Here are universities, a free trade zone, international airports, hotel chains, international banks, foreign consulates, major manufacturing industries, and so on.

During the active pioneer agricultural and relict frontier phases, urban settlements undergo structural alterations in response to their selective integration into the national and global economy. Labor, needed in resource extraction and family farming, becomes less important as extractive resources are exhausted and as young unremunerated household members grow up and leave the farm. Settlements that survive do so because they either enjoy multiple resource bases or successfully subdivide from existing *municípios*. In the latter case, government services usually provided by the state fill the economic gaps left by the retreating extractive industry, commerce, and failed agriculture.

Most rainforest cities do not reach the status of *município* service center. But even those that do grow to this level face an uncertain future once the local natural resource base has been depleted or its extensive market area has been claimed by smaller, more remote service centers. Stagnation, then, is the probable finality most rainforest cities move toward on the agrarian frontier. But for a brief time frontier towns provide a useful service to their inhabitants, to the government institutions that sponsored their creation, and to the extractive enterprises at national and global economic levels that benefit from the support services provided. We speculate that over the long haul the agrarian frontier spawns new forms of urban primacy and regional polarization, leaving relicts of urban decentralization in their wake.

Urban Change on the Corporatist Frontier

The second pattern of urban system formation arises from a vastly different set of circumstances. It originates directly from transnational or state capital, which imposes a blueprint for planned urbanization onto the rainforest landscape, setting in motion a different process of regional change. In contrast to the populist frontier, propelled largely by autonomous labor (e.g., small-scale peasant farmers or miners) and acting in rational response to incentives created by specific state institutions (Collins 1986; Schmink 1987), in many Amazon locations the central state has decisively favored the development of a frontier dominated by corporate capital (Bunker 1985; Butler 1985; Pompermayer 1979; Roberts 1991; Schmink and Wood 1992). The corporatist frontier is a visible spatial manifestation on the national periphery of what Evans (1979) called the "triple alliance" of state, foreign, and national capital, the development strategy favored by Brazil's military-authoritarian regime especially in the periods from 1964 to 1970 and from 1975 to 1985.

The corporatist frontier is more likely to be driven by a large-scale centralized operation, dependent on domestically subsidized transnational investment, centralized planning, and state approval and financial co-sponsorship. The specific activities occurring on the corporatist frontier vary widely, but, even to a casual visitor, they are clearly different from those on the populist frontier. They include huge cattle ranches with private airports, large industrial mining operations, industrial tree farms, mechanized soybean estates with direct satellite hookups to world commodity exchanges, behemoth hydroelectric development projects along the Amazon's vast waterways, state-of-the-art industrial wood processing complexes, computer microchip plants, and genetic engineering laboratories. These corporatist big projects, many located deep in the rainforest, produce a distinctive form of associated urbanization.

Although the corporatist frontier assumes centralized project planning, its full implementation requires the participation of workers, both formally as project employees and informally as subcontracted laborers and service providers in nearby settlements. As the artifact of massive capital investments, corporatist frontiers attract hoards of underemployed migrants who are forcibly excluded from entering and squatting on land in the project areas. The corporatist frontier depends upon centralized control, enclosed spaces, and the military-like repression of populist incursions through guards and fortress-style gates.[23] Populist workers gather outside the project gates, serving as a source of cheap labor, often sub-contracted, in flexible production schemes (Roberts 1991). Although these masses may be subordinated to the interests of the locally dominant corporation, popular resistance often arises in the form of local land invasions and organized protests (Foweraker 1981; Hall 1989; Schmink and Wood 1992).

Generally, the development corporation constructs a neatly planned company town at its project site with all the amenities of an upper-income planned urban development, where managers, higher-level company technicians, and their families safely reside. In contrast to the highly regulated social space of the "bungalow city," with its swimming pool and satellite dishes, most construction workers, domestic servants, and an array of itinerant urban service providers gather in unplanned, self-built shantytowns erected on the periphery of the corporatist project (Butler 1985; Hall 1989; Roberts 1991).

One example of a corporatist settlement is the infamous Jarí project. In 1967 the North American billionaire shipping magnate Daniel Ludwig bought up 7,400 square kilometers of Amazonia, a chunk of rainforest the size of the state of Connecticut, for U.S. $3 million to build the world's largest pulp processing project. Fast-growing soft-wood plantations (*Gmelina* spp.) would replace thousands of hectares of natural tropical rainforests. A state-of-the-art U.S. $275 million floating pulp mill was barged from Japan to cruise the Rio Jarí and produce cel-lulose in an integrated mobile production system. Channeled through his Amazon company, Jarí Forest Products, Ludwig's personal invest-ment was conservatively estimated to range from U.S. $200 to U.S. $500 million (*Time* 1976). Total corporate investment probably exceeded U.S. $1 billion before the project was nationalized. This bub-ble of corporatist expansion in Amazonia required an urban nucleus to ground the project's command and control activities. Thus was born the bungalow city of Monte Dourado:

> The capital of this jungle kingdom is Monte Dourado [golden mountain] (present population 3,500) a sprawling new community of attractive bungalows, town houses and apartments. A Jarí-built hospital staffed by seven doctors cares for the sick, and a Jarí school educates the employees' children. A giant service depot stocks nearly $6 million worth of spare parts and equipment so that a force of 266 mechanics can keep heavy-duty machines busy build-ing more roads, more industrial sites and ports, and even a roadbed for a 43-mile private railroad. (*Time* 1976).

Three years later journalists from the same periodical reported the fol-lowing:

> Ludwig has built barracks for ordinary laborers as well as fancier bungalows for the technical and managerial staff. But he cut back substantially on plans for additional housing, especially for the low-est-paid workers. Result: squalid slum towns inhabited partly by whores and thieves, have sprung up near the sites, and many work-ers live in unsanitary and unsavory conditions. . . . The lack of sat-isfactory living conditions in the rough territory has contributed to a debilitating rate of turnover, more than 50% a year. This adds enormously to the cost of training programs that Ludwig's man-

agers must conduct in order to acquaint the largely backwoods work force with modern machines. (*Time* 1979)

A ring of shantytowns emerged on this corporatist frontier landscape, defying any notion of orderly central place hierarchies. Other examples of the hybrid nature of the corporatist frontier, uneasily combining the planned company town with the spontaneous popular settlement, can be found in the huge Tucuruí hydroelectric project and the Carajás mining project in southern Pará (chapters 7 and 10).

State-sponsored megaprojects also engender corporatist urban enclaves such as those scattered throughout the Amazon near various hydroelectric development sites of ELETRONORTE, the federal government's power utility subsidiary for the Amazon. Fearnside (1989:405) describes how life in the *barrageiro* (dam builders) colony of the Balbina hydroelectric project outside of Manaus forges social class identity within Brazil's corporate elite:

> The *barrageiros* move from project to project living in comfortable but remote colonies built at each site. Life in the colony can appear idyllically free of the social problems of the rest of Brazil—a situation maintained by armed guards who prevent any laborers from entering the "class A" residential areas. . . . Separate social clubs divide engineers from other categories: mere scientists are not allowed in the engineers' club. This isolation has encouraged formation of a strong interest group that battles furiously any who question the wisdom of Balbina [or hydroelectric dams in general]—any doubt is perceived as a threat to the *barrageiro* way of life.

But like other resource-extractive rainforest cities on the corporatist frontier, the corporatist presence at the "Vila Eletronortes" around the Amazon contracts over time, representing a different dynamic of regional change than is found in the populist frontier. The pattern of corporate urbanization follows roughly the following sequence (fig. 4.3). Initially, corporate entities, usually privileged by government subsidies, build a luxury bedroom city (*a*, fig. 4.3A) to house their permanent professional employees in a "hardship post" environment (e.g., an industrial bauxite extraction project). The corporation, often a parastatal enterprise, may build a much less commodious dormitory city (settlement *b*) for its regular nonprofessional workers. The sheer mass of the capital investment in the urban-industrial project draws workers from far and wide. Makeshift shantytowns (*g1*, *g2*, and *g3*, fig. 4.3B) appear seemingly overnight, often created

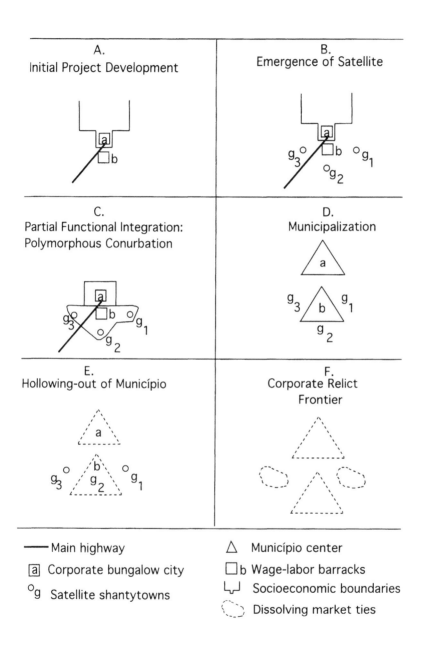

A.	B.
Initial Project Development	Emergence of Satellite

C.	D.
Partial Functional Integration: Polymorphous Conurbation	Municipalization

E.	F.
Hollowing-out of Município	Corporate Relict Frontier

—— Main highway

$\boxed{a}$ Corporate bungalow city

^{o}g Satellite shantytowns

$\triangle$ Município center

$\square$b   Wage-labor barracks

$\llcorner\lrcorner$ Socioeconomic boundaries

$\langle\cdots\rangle$ Dissolving market ties

Figure 4.3. Urban system change on corporatist frontiers

by invasion, to house in abysmal conditions the droves of aspiring contract workers that arrive daily. A polymorphic urban pattern appears as shanty-towns begin to establish social and economic linkages with each other (fig. 4.3C). Still, the corporate development project is the economic hub of the emerging urban complex. Although sometimes the entire conurbation becomes incorporated as one municipal jurisdiction, often the corporate-shantytown polarization becomes institutionalized in separate municipal jurisdictions (fig. 4.3D). Even in the case of two separate municipalities, the original corporate settlement, settlement *a*, becomes the upper-income neighborhood in a larger conurbation, as in the case of Tucumã-Ourilândia (chapter 7). Eventually, the urban core of the settlement complex begins to hollow out when the corporation withdraws following the depletion of the resource it was created to exploit (fig. 4.3E). The system, originally composed of carefully segregated elements, now dissolves. In the final phase, the old system reverts to a new system based on small isolated and marginally productive settlements that subsist on government services and petty commodity exchange (fig. 4.3F). Unlike its populist counterpart, the corporatist frontier does not evolve toward increasingly complex forms of urbanization; it begins urbanized and then degenerates into a nonnodal conurbation.

Like the pattern of urbanization on our "agrarian" frontier, regularity in the structure of the urban networks oriented to corporate development projects is susceptible to numerous distortions. In the corporate frontier, the intrusion of migrants (overwhelming the barracks and bedroom communities), the discovery of more lucrative opportunities for labor (vacating the planned development of its labor force), or shifts in government or transnational financing (i.e., cutting off the project and its town from further subsidy) all contribute unpredictable emergent properties to the dynamic of urban change on the corporate frontier.

CONCLUSIONS

The processes of urbanization in Brazilian Amazonia do not easily fit into any one of the theoretical frameworks reviewed in chapter 2. We therefore suggest a pluralistic perspective, which allows for the possibility that different frameworks explain different aspects of urbanization simultaneously. The concept of disarticulated urbanization is a useful conceptual scaffold in this regard. As selective zones of Amazonia become incorporated into national and international economies, regional urban centers become differentially articulated to different levels of the global urban hierarchy. New linkages to distant centers erode traditional regional ties. Patterns of regional urbanization thus become variable in both spatial and temporal terms.

Two generalized patterns of urbanization, each associated with a distinctive mode of frontier expansion, can be discerned in contemporary Amazonia. Any general typology is susceptible to the criticism of oversimplification. We acknowledge that both the populist and corporatist frontiers presented here constitute ideal types. In this sense, the two frontier models are distinctive conceptual points on a continuum of social spaces that inevitably interact and overlap in patterned but locally variable ways. These types of frontiers are defined less by specific activities than by their functional orientation (to localized or global processes of development), class control (populist or elitist), political-economic linkages (participatory or centralized), and specific institutional orientations and associated social histories. The aggregated effect of these incongruous processes of urban network formation in the frontier is a highly contingent pattern of disarticulated urbanization at the regional level.

Notes

1. We will use the term *village* to denote settlements of less than two thousand inhabitants; *towns* are defined as ranging from two thousand to twenty thousand, and *urban centers* or *cities* as exceeding twenty thousand.

2. Admittedly, this is an eclectic array of source material, modernist and postmodernist alike. Our aim, however, is not to achieve intellectual purity but rather to identify important strands of contemporary theorizing, however disparate, that provide a transdisciplinary basis for conceptual pluralism.

3. Our usage of the term *systems*, with its broad allusions to rational scientism, will undoubtedly perturb ardent postmodernist readers. Like many contemporary writers of the postmodernist genre, however, our usage is mainly metaphorical rather than analytical, and we do not wish to imply that our position arises from an application of general systems theory.

4. Readers interested in the Chicago school of human ecology are referred to the chapters by Robert Park, Louis Wirth, and Robert Redfield in Richard Sennett's book, *Classic Essays on the Culture of Cities* (1969). Herbert Spencer (1972) is perhaps best known for linking sociological explanations of societal change with biology, and the observed tendency of organisms (like societies) to evolve toward more highly differentiated, complex forms. Spencer's ideas later contributed to the long-standing sociological discourse on the relative importance of structure versus agency in positivist explanation, a discourse that has helped frame contemporary debates over structuralist and postmodernist frameworks.

5. Our discussion of ecological rationality draws upon John S. Dryzek's *Rational Ecology: Environment and Political Economy* (1987).

6. We would not be inclined to draw the analogy with ecological rationality much further on this point. Dryzek (1987) adopts a successional framework common among ecologists to explain the process of differentiation toward more complex systems. While we concur that

pioneer urban systems are generally more simplified, it is difficult to imagine the urban corollary of a "climax" ecosystem.

7. A relatively recent movement afoot in the life sciences, dubbed the "New Biology" rejects the organism-environment dichotomy and argues that one is integral to the other (Dickens 1992:144).

8. We will not attempt a thorough review of these perspectives here, pointing interested readers instead to the excellent summary by Cloke et al. (1991), from which much of the material presented here is taken.

9. We acknowledge that the term "conceptual pluralism" has diverse transdisciplinary connotations. In this regard, we also note the pioneering work of economist Richard Norgaard (1994) in positioning pluralism as an emerging perspective in the social sciences. Norgaard's exposition on contingency and contextuality in explanation is tied to the notion of coevolution of interacting ecological and social systems, rather than to critical realism, reflecting the different forms that pluralism may take in different disciplines.

10. While pointing our readers to the works of British scholars Roy Bhaskar, Derek Gregory, Andrew Sayer, we distinguish this recent literature from its forerunner, "critical realism," that emerged in the United States around World War I (e.g., Sellars 1969 [1916]; Drake et al. 1968 [1920]). We also need to distinguish critical realism, situated as it is in epistemology, from "metaphysical realism" although important linkages between the two and conceptual pluralism exist. For example, the distinguished philosopher Hilary Putnam (1994:352–353) writes: "The metaphysical realist views truth as radically non-epistemic. . . He [sic] admits that it is logically possible that we will have to live with a number of empirically equivalent systems of the world, or theories, which equally well satisfy all our operational and theoretical constraints even as scientific inquiry continues indefinitely."

11. David Simon (1992, 1995), for example, proposes a conceptual framework for urbanization, grounded in political economy but sensitive to human agency and social and cultural expression that integrates data from five scales of analysis: global, continental regional, national, urban, and intraurban.

12. Ruy Moreira (1985) proposes similar spatial categories, which he calls "molecular space" and "monopoly space." While he and other neo-Marxists adopt this typology to characterize national space in different historical periods (in this case, the periods of merchant capital hegemony and industrial capital, respectively), we suggest that these spatial characterizations occur contemporaneously in the frontier creating a mosaic of fragmented subregions, each articulated to different underlying logics of spatial organization.

13. Many frontier towns of southern Pará, for example, have come to depend in socioeconomic terms more on interactions with Goiânia, Brasília, and industrial centers in the Brazilian Southeast than with Belém (see chapter 7). In Rondônia, most of the inland cities on the BR-364 highway, especially the state's second largest city, Ji-Paraná, have greater economic linkages to Cuiaba, Goiânia, and São Paulo than to Porto Velho (see chapter 6).

14. That the Third World's rural poor are not homogeneous masses of people but rather are highly differentiated groups engaged in diverse production systems seems to have become apparent to First World researchers during the 1970s when international development assistance was oriented toward the reduction of rural poverty (Mintz 1976; Clay 1979).

15. As Cleary points out "[a] revised political economy would also need a newly sanitized analytical vocabulary. It is wrong to use evolutionist terminology such as 'pre-capitalist' to describe Amazonian economies or social formations" (Cleary 1993:346).

16. Examples of emergent secondary cities in the Amazon region include Santarém, Marabá, and Conceição de Araguaia (all in Pará), Itacoatiara (in Amazonas), and Ji-Paraná (in Rondônia), all with populations exceeding fifty thousand inhabitants.

17. In Rondônia, for instance, the secondary city of Jí-Paraná serves several economic functions, most notably as a statewide wholesale distribution point, a function that, according to central place theory, likely would be provided by the largest city, Porto Velho. The capital city, Porto Velho, a creation of the Amazon rubber boom, is largely irrelevant to the state's current economy, given its anachronistic location on the periphery of the state's settlement frontier. But because it provides mainly administrative functions and is thus the main arena for institutional mediation of frontier land use in the state, its importance as a quasi-metropolitan center in the region's political economy is maintained or even heightened with the recent participation of global capital (i.e., World Bank programs) in the state's economy.

18. The incorporation of Third World regions into the world economy has been neither unilinear nor permanent. Incorporated populations, while producing items for extra-local markets, rarely produce commodities through purely capitalistic modes of production. The diversity of production strategies allows for the development of interdependent or symbiotic relationships between individuals or groups that, in turn, provide the basis for a multitude of noncapitalist modes of production found throughout the world (Clay 1979; Greenwood 1973; Mintz 1976).

19. For instance, the influence of SUDAM and GETAT on the functions urban areas acquired in southern Pará was vastly greater than that of INCRA. Yet, "without doubt, the predominance of INCRA. in the State of Rondônia after 1968 was one of the components that would contribute to the character of the occupation of Rondônia" (SET 1983:n.p.).

20. Populist political movements in Latin America are associated with rapid urban expansion, wage pressures on middle-income classes, and the growth in popular expectations. For a brief summary of selected historical examples see Outhwaite and Bottomore (1993:494). A more detailed recent treatment may be found in di Tella (1989).

21. Our definition of *corporatism* is roughly consistent with common contemporary usage, although it diverges from classical Taylorian notions based on a mechanistic conception of the corporation as an apolitical production function. For example, Outhwaite and Bottomore (1993:116) define corporatism as "a specific sociopolitical process in which organizations representing monopolistic functional interests engage in political exchange with state agencies over public policy outputs." See also Hoppe and Peterse (1993), Cawson (1986), and Malloy (1977). For an interesting review of different conceptual models of corporate control, see Faris (1991).

22. More successful boomtowns, favored by locational advantages of transportation, commerce, and industry, may retain their competitive position relative to younger centers in the interior. Those less favored decline in importance or even disappear entirely.

23. In an interesting variant of this pattern, during the early 1980s a private Brazilian logging corporation with direct contractual ties to a North American hardwood importer illegally invaded a mahogany-rich portion of the Guaporé Biological Reserve in Rondônia. To discourage interlopers it established an armed guard in a house at the entrance to its "territory" and recruited a small number of colonists to occupy the enclosed area, thereby legitimizing its own illegal occupation of the reserve (Browder 1986). Interestingly enough, this corporatist social formation, although clearly of an expeditionary nature, was situated at the cutting edge of the populist frontier, illustrating the complexity and fluidity of these social spaces.

5

Metropolitan Centers in Amazonia:
Regional Cities and Urban Primacy

❧❧

Amazonia's contemporary urban transition, promoted by state-sponsored regional development programs since the 1960s, has taken several
discernible forms. One obvious demographic change is the increasing
proportion of the regional population living in cities. The Brazilian census of 1980 showed for the first time that just over half the region's population lived in urban centers. By 1991 nearly 58 percent of the
Amazonian population resided in cities. A second significant aspect of the
regional transformation has been the appearance of new inland settlements and the elevation of existing settlements to the status of municipality or urban center.[1] A third distinctive feature has been the declining
regional primacy of the major metropolitan centers, Belém and Manaus.
Small and medium-sized settlements, now directly linked to national
metropolises beyond the regional central places, have grown at faster
rates than the largest cities of Amazonia. Nevertheless, the traditional primate cities have witnessed massive in-migration, squatter settlement, and
activity of the informal sector on the urban periphery. Finally, and not
surprisingly, the functions of the metropolitan centers have changed,
becoming more multifaceted than they had been during the days of the
rubber boom. Unlike the unidimensional riverine entrepôt system of the
past, contemporary patterns of urbanization reflect the growing disartic-

ulation of the regional city system, combined with dramatically enhanced spatial extension and functional complexity.

This chapter examines the changing historical patterns, spatial structures, and regional political and economic roles of Amazonia's three largest cities: Belém, Manaus, and Porto Velho. The largest two metropolitan areas, Belém and Manaus, now both have more than one million inhabitants. These regional metropolises have long served as Amazonia's main gateway cities. As the upper level of the city-system of northern Brazil, these populous centers continue to play crucial intermediary roles in the regional interface among the Brazilian government, externally based corporations, and the larger world system. Porto Velho, the largest of the North Region's intermediate cities, now exceeds a quarter of a million in population (half a million in the larger metropolitan area) and rapidly approaches the size of a regional metropolis. In fact, the capital of Rondônia increased in population by roughly 7.6 percent annually between 1970 and 1991, the fastest rate of the three largest regional cities (IBGE 1991).

As state capitals, Belém, Manaus, and Porto Velho have a profound political impact at the local level. As the seats of national and state agencies, these regional cities disburse central budgets, administer fiscal incentives, and are the legal and bureaucratic centers of their states. In their regional economic roles, however, the metropolitan centers now confront new challenges. Once primarily commercial entrepôts for the export of Amazonia's abundant natural resources, the large metropolitan centers no longer enjoy the unrivaled urban primacy they had because of their regional isolation.

HISTORICAL PATTERNS OF URBAN PRIMACY

Urban life in Amazonia has long radiated outward from the historic gateway cities of Belém and Manaus. Colonial urbanism emphasized the Portuguese founding of Belém and many small settlements; the rubber trade of the late nineteenth century, administered through the *aviamento* system's regional hierarchy of places, allowed the two primate cities to grow from rudimentary river towns to prosperous metropolitan centers (see chapter 3). During the last century these two urban rivals have competed for leadership as preeminent regional poles of commercial activity, technological progress, and imported civilization. Both cities now boast carefully preserved opera houses, municipal markets prefabricated in Europe, impressive beaux-arts architecture, grand neoclassical parks, and tree-lined boulevards. Such civic monuments were built as turn-of-the-century status symbols during the apogee of the rubber boom. The Teatro Amazonas in Manaus and the Teatro da Paz in Belém, the grand regional opera houses, symbolized with architectural exuberance the local elites' conspicuous consumption during the Amazonian rubber boom. The renovated gold dome of the Teatro Amazonas, one of Brazil's most widely known landmarks, still dominates the urban skyline of Manaus.

Belém and Manaus retained their metropolitan dominance in the far-flung urban system of Amazonia even after the end of the rubber boom in the early twentieth century. As late as 1960 the entire Brazilian North Region had only twenty-two cities exceeding populations of five thousand; of these urban centers, only four had populations of more than twenty thousand, and only Belém and Manaus had more than one hundred thousand residents (table 1.2). Yet Amazonia's regional city-system changed dramatically after 1970, largely because of government-sponsored programs of regional integration. The number of urban centers with more than five thousand residents roughly doubled from 33 cities in 1970 to 67 in 1980; by 1991 the total number had doubled again to 133 cities in Amazonia (IBGE 1992).

As recounted in chapter 3, regional programs of national integration began in earnest with the completion of the Belém-Brasília Highway in the 1960s. The Transamazon Highway, cut across the breadth of the Amazon Basin in the early 1970s, failed to achieve the grandiose original goals of the directed colonization program, but it served to open new areas of the *terra firme* uplands to spontaneous in-migration. A series of other new highways followed the Transamazon Highway, as did new regional airports, improved telephone communications, and national television and radio networks. The POLAMAZONIA program of the late 1970s, which emphasized large-scale cattle ranching, mining, and timber projects, also induced migrants and corporations to move into the region. The POLONOROESTE regional development program of the early 1980s again promoted colonization, this time in Rondônia, with disastrous ensuing impacts on the rainforest.

These contemporary programs of regional development tended to promote a common geopolitical strategy: to direct the expansion of economic activity into specific frontier areas (southern Pará, Rondônia, Acre, northern Mato Grosso) deemed vital to national interests and to link those areas directly to the populous economic heartland of the Brazilian Southeast. The national metropolitan centers of Brazil, once marginal to the extractive-mercantile economy of Amazonia, have come to play more central political-economic roles in regional affairs under the state-sponsored regime of national integration. Conversely, previously isolated areas of the Amazon periphery were thrust into the prevalent national currents of commerce, migration, and popular culture. No longer so dependent on circuitous river transport to the regional metropolitan centers of Amazonia, residents of the frontier towns could travel to and communicate with national metropolitan centers directly.

Along with the proliferation of new settlements in the region, the degree of metropolitan primacy in Amazonia has steadily declined. Although the large metropolitan centers have grown to unprecedented sizes, in relative terms their population shares have steadily declined. In 1950, 67.9 percent of the total urban population of the Brazilian North Region resided in the municipalities of Belém and Manaus. By 1991 only 41.8 percent of Amazonia's urbanites lived in the two primate municipalities (table 5.1).

Table 5.1 Municipal populations of Belém and Manaus, as percentages of the total urban population of the North Region, 1950—-1991

	1950	1960	1970	1980	1991
Belém	43.9%	41.7%	38.9%	30.7%	23.0%
Manaus	24.0	18.1	19.1	20.8	18.7
Belém and Manaus	67.9	59.8	58.1	51.6	41.8
Other urban centers	32.1	40.2	41.9	48.4	58.2

Sources: IBGE, *Anuario estatistico do Brasil: 1984,* Rio de Janeiro, 1985; and IBGE, *Sinopse preliminar do censo demográfico: 1991,* Rio de Janeiro, 1991.

Another dramatic urban change in Amazonia has been the relative decline of Belém vis-à-vis the recent rise of Manaus. The urban concentration in Belém has fallen by half in recent decades: in 1950 the city claimed 43.9 percent of the North Region's urban population, compared to 23.0 percent of the regional city-dwellers in 1991. Manaus, on the other hand, has witnessed spectacular growth due to the commercial and industrial development of its free trade zone. The local elites of Belém simply have not been able to compete with the national and international capital pouring into Manaus since the 1960s in response to federally subsidized growth-pole policies of the Zona Franca de Manaus. This free trade zone, which emphasized industrial employment in off-shore assembly plants of multinational corporations, in turn attracted massive interregional migrations to the upper Amazon region (Correa 1991; Despres

1991). Since 1970 the population of Manaus has burgeoned from 286,000 to just over 1 million inhabitants in 1991 (table 5.2). During the 1980s, however, Manaus also experienced a declining concentration of the urban population of the Brazilian North Region due to the growth of the smaller inland cities. In 1991, 18.7 percent of Amazonia's urban residents resided in the capital of Amazonas (table 5.1).

Changing patterns of urban primacy are also reflected in the more equalized demographic profiles of regional cities. Belém do Pará, long the preeminent regional metropolis, as recently as 1970 had a metropolitan population of over twice the size as that of Manaus. In fact, Belém's population in 1970 exceeded the combined total of the rest of the fifteen most populous municipalities in the Brazilian North Region. The erosion of Belém's position has been evident more in relative than in absolute terms, since the capital of the state of Pará has continued to grow steadily throughout the postwar period. Metropolitan Belém grew by a healthy 3.69 percent annually between 1970 to 1991. However, this urban growth rate, high by North American or European standards, was the lowest of any of the fifteen most populous municipalities in Amazonia. Manaus nearly doubled this growth rate, averaging a vigorous 6.17 percent of annual increase during the last two decades. Manaus has closed the former gap with Belém. If recent rates of growth continue, Manaus eventually will be the largest regional metropolis (table 5.2).

Despite the continued growth of the large metropolitan areas, particularly Manaus, the intermediate cities now grow even more rapidly. For example, the capital of the land-rush state of Rondônia, Porto Velho, officially grew from just 48,839 in 1970 to 229,410 in 1991. With an average annual population growth rate of 7.64 percent over the last two decades, Porto Velho surpassed both Santarém, Pará, and Macapá, Amapá, to become Amazonia's third largest metropolitan area.[2] Rio Branco, capital of the neighboring state of Acre, experienced a 7.58 percent annual growth rate from 1970 to 1991. Remote Boa Vista, capital of gold-rich Roraima, grew by a dizzying 9.66 percent a year over the recent twenty-one-year period (table 5.2).

The inland cities located in areas of active frontier settlement experienced the highest rates of demographic growth among the fifteen largest urban centers of the North Region between 1970 and 1991. The cities of southern Pará, recipients of massive in-migrations from adjacent states, grew the fastest. Marabá, the emerging regional metropolis, saw its population increase annually by 9.08 percent. The town of Altamira, along the Transamazon Highway in Pará, expanded by 11.03 percent annually. Itaituba, located near the gold camps in the vicinity of the Tapajós River in Pará, grew by an astronomical 16.00 percent a year. The highest growth rate of all occurred in a cattle boomtown of southern Pará, Paragominas. This town, named for the many migrants from the states of Pará, Goiás, and Minas Gerais, emerged near the Belém-Brasília Highway in the 1960s. Paragominas added an average annual increase of 16.92 percent to its population between 1970 and 1991 (table 5.2).

Table 5.2 Fifteen largest urban centers, by municipal urban populations, North Region of Brazil, 1970—-1991
1991

RANK	MUNICIPALITY & STATE	1970 URBAN POPULATION	1991 URBAN POPULATION	ANNUAL GROWTH RATE
1	Belém, Pará[a]	611,497	1,309,517	3.69%
2	Manaus, Amazonas	286,083	1,005,634	6.17
3	Porto Velho, Rondônia	48,839	229,410	7.64
4	Santarém, Pará	70,021	179,759	4.59
5	Rio Branco, Acre	36,095	167,457	7.58
6	Macapá, Amapá	55,915	153,556	4.93
7	Boa Vista, Roraima	17,154	118,928	9.66
8	Marabá, Pará	16,486	102,364	9.08
9	Castanhal, Pará	26,192	92,740	6.21
10	Jí-Paraná, Rondônia[b]	——-	75,384	——-
11	Itaituba, Pará	2,760	62,278	16.00
12	Abaetetuba, Pará	19,108	56,400	5.29
13	Altamira, Pará	5,734	51,587	11.03
14	Bragança, Pará	16,606	49,537	5.34
15	Paragominas, Pará	1,747	46,584	16.92

Source: Instituto Brasileiro de Geografia e Estatística, *Censo demográfico,* various years.

[a]*Includes the municipalities of Belém and adjacent Ananindeua, which jointly constitute the metropolitan area. The 1991 population includes suburban and exurban "agglomerations of urban extension" in the rural zones.*
[b]*The municipality of Jí-Paraná, Rondônia, had not yet been created in 1970.*

The contemporary regional decline in overall metropolitan primacy should not obscure the continuing importance of Amazonia's historic primate cities. Though the smaller, more remote urban centers generally have grown the fastest in recent years, the largest metropolises continue to experience rapid growth, too. Indeed, the average annual 6.17 percent increase in the population of Manaus over the last two decades is high by any conventional standard, though it ranks ninth among the fifteen largest urban centers of the North Region. Even the recent growth rate of

Belém, 3.69 percent annually, which is relatively modest for the region, far exceeds that of most cities in the developed world. For example, most older North American central cities have declined in population since 1950.[3] The contemporary growth of massive regional metropolises in Amazonia reflects a complex interaction of global and local forces to which we now turn.

REGIONAL CITIES IN THE GLOBAL SYSTEM

The urban transformation of Amazonia, long a peripheral region in the global economy, provides empirical evidence for evaluating the contemporary "world cities" hypothesis, which is summarized in chapter 2. This perspective stresses the development of a global hierarchy of cities, based on surplus extraction from the resource- and labor-extractive peripheries, benefiting the larger metropolitan centers, a theory we test in chapter 9. Armstrong and McGee (1985), for example, outline four general levels in the global political-economic hierarchy of places: global centers, national metropolises, regional centers, and small cities and rural areas (figure 2.4). In this model of global interactions, the higher levels of the urban hierarchy centralize strategic decision making, exercise technological and financial dominance, promote consumer values through the media, and accumulate capital through surplus transfers. The lower levels of the urban hierarchy, the staging grounds of capitalist expansion, supply an essential reservoir of raw materials, human energies, and internal markets for global economic expansion. The world cities hypothesis explicitly adopts a hierarchical form of spatial linkages.

The most important contemporary global centers are New York, London, and Tokyo, from which the most important international banks, financial markets, and transnational corporations operate. Continental cities perform such financial and corporate roles on a more regional scale: Singapore and Hong Kong act as primary centers of accumulation in eastern Asia, while São Paulo and Buenos Aires do so in South America. National metropolises are centers of political control, cultural influence, and corporate-financial headquarters at the country level. In Brazil, in addition to the preeminent financial-industrial center of São Paulo, the national capital of Brasília and the former capital of Rio de Janeiro also exercise important political, socioeconomic, and cultural functions of national cities. These international and national centers, which progressively accumulate capital and attract migrants from the hinterlands, are the nerve centers and the principal beneficiaries of economic globalization.

Dynamic regional cities may also serve as important staging areas of capital accumulation in the world-city system. Belém and Manaus, the dominant regional cities of Amazonia, have long maintained strategic political-economic links both to external centers of power and to the lower-order settlements. Historic patterns of mercantile trade favored urban primacy during Amazonia's past boom-and-bust cycles in natural resources. From the large metropolitan centers, commercial elites, some directly linked to foreign capital, dominated the local economies of vast rainforest hinterlands. Yet, contemporary national integration, linking Amazonia directly to the economic heartland of Southeast Brazil, has selectively and differentially integrated two of the Amazon's regional metropolises (Belém and Manaus) into the national economy. After thirty years of national integration under the military regime, many of the intermediate functions (e.g., finance, management, information services) provided by these two regional metropolises have been usurped by national metropolises in the economic core region in the Southeast. In contrast to Manaus, which paradoxically is more favorably positioned in relation to the transnational level of production, Porto Velho has become effectively marginalized in both the national and global economic spheres.

These important shifts in geopolitical and economic influence suggest that it would be a mistake to collapse the complex political, economic, and demographic changes of the contemporary period into a single world cities hypothesis. Numerous idiosyncrasies emerge in the world cities analysis of Amazonia. For example, while Belém, once the Amazon region's gateway to the world, has been relatively insulated from transnational activity, it retains a significant role as the administrative capital of the Legal Amazon region, disbursing SUDAM fiscal incentives throughout the region. Manaus, meanwhile, is poised to assume a very different role in the world-city system, possibly as a high-technology biodiversity center or a regional "technopole" (Castells and Hall 1994). Other state capitals have seen their direct market relations with local centers in the frontier areas decline in relative importance. New transportation and communications connections now link the inland centers directly to the dominant economic centers of southern and central Brazil, defying the hierarchical spatial framework accepted in some world cities formulations (Armstrong and McGee 1985). Hence, it is necessary to consider factors of spatial change originating not solely from the global forces of capital accumulation but also from distinctive national, regional, and local sources as well.

THE INFORMAL SECTOR AND
PERI-URBAN EXPANSION

As Amazonia's city-system has become more fully integrated into international currents of trade, investment, and consumption, the region's social structure has simultaneously become internally differentiated and stratified. The increasing internal differentiation of Amazonia is vividly apparent in the expansion of peripheral squatter settlements, that is, in the process of "peri-urbanization." Before turning to case studies of Belém, Manaus, and Porto Velho, we briefly explore the burgeoning shantytowns and informal sector economies of regional cities.

The rapid urbanization of the metropolitan fringe has been a largely unplanned by-product of the government's official growth-pole development strategy of the post–World War II era. Throughout Amazonia, expansion of the informal sector of less capitalized, labor-intensive, officially unregulated economic activities has found its most visible expression in peri-urban expansion: the spontaneous emergence of peripheral shantytowns. These shantytowns of self-constructed housing, often located illegally on unoccupied landholdings, go by a variety of regional names: *favela* is the most common term for informal urban housing but generally denotes a hillside slum residence; the *baixada,* a popular neighborhood located on low-lying, marshy terrain, predominates in Amazonia.[4]

Whatever the local terminology, the peripheral shantytowns exhibit abundant social problems, public health dangers, and deficiencies of physical infrastructure. Yet the squatter settlements in Amazonia do not conform to many of the negative stereotypes associated with marginality theories, as Perlman (1976) argues in her classic study of the favelas of Rio de Janeiro (also see Bienen 1984; Lobo 1982). The shantytown dwellers may be geographically peripheral and politically marginalized, but in socioeconomic terms they participate actively in urban life. The dynamic informal sector of small-scale activities also provides much of the manual labor for the urban economy.

Without a doubt the general process of shantytown development, often called favelization, has greatly fueled the Amazon's urban explosion. Yet it is difficult, if not impossible, to know the precise extent of the phenomenon as informal housing evolves toward more regularized forms over time. On the one hand, official figures generally understate the pervasiveness of the informal sector. The Brazilian Institute of Geography and Statistics, for example, now publishes statistics on the number and size of shantytowns in Brazilian cities.[5] According to official criteria, authorities counted twenty-seven squatter settlements in the metropolitan area of Belém, encompassing 42,096 housing units or 15.3 percent of the housing stock in the metropolis (IBGE 1992). Conservatively assuming five residents per household, official statistics would indicate a population of more than two hundred thousand in Belém's squatter settlements.

Academic and journalistic accounts, on the other hand, often arrive at higher estimates of informal housing units. For example, a survey carried out by social scientists at the Federal University of Pará, located next to one of Belém's largest shanties at Guamá, found that 60 percent of the metropolitan area consisted of spontaneous, unregulated settlements. Low-lying shantytowns subject to frequent flooding, the vast *baixadas* encompassed an area of over 20 square kilometers and housed more than 450,000 residents of the city (plate 5.1). It has been asserted that this informal urbanization represents a larger regional process of "passive proletarianization, dissolving traditional forms of (re)production, which

Plate 5.1. View of Vila de Barca, a low-lying *baixada* shantytown of Belém, 1990   *Photo: Brian J. Godfrey*

for the great majority of direct producers does not translate into a salaried position in the formal labor market" (Mitschein et al. 1989:23).

In contrast to Belém, Manaus exhibits a more pervasive formal sector of industrial wage-labor and associated urban services. Due to investments generated by the free trade zone, the capital of Amazonas has witnessed the development of an industrial complex remarkable both for its massive scale and remote regional location. Yet the industrialization associated with the Zona Franca de Manaus, favored by the massive fiscal incentives of the federal government, remains anomalous among the larger metropolitan areas of Amazonia. This kind of intraregional idiosyncrasy, however, indicates the need to ground our comparative analysis in the finer points of particular cities. The succeeding case studies of the region's three principal cities illustrate in greater detail the general patterns and local variations in Amazonia's metropolitan growth.

Belém: Primate City of the Lower Amazon

The evolution of Belém do Pará, long the preeminent regional metropolis, illustrates two contrasting historical phases of urban growth in Amazonia. First, during the Portuguese colonial period and for more than a century of Brazilian national life, the city served as a strategic point of geopolitical control and as a center for extractive-mercantile trade during the various boom-and-bust cycles in natural resources. Second, during the postwar period of national political and economic integration, Belém became the regional administrative center for development programs in the Lower Amazon Basin. Although Belém cannot rival the extensive modern industrial plant of Manaus, the capital of Pará has become an important regional development pole in its own right. The city's recent growth also has uneasily juxtaposed modern and informal sectors. Sleek, tall, downtown skyscrapers now contrast with spontaneous shantytowns on the vast urban peripheries (plate 5.2).

Belém's historic importance as a regional center was due to its privileged geographic situation as an accessible gateway port located at the entrance to the vast Amazon River system. In 1616 the Portuguese founded Nossa Senhora do Belém do Grão Pará, known popularly as Belém do Pará, on the navigable estuary south of Marajó Island leading to the main channel of the Amazon River. Near the tip of a low-lying peninsula jutting into the confluence of the Guamá River and Guajará Bay, the Portuguese established a fort, cathedral, and residential square on a

Plate 5.2. View of the downtown skyline of Belém, 1990. Modern sky-scrapers loom over the older, low-lying structures. In the foreground are the Victorian towers of the nineteenth-century Palacete Bologna, once the mansion of a local rubber baron. *Photo: Brian J. Godfrey*

defensible bluff overlooking a sheltered harbor. This upper city (called simply *cidade*) became the administrative core of colonial Belém, over-looking the harbor, by which a lower city of commerce and residence (called *campina*) grew up on marshy ground. During a series of economic cycles in primary products Belém controlled the mercantile functions required for interregional trade and therefore became the main urban center in eastern Amazonia (Penteado 1968).

The most notable phase of resource-based growth occurred with the rubber boom (circa 1870–1920). As the main regional commercial cen-ter, Belém served as the headquarters for Amazonia's most important import-export houses and commercial enterprises at the turn of the cen-tury. Prosperity from the rubber trade permitted urban expansion, and a cosmopolitan flair was reflected in Belém's elegant opera house, impres-sive mansions, expanding streetcar lines, and bustling commerce. Belém's population grew from 61,997 in 1872 to 236,402 in 1920—a 281 percent increase. The capital's growth outpaced that of the state of Pará, which grew from 275,237 in 1872 to 983,507 in 1920—a 257 per-cent increase over a period of roughly fifty years (IBGE 1985).

The mercantile phase of metropolitan development reached its apogee in turn-of-the-century Belém do Pará. The wealth from the rubber trade permitted conspicuous consumption on the part of the rich and an impressive municipal modernization of the urban infrastructure, rivaled regionally only by the similar transformation of Manaus. From 1897 to 1911 a powerful mayor, Antonio Lemos, launched ambitious public works projects to beautify and sanitize Belém with grand public squares and parks, new municipal buildings and monuments, paved and tree-lined avenues, a new public market (the Mercado "Ver-o-Peso"), electric streetcars, and a sewage system still in use. When Lemos was criticized for his authoritarian methods of urban renewal, he cited the inspiring example of Baron Haussmann, who in the late nineteenth century had demolished and rebuilt the medieval core of Paris. As a result of rising popular discontent, fed by vitriolic political reports in a rival newspaper, violent mobs took to the streets and overthrew the Lemos administration in 1912, shortly before the prices for natural rubber plummeted on the world market. Belém's boom years, which had lasted for several decades, thus abruptly ended with the collapse both of the reigning political machine and of the Brazilian rubber trade (Rocque 1973:447).

After the end of the rubber boom, Belém entered a prolonged period of economic stagnation in the early twentieth century. The population of the city actually declined from 236,402 in 1920 to 206,331 in 1940. Only after World War II, as the city's rejuvenated role as an Amazonian development pole encouraged new in-migration, did the metropolitan population began to grow again. By 1950 Belém's municipal population of 254,949 finally surpassed that of the census of 1920. In 1980 the population of Belém had reached 933,287 (IBGE 1985:81). In 1991, the year of the last decennial census, Belém's municipal population exceeded 1.2 million, and the metropolitan population had grown to more than 1.3 million (IBGE 1992).

Against the backdrop of this growing population, contemporary urban expansion has been constrained by the low-lying peninsular site of Belém. At least 40 percent of the metropolitan area lies on terrain less than two meters above sea level, making it subject to flooding as the rivers rise during the rainy season (Santos et al. 1991). Historically, as Belém expanded outward from the central area near the port, the fashionable neighborhoods took the higher topographic positions along the elevated central spine of the peninsula, away from the swampy riverfronts, which are vulnerable to flooding, insects, and diseases such as malaria and hepatitis. The wealthy neighborhoods of Batista Campos, Nazaré, and São Brás, for example, occupy the central residential axis of the city. These neighborhoods have witnessed the conversion of decaying mansions, once homes of wealthy rubber barons, to higher-density residential towers since World War II. Planned suburban developments on the metropolitan periphery also have become popular among the upper-middle classes in recent years. However, most new housing in this circumscribed riverine city has come in the form of spontaneous settlement in peripheral low-lying areas, the *baixadas,* which are unlikely to have piped water, sewerage, public transportation, and other basic urban services (fig. 5.2). Roughly half the metropolitan population now lives in the shantytowns of the swampy *baixadas* (Mitschein et al. 1989).

Despite Belém's continued growth, the proportion of the state's population residing in the capital region has declined in recent years.[6] In 1950 about a quarter of the state's population lived in the immediate vicinity of the Belém. Improved regional transportation and resultant interregional migrations initially favored the capital's urban primacy in Pará. The opening of the Belém-Brasília Highway in 1960 broke the state's long terrestrial isolation from the other major cities of Brazil. Subsequently, the Belém microregion's percentage share of Pará's total population grew, reaching a peak of nearly one-third of the state's total in 1970. The construction of the Transamazon Highway and other frontier roads after 1970, however, led to decentralizing tendencies in Pará. With the growth of the state's inland cities, the relative demographic concentration in Belém has declined somewhat to about 28 percent of the total population of Pará (table 5.3).

Pará's recent tendency toward urban decentralization should be put into a wider regional perspective. After all, the accessible areas of eastern Amazonia have been among the regions most affected by contemporary frontier urbanization. Large-scale migration to southeastern Pará has expanded the settlement frontier, peopled new inland boomtowns, and gave them strong ties to adjacent states. The weakened socioeconomic links to Belém, along with the distinct regional origins of the largely migrant population, have given southern Pará a strong separatist movement in favor of independent statehood (see chapter 7). The massive interregional migratory influx also helps explain an apparent paradox. On the one hand, only 50.3 percent of the population of Pará was urbanized in 1991, falling well below the regional average of approximately 58 percent urban residents (IBGE 1991). On the other hand, nine of the fifteen largest urban centers in the entire North Region, as measured by municipal urban population, were located in Pará (table 5.2).

Table 5.3 Population concentration, Belém, Manaus, and Porto Velho, 1950—1991

	PARÁ		AMAZONAS		RONDÔNIA	
Year	Total State Pop. (000)	% of total pop. in Belém metro area[a]	Total State Pop. (000)	% of total pop. in Manaus metro area[b]	Total State Pop. (000)	% of total pop. in Porto Velho metro area[c]
1950	1,123	25.1	514	32.5	37	73.8
1960	1,538	28.2	715	29.0	70	72.2
1970	2,197	31.8	961	44.0	111	75.7
1980	3,403	30.0	1,430	52.1	491	29.2
1991	5,182	27.9	2,103	55.4	1,131	26.0

Source: Instituto Brasileiro de Geografia e Estatística (IBGE), *Censo demográfico,* various years.

[a]The Belém metropolitan area (microregion), as defined by the IBGE (1991), includes the municipalities of Ananindeua, Barcarena, Belém, and Benevides.

[b]The Manaus metropolitan area (microregion) includes the municipalities of Autazes, Careiro, Careiro de Várzea, Iranduba, Manacapuru, Nanaquiri, and Manaus (IBGE 1991).

[c]The Porto Velho metropolitan area (microcregion) includes the municipalities of Porto Velho and Vila Nova de Mamoré (IBGE 1991).

Belém's declining primacy in the state of Pará has been apparent in economic as well as in demographic terms. Once the dominant industrial, commercial, and service center in Pará, the capital region has experienced a steady decline in its proportion of statewide economic output since about 1970. By 1985 the capital's position was strongest in commerce and services. The large-scale development projects implanted in the state's interior in recent years have simply outstripped Belém's industrial base, which was mainly dedicated to wood products (e.g., plywood and paper pulp) and food processing (e.g., palm oil, agricultural products, and seafood). Future economic prospects for Belém depend largely on strengthening the city's role as a regional business headquarters, governmental administrative base, commercial entrepôt, and service center (IDESP 1991). The recent experience of urban deconcentration in lower Amazonia contrasts sharply with the contemporary situation in the Upper Amazon region. In the other giant state of the North Region, Amazonas, metropolitan primacy remains very much intact.

Manaus: Development Technopole of the Upper Amazon

The Portuguese originally founded the settlement at Manaus in 1669 when they christened a fortress, São José da Barra do Rio Negro, located 18 kilometers above the confluence with the main channel of the Amazon River.[7] This defensive outpost was strategically set on a gently rising peninsula jutting into the Rio Negro and bounded on several sides by small streams *(igarapés)*. Portugal's motives for establishing this military post and trading center in the Upper Amazon Basin, roughly 1,200 kilometers inland from the mouth of the Amazon River system, were clearly geopolitical. Locked in imperialistic competition with Spain over control of the disputed colonial frontiers of South America, Portugal intended to win control of the Amazon Basin far beyond the line of demarcation set by the earlier Treaty of Tordesillas. Physical geography and colonization history favored Portuguese aspirations: as opposed to the Spanish, faced with difficult overland routes across the Andes, Brazilian settlements on the Atlantic coast permitted relatively easy regional access via the Amazon River system. The Portuguese territorial ambitions ultimately succeeded and were formally ratified by the Treaty of Madrid, signed in 1750, which officially granted the bulk of the Amazon Basin to the Luso-Brazilians (Benchimol 1977).

The colonial settlement of Manaus, then known as Vila da Barra, centered on a fledgling commercial district, located near the fort and river port. Although then still a small town, Manaus already had a discernible social ordering of space in the late eighteenth century. The white settlers occupied the better houses on two streets close to the fort and the church, the two most important local institutions. The Indians, low in social status, lived in huts farther from the town center. At the public market, located on the shores of the Rio Negro, the local population of Indians and *caboclos* sold agricultural produce and forest commodities and purchased diverse manufactured commodities. As the small settlement grew to a metropolis during the nineteenth and twentieth centuries, this basic social stratification of space persisted: the upper classes preferred the central areas of Manaus, and those of lower status were relegated to the urban periphery (Benchimol 1977). Even today residential patterns favor metropolitan centrality along a dominant commercial spine leading outward from the central business district near the waterfront. The contemporary elite district of Adrianópolis lies in an affluent central axis of residential expansion, where the most elegant shops, the Amazonas Shopping Center, country clubs, and other urban amenities are now to be found.

During the rubber boom from roughly 1870 to 1920 migrants from other regions of Brazil poured into Amazonia. According to official Brazilian census data, the population of the state of Amazonas exploded from 57,610 in 1872 to 363,166 in 1920—a 530 percent increase over the period. The demographic growth of remote Amazonas between 1872 and 1920 outpaced that of more populous Pará, which grew by 281 per-cent. Nevertheless, the population growth in the capital of Amazonas actually lagged behind rural areas during the boom years: the city of Manaus increased from 29,334 residents in 1872 to 75,704 in 1920—a 158 percent increase. As noted earlier, Belém witnessed a 281 percent increase in its population between 1872 and 1920 (IBGE 1985).

Even if the demographic growth of Manaus did not keep pace with that of Belém or the rural zones of production during the boom years, the regional benefits of the Upper Amazon rubber trade transformed the capital of Amazonas. The prosperity of the boom years attracted new groups to Manaus, giving the city a vibrant and cosmopolitan air. English investors and engineers entered the city to build the physical infrastruc-ture of the port, trolley lines, and a power plant. Portuguese merchants ran the most important import-export houses. Syrian and Lebanese ped-dlers moved into small-scale commerce of a more informal nature. Migrants from the northeastern state of Ceará, then in the midst of a pro-longed drought, migrated to Amazonas in large numbers, joining the ranks of both *seringueiros* and *seringalistas*. Successful immigrants mingled with the local commercial and governmental elites in exclusive social clubs. In addition, it became fashionable to patronize the arts, which gave the city a patina of European culture evident in the opera performances at the Teatro Amazonas. Although the commercial expansion of the down-town area gradually subsumed many residential areas, the high-income groups moved along an affluent axis of central districts, served by new trolley lines, streetlights, and other urban amenities, while the poorer classes remained on the urban peripheries. Although many foreign immi-grants eventually left and the conspicuous consumption by elites virtually vanished after the collapse of the rubber trade in the early twentieth cen-tury, Manaus remained the undisputed urban center of the Upper Amazon Basin (Benchimol 1977; Despres 1991).

Until the implementation of the free trade zone in the 1960s, the economy of Manaus expanded and contracted with the boom-and-bust cycles of resources extracted from the vast forests and rivers of the Upper Amazon Basin. The contemporary urban-industrial growth of Manaus is all the more remarkable in the historical-geographical context of a peripheral extractive economy, stunted by long-term regional isolation and intractable economic dependence on resource extraction. For centuries Manaus has served as a peripheral mercantile entrepôt for trade in extractive commodities, such as fish, rubber latex, and other forest products, on the distant margins of the world economy. The growth-pole policy of urban-industrial development in western Amazonia, initiated by the Brazilian military after the 1964 coup d'état, followed a geopolitical rather than purely economic rationale. The remote location of Manaus raised transport costs for regional construction, manufacturing, and commerce. Yet the Brazilian generals intended to complete a process of territorial expansion begun by the Portuguese colonizers three centuries before: modern industrial and commercial development would finally conquer the vast national interior.

The industrialization of Manaus followed the military's promulgation of Decree Law 228 on February 18, 1967, which established the Manaus Free Trade Zone, known in Portuguese as the Zona Franca de Manaus (ZFM), along with a supervisory agency, the Superintendência da Zona Franca de Manaus (SUFRAMA). The ZFM provided an export enclave for industrial growth, based mainly on the development of assembly plants manufacturing imported component parts (plate 5.3). The public sector was able to attract national and foreign capital to the remote Upper Amazon by improving the physical infrastructure and providing generous fiscal incentives for investors. Hundreds of kilometers of new roads were built in Amazonas, and a new international airport received more than 203,000 passengers in 1975 (SEPLAN 1976). In addition, SUFRAMA completed construction of a vast industrial district, located adjacent to the new international airport about five kilometers from the center of Manaus. Approved development projects were to be exempt from import and export duties and from the manufacturers' sales tax (Despres 1991).

These inducements were sufficient to attract many national and foreign firms to the ZFM By 1975 less than a decade after the inauguration of the free trade zone, SUFRAMA had approved 140 development projects. By 1991, 811 projects had been approved by SUFRAMA. Of the 661 projects already in progress by 1991, 184 (27.8 percent) were in the Industrial District, 330 (49.9 percent) elsewhere in Manaus, and 147 (22.2 percent) outside of the capital of Amazonas. The electronics industry has economically dominated the SUFRAMA-supported projects, despite the formidable locational disadvantages of Manaus for a sector dependent on imported components and external markets. In terms of the investment value of the industrial operations established by 1991, nearly half (49.4 percent) of the capital was dedicated to electronic and communications assembly plants, 15.2 percent to transport materials, and 7.1 percent to mechanical industries (Salazar 1992).

Although the precise proportion of foreign capital invested in SUFRAMA projects is hard to ascertain, local affiliates of Japanese, American, and European multinational corporations clearly have led the hegemonic electronics sector. Except for one large Brazilian firm, Gradiente, such multinationals as Phillips, Philco, Toshiba, and Sony have predominated. Of the forty-three electronics firms operating in 1983, twenty-two multinational affiliates stood out. These multinationals employed 65 percent of the 21,860 workers in the electronics industry of Manaus (Despres 1991:48–49).

These investments created a new urban industrial labor force in Manaus. By 1991 the state of Amazonas registered 151,432 industrial sector employees; 46.8 percent worked at SUFRAMA-approved projects in

Plate 5.3. View of factory in the Zona Franca de Manaus, 1993. In the foreground potential workers line up to fill out job applications.
Photo: Brian J. Godfrey

the Industrial District, 47.3 percent worked elsewhere in Manaus, and 5.9 percent worked in the interior of the state (Salazar 1992). Given the high percentage of the labor force employed in industry, the informal sector of Manaus appears to be disproportionately smaller than in either Belém or Porto Velho. Although thousands of residents live precariously in stilt houses *(palafitas)* over the river and the swamps on the periphery of Manaus, the proportion probably does not match that of the *baixada* shantytown dwellers of Belém.

A large-scale survey of 6,500 families in Manaus during 1984 found that 94.2 percent of the population lived in single-family homes on firm land *(terra firme)*. The provision of urban services was relatively high for a rapidly urbanizing center. Most of Manaus's population was connected to the public water supply (96.7 percent) and to the public electric lines (98.7 percent). More problematic was the sanitary situation. Only 8.6 percent of the survey respondents were connected to the city's sewers, fully 50 percent depended on septic tanks, and the rest of the respondents appeared to have precarious patterns of waste disposal (CODEMA 1984).

The contemporary industrial boom of Manaus has transformed the city and its regional situation. As opposed to the previous extractive-mercantile patterns, current industrialization is state-sponsored and urban-based, unrelated to the traditional forest economy of the region. The ZFM bypasses the traditional patron-client exchange network of the *aviamento* system, replacing it with a dependence on external investments by foreign and multinational firms. These structural changes are profound and far-reaching, as noted by Despres (1991:243):

> With the creation of the Free Trade Zone and the industrialization of Manaus, the social formation that came into being with the *aviamento* system and the collection and trade of forest products gave way to one in which the capitalist production of consumer goods by national and multinational firms assumed a hegemonic position relative to traditional mercantilist interests, petty commodity producers, and the like. The hegemony of these exogenous firms is substantiated by the position they occupy in the urban economy as a consequence of the capital they control, the employment they provide, the fiscal incentives they receive from the federal government, and the investment of public funds in infrastructural projects that is made on their behalf.

Whatever its merits, the growth fostered by the ZFM has exacerbated the problems of uneven regional development in the Upper Amazon Basin. Lacking the expansive agrarian frontiers of Pará and Rondônia, the state of Amazonas has grown mainly around the industrialization and commercial expansion in the export enclave of Manaus. This kind of industry-led growth in turn has led to a proliferation of local service-sector activities in support of both the leading businesses and the growing local consumer market. In 1985 the capital city claimed 91 percent of the state's industrial gross product, 86.6 percent of the state's commercial product, and 85.8 percent of the state's service-sector gross product. Although the statewide economic hegemony of Manaus has slipped somewhat in recent years, the concentration in the capital of Amazonas still is overwhelming and roughly twice that of both Pará and Rondônia.

As a consequence of the ZFM, the demographic dominance of Manaus within the huge state of Amazonas has steadily increased in recent years, in contrast to Belém's decline in significance. Interregional migrations, from the interior of Amazonas and other Brazilian states, have swollen the population of Manaus. In fact, about half the state's population now resides in the capital. The growth of Manaus presents another paradox: although Amazonas remains a scantily populated state, with an average population density of only 1.3 persons per square kilometer, it is now more than 72 percent urbanized (IBGE 1991).

In significant ways the states of Pará and Amazonas have pursued divergent paths of urban and regional development in recent years. In Pará, the demographic gap between Belém and other urban centers has steadily diminished in recent decades. In 1960 the city of Belém had a population of 359,988, a figure 14.4 times the size of the state's second largest city, Santarém, which had a population of only 24,924. By 1991 the difference in proportions between the populations of Belém (952,170) and Santarém (222,708) had shrunk to a factor of 4.2. Contemporary growth-pole policies in Amazonas have had the opposite effect. In 1960 Manaus had a population of 154,040, that is 16.9 times the population of the next largest city in the state, Parintins, which had a population of only 9,068. By 1991 the difference between the cities of Manaus (1,005,634) and Parintins (40,457) had widened by a factor of 24.8 (IBGE 1992).

The lack of any intermediate-sized cities between Manaus and Parintins indicates a high degree of metropolitan primacy in the asymmetrical urban system of the state of Amazonas, in contrast to the more evenly tiered city-system emerging in Pará. Since the Second World War the concentration of the state population in metropolitan Manaus has risen steadily as the capital region has increasingly polarized the entire territory of Amazonas. The microregion of Manaus, including the capital and six nearby municipalities, has received a heavy migratory influx from other regions of the state and the nation.[8] In 1950 only 32.5 percent of the state's population lived in the Manaus microregion, while 55.4 percent did so in 1991. Conversely, the percentage of the state's population living in the interior of Amazonas has steadily fallen from 67.5 percent in 1950 to 44.6 percent in 1991 (table 5.3).

In short, Manaus has become the urban-industrial showcase of Amazonia, surpassing in many ways its chief regional rival, Belém. With an urban population exceeding 1.0 million, Manaus now closely approaches in size the other metropolis on the Lower Amazon with its 1.3 million residents. In addition to narrowing the demographic gap between the two cities, Manaus has taken the leading regional role in advanced technology and industrial development. Its location in the heart of the Amazon's biologically rich rainforests, from which one quarter of all pharmaceutical drugs are derived, presents unique opportunities for its expansion as a secondary global technopole. As a result of the federal policy of fiscal incentives to promote the Free Trade Zone of Manaus, the capital city of Amazonas now outdoes Belém in several conspicuous symbols of modernization: the ultramodern airport of Manaus serves as a major international gateway to Brazil and South America; the city's glittering shopping center, "Shopping Amazonas," boasts the latest in modern consumerism; and the immense industrial park at Manaus is a model of export-enclave planning in the developing world. However, recent cutbacks in federal subsidies for the ZFM, begun during the Collor administration in the late 1980s, make investments in the SUFRAMA-sponsored assembly plants less attractive to foreign and national capital. Given a growing rate of urban unemployment in Manaus by the early 1990s, ecological tourism has received growing attention as an alternative form of regional development—one at odds with the urban-industrial emphasis. In any case, it remains to be seen whether contemporary state-sponsored, import-substitution industrialization can provide sustainable economic growth for the burgeoning population of Manaus.

Porto Velho: The Disarticulated Growth of an Intermediate Regional City

Porto Velho has been shaped by many of the same historic forces that influenced Manaus and Belém, most notably the regional trading boom in rubber latex during the late nineteenth and early twentieth centuries. Yet the contemporary urban growth of Porto Velho generally is more disarticulated from national and global currents than that of the two larger metropolitan centers of Amazonia. Unlike Manaus, a bustling export enclave with growing hegemonic primacy over the urban network of Upper Amazonia, Porto Velho's statewide urban primacy has declined with the recent colonization of the interior of Rondônia. Porto Velho's relative demographic importance within Rondônia has plummeted more dramatically than has been the case in Belém. Although Porto Velho itself has experienced rapid population increase in absolute terms, it has been overshadowed by, and even isolated from, the growth of Rondônia's inland settlement frontiers.

Despite the steady contemporary growth of Porto Velho's population, the capital's relative demographic position in the state has declined precipitously in recent decades. After the end of the rubber boom in the early twentieth century, most of Rondônia's population clustered in the capital. The Brazilian censuses of 1950, 1960, and 1970 found roughly three-quarters of Rondônia's population living in Porto Velho. Although with the massive migratory influx to the inland agrar-

Plate 5.4. View of downtown skyline and Rio Madeira, Porto Velho, Rondônia, 1994   *Photo: John O. Browder*

ian frontier the microregion of Porto Velho grew in population from 84,000 to 134,000 during the 1970s, the capital's proportion of the state's population fell to just 27 percent in 1980. By 1991 the Porto Velho microregion accounted for only about 26 percent of the state's population (table 5.3).

The case of contemporary Porto Velho illustrates a disjunction between several political and socioeconomic forces. Burgeoning Porto Velho serves two main geopolitical functions. First, it functions as Rondônia's political capital, which provides a central bureaucracy, tax revenues, and some leverage over local affairs in the state. Second, the city continues to have a nominal strategic significance to the Brazilian military establishment as an outpost on the Madeira River near the Bolivian border. On the other hand, Porto Velho has little economic importance in Rondônia's largely agricultural and mining economy; moreover, the capital has relatively minimal commercial linkages with the rest of the state. In addition, the demographic and migratory patterns of the Porto Velho population are significantly different from those characterizing the inland settlements. Whereas settlers in the interior generally maintain linkages to their origins in the South and Southeast, Porto Velho's orientation is more toward the traditional riverine populations and the mining sectors of the northwestern corner of the state. To a greater degree than either Manaus or Belém, Porto Velho is functionally disarticulated within the state's urban system.

Porto Velho's origins were both messianic and strategic in nature. The earliest settlement established in the vicinity of present-day Porto Velho was the Jesuit mission village, Santo Antonio, organized by Padre João San Payo in 1723.[9] Repeatedly abandoned and reoccupied, the location acquired both economic and strategic significance as a Brazilian military garrison protecting this rubber-producing region from suspected Bolivian aggressions during the War of the Triple Alliance, which pitted Brazil, Uruguay, and Argentina against Paraguay from 1865 to 1870. The rubber trade that brought such prosperity to Manaus and Belém began to reach the Guaporé Valley in earnest in the 1870s. The bend in the Madeira River near Santo Antonio became a small but bustling center for an unlikely conglomeration of rustic northeastern rubber tappers and sophisticated European and North American railroad construction engineers and bankers.[10] The pivotal event in the early history of Porto Velho was the 1907 decision by an international consortium of business interests to locate the terminus of the infamous Madeira-Mamoré Railroad at a location seven kilometers south of the site of Santo Antonio (Gomes da Silva 1984, 1991; Hardman 1988).

The story of the ill-fated Madeira-Mamoré Railroad provides a prophetic account of the forces that have persistently disfigured the spatial organization of Amazonia. While the vision of building a railroad across the Madeira and Mamoré River interfluve is widely thought to have been the brainchild of the North American Percival Farguhar in the first decade of the twentieth century, the idea was first articulated by the Bolivian general Quentin Quevedo in 1861 and reaffirmed shortly thereafter by the Brazilian engineer João Martins da Silva Coutinho (Gomes da Silva 1984:45–46). In its original Bolivian formulation, the railroad was intended not just to move rubber latex more quickly from the jungles of the Guaporé and Beni Valleys to Santo Antonio, and from there to Manaus, but rather to give landlocked Bolivia access to the sea lanes of the Atlantic Ocean. From Brazil's perspective, the railroad would open Andean gold and silver as well as Bolivian rubber resources to Brazilian exploitation.

Beginning in 1872, with a cooperation treaty in hand, the Brazilian government with its own and then North American contractors launched two ill-fated attempts to construct the railroad in segments. Bad organization, incompetent management, tropical diseases, and almost daily Indian attacks on the gandy dancers foiled the effort (Gomes da Silva 1984). In 1879 another construction attempt was abandoned after hostile Indians wounded its principal builder, the North American engineer Thomas Collins, forcing him to return to the United States, where his personal tragedy was completed when his wife died in an insane asylum. Left behind in Brazil were one locomotive, three freight cars, and hundreds of coffins containing the corpses of North American, Irish, Italian, and Brazilian workers. Just seven kilometers of railroad had been constructed (Gomes da Silva 1984:46).

Nearly thirty years later, in 1907, the North American company of May, Jekill, and Randolph was contracted by Farguhar to reconstruct the ruins of Collins's work. To expedite rail-to-barge transshipment, the terminus of the railroad was moved from Santo Antonio to a site on the banks of the Madeira River seven kilometers away. This point had been named Porto Velho (Old Port) by the Brazilian military fifty years before. The reconstructed roadbed extended from Porto Velho to the Brazilian port town of Guajará Mirim on the Mamoré River border with Bolivia. Yet the timing of this reinvigorated project could not have been more inauspicious. Rail service was finally inaugurated between Porto Velho and Guajará Mirim in 1912, just as the South American trade in rubber latex collapsed, never to revive fully, under pressure from more productive Asian plantations.

During the Farguhar years, lasting roughly from 1907 to 1912, the growing profits of the rubber trade, controlled largely by foreign syndicates, gave international significance to an essentially bilateral treaty between the two neighboring South American nations. The introduction of multinational capital into the Guaporé Valley to underwrite the infamous "Railroad of Doom," renowned as such for its colossal human costs, gave rise to the city of Porto Velho. In comparative historical terms, developments of the Farguhar era in Rondônia provide an interesting precursor to the contemporary corporatist mode of urban development outlined in chapter 4. The influx of foreign capital reduced the historic settlement at Santo Antonio to a squalid barracks community for would-be day workers on the railroad. Meanwhile, the new terminus of the railroad, Porto Velho, became an orderly and clean company town. This social contrast between the neat planned town and the chaotic informal settlement anticipated the corporatist scenarios of urbanization in contemporary Amazonia at Carajás, Tucuruí, and Jarí. A description of the contrast between Santo Antonio and Porto Velho in the early twentieth century might well fit numerous present-day Amazon settlements:

> The urban scenes encountered in Santo Antonio and Porto Velho were completely different: Santo Antonio was favored by braggarts, hucksters, gunslingers. Wine, women, and song were sold for weight in gold. . . . The scenario found in Porto Velho was [sic] opposite of Santo Antonio; there were signs of progress: symmetrical street designs, sewage service, running water systems, a dry cleaner, a great ice factory . . . , a luxury hotel. . . . This contrast was born of the development of Porto Velho, after 1907, in the nature of a headquarters of an international private corporation. Just seven kilometers away, Santo Antonio remained as degraded public property, place of strangers without names or law, the reverse image of urbanization based on the production and commerce. (Hardman 1988, cited in Perdigão and Bassegio 1992:157—-158)

With the collapse of the rubber boom during the second decade of the twentieth century, Porto Velho quickly fell into decline and isolation. Rail service on the Madeira-Mamoré line struggled on, unprofitably, for decades. British interests took over the line from 1919 to 1932, when the Brazilian government assumed control of the indebted railroad. Official attempts to revive the moribund Rondônia economy continued with mixed results. In 1943 Brazilian President Getúlio Vargas established the Federal Territory of Guaporé and named Porto Velho the administrative capital. With the discovery of cassiterite ("tin stone") in the 1950s, Porto Velho began to grow again with a frenzied influx of miners (SET 1983). In 1972 the Brazilian Government finally abandoned the decrepit Madeira-Mamoré Railroad in favor of a new highway project, the BR-364 (Hecht and Cockburn 1989:82). In 1981 the Territory was elevated to the State of Rondônia, with Porto Velho as its capital city. Today the infamous Madeira-Mamoré Railroad occasionally carries tourists on day excursions between Porto Velho and Santo Antonio.

The most important phase in the contemporary development of Rondônia began with large-scale programs to colonize the interior in the early 1970s, as discussed in chapter 6. Although farmlands close to the capital city enjoyed a locational advantage in regard to a major consumer market, soils in Porto Velho *município* were considered unsuitable for intensive cultivation. Consequently, the Rondônia government did not promote agricultural development in the vicinity of the capital, as it did in the interior of the state. Rondônia's contemporary rural settlement focused on opening the state's inland *municípios,* thus diminishing Porto Velho's demographic and economic hegemony (Becker 1987; Coy 1987; Fearnside 1986; Martine 1990; Millikan 1988).

The contemporary extension of Brazil's highway network into Rondônia has eroded the capital's commercial preeminence, which was based historically on Porto Velho's inland river port. With the opening of the BR-319 from Porto Velho to Manaus in the late 1970s and of the BR-364 highway from Cuiabá to Porto Velho in the early 1980s, direct overland road transport between Manaus and Brazil's southeastern industrial core prevailed for the first time over river navigation. Although the volume of freight moving through the port of Porto Velho increased as imports of construction materials, machines, food, and fuel exceeded exports of agricultural and extractive products, Porto Velho's commercial function as a river entrepôt declined relative to overland transport. The capital's commercial sector remains important to the local economy, but it represents a declining share of the state's total commerce. In 1980, 59.6 percent of the value of the statewide commercial output was generated in Porto Velho; in 1985 the capital generated only 52.4 percent of the state's commercial output. In terms of the total number of commercial establishments, only 37.4 percent were located in Porto Velho in 1985, down from 40.5 percent in 1980. The businesses of Porto Velho employed 35.6 percent of all commercial sector workers in the state in 1985, compared to 47.2 in 1980. The expansion of the state's city-system, due to the proliferation of new settlements in the interior, has called into question the capital's former commercial preeminence (IBGE, various years).

Industry is comparatively insignificant in Porto Velho, unlike in Manaus. In 1985 Manaus accounted for 91 percent of the state's industrial output, 66.3 percent of the private industrial firms, and 88.6 percent of the industrial workers. In Rondônia the relative contribution of Porto Velho to the state's industrial product dropped from 54.6 percent to 37.9 percent between 1980 and 1985. During the same period, Porto Velho's share of the state's industrial establishments dropped from 24.7 percent to 22.6 percent. The number of industrial workers in Porto Velho fell from 47.4 percent of the state total in 1980 to 33.1 percent in 1985. This industrial decline results, in part, from the vicissitudes of the gold-mining sector of the *garimpeiros.*

The relative contribution of the mineral extraction sector to Porto Velho's economic output has also declined in recent years, from 86.7 percent in 1980 to 50.1 percent in 1985 (IBGE 1985). Gold mining gained new momentum between 1985 and 1990 although activities were temporarily suspended by the state government in 1989.

Besides the local *garimpo,* Porto Velho's growth has depended on public employment and the service sector. Porto Velho lives largely on public employment: 18,000 federal and 3,000 state workers and 15,000 military jobs (Gomes da Silva 1993[11]). Services also remain essential to the local economy of Porto Velho, even as the capital's overall proportion of statewide service-sector employees has declined in recent years. Overall, the demographic and economic indicators point to a pattern of increasing urban decentralization in Rondônia. Porto Velho also plays a declining role in the state's industrial, commercial, and service-sector activities. Only about a quarter of the state's population now resides in the capital, compared to fully three-quarters as late as 1970 (table 5.3). In short, the contemporary growth of the capital city has been steadily disarticulated from that of its interior hinterland.

PERI-URBANIZATION OF PORTO VELHO

As in the other metropolitan centers of Amazonia, the urban expansion of Porto Velho has occurred mainly on the metropolitan fringe with the rapid growth of shantytowns (fig. 5.1). Since 1990 an estimated 80,000 to 90,000 people have moved to some thirty new peri-urban settlements, some as far as 25 kilometers away from the central business district.[12] According to local authorities, between 60 and 70 percent of the total urban population of Porto Velho now lives in the new peripheral neighborhoods.[13]

As elsewhere in Brazil and Latin America, the peri-urban settlements of Porto Velho often are depicted as housing marginal populations, peopled in this case largely by drug dealers, prostitutes, gold miners, and vagrants. Yet a 1993 survey of randomly selected households in five different peri-urban neighborhoods of Porto Velho refutes such stereotypes. Instead, what emerged was a snapshot of the socioeconomic heterogeneity of Porto Velho's urban fringe. The 298 households surveyed in 1993 were located in five new *bairros* on the edge of the capital: Jardim Eldorado, Esperança da Comunidade, São Sebastião, Ulisses Guimarães, and União de Vitória. In terms of general demographic characteristics, the households on the average consisted of five members; the mean age of the household head was thirty-seven years. With the exception of one *bairro*, São Sebastião, the population was predominantly male. Overall, males represented 56 percent and females 44 percent of the total population of the households surveyed in the five neighborhoods.

Porto Velho's peri-urban population consists mainly of migrants from Amazonia, in contrast to the population of the interior of Rondônia, where recent migrants from the South, Southeast, and Center-West regions of Brazil predominate. More than half of the Porto Velho population is native to other states of the North Region, and 20.5 percent was born in Rondônia itself. This finding suggests that Porto Velho does not serve as a magnet for displaced or failed farmers from the recent colonization areas of the interior. Rather, Porto Velho's population is drawn mainly from traditional Amazonian settings along the regional rivers.

Porto Velho's peri-urban residents, despite their predominant origins in Amazonia, are relative newcomers to the state capital. In the overall sample, the mean number of years of residence in the Porto

Velho metropolitan area was ten years. Of these ten years, only the last three, on average, had been spent living in the current peri-urban district. Over thirty different locations were cited as last residences of the heads of household surveyed. Most of these previous residences were located in other peripheral districts in the Porto Velho *município*. Only seven of 298 household heads (3.1 percent) reported a rural-urban move from the last residence. The dominant mode of local migration was urban-urban. Thus, Porto Velho's peri-urban population is composed mainly of migrants from within the urban area of the Porto Velho *município* itself, rather than of peasants recently arrived from the countryside.

The 1993 survey also examined the distribution of the labor force by sector, type of employment position, and location of workplace. In terms of its sector distribution, employment was heavily concentrated in private services. Overall, 70.35 percent of the survey respondents reported their primary economic activity to be in the private service sector. Unlicensed or "informal" activities, often not reported in official census surveys, were particularly important. Indeed, the largest single employment sector was "unskilled services," which contributed 31.4 percent to the overall employment of the five neighborhoods; in Esperança da Comunidade, the figure rose to 54 percent. These findings indicate that a high proportion of the peri-urban population is actively employed in informal services. Far from being idle or inactive relative to the larger population, the labor participation of peri-urban residents is quite high, especially if informal activities are included (table 5.4).

Of private services, the most common type of employment was civil construction, which accounted for 5.7 percent of the total jobs reported. Construction employment varied from 16.7 percent of the labor force in Jardim Eldorado to 3.3 percent in Ulisses Guimarães. This considerable variation among the neighborhoods in the construction workforce reflected the significance of new building found in peri-urban settlements. Also notable are the high rates of employment in the public service sector in São Sebastião and União Vitória. Over a quarter of the economically active population in these neighborhoods worked in public services, which suggests more middle-class residents than in the other *bairros*. These patterns of employment in both civil construction and public services emphasize the heterogeneity of the peri-urban population of Porto Velho.

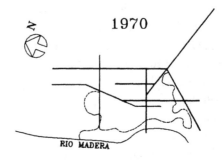

1970

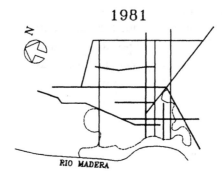

1981

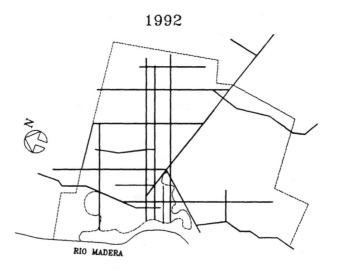

1992

Fig. 5.1 Urban area expansion Porto Velho, 1970–1992

Table 5.4 Distribution of Porto Velho's peri-urban labor force by sector and neighborhood

OCCUPATION	OVERALL	JE	EC	SS	UG	UV
Extraction	4.4	12.5	5.7	3.6	1.6	0
Agriculture	2.5	0	0	7.1	3.3	0
Industry	3.1	0	0	3.6	6.6	0
Private Services	70.3	50.0	83.0	50.1	80.3	72.8
Public Services	6.9	0	0	28.6	0	27.3
Other	12.5	37.6	11.5	7.2	8.2	0
Openly Unemployed	1.25	0	2.9	3.6	0	0

Source: Browder, unpublished survey, 1993.

Note:

JE = Jardim Eldorado

EC = Esperança da Comunidade

SS = São Sebastião

UG = Ulisses Guimarães

UV = União Vitória

The low rates of employment in primary (extraction and agriculture) and secondary (industry) sectors are also noteworthy. Only 3.8 percent of the sample reported working principally in mineral extraction, which refutes a popular misconception that Porto Velho's peripheral favelas are teeming with *garimpeiros*. No one worked in timber extraction, and only one respondent reported working in nonwood forest product extraction. Only 3.1 percent of the sample overall reported working in industry, mainly in manufacturing wood products or in food processing. Finally, agricultural employment represented merely 2.5 percent of the total sample, ranging from 0 percent in Jardim Esperança to 7.1 percent in Ulisses Guimarães. The peri-urban population, often believed to work in the rural sector, actually is almost entirely urban in its economic orientation. The peri-urban neighborhoods of Porto Velho definitely are not reservoirs of peasants and agricultural day laborers.

The 1993 survey of peri-urban neighborhoods in Porto Velho also classified the respondents according to the following types of employment positions: business owner, self-employed, permanent employee, temporary employee, unpaid family worker, and out of work (table 5.5). The most important employment category was permanent jobholder, representing 35 percent of the sample. This finding indicates that a substantial portion of the peri-urban labor force is formally or legally employed. The second most important type of employment position, however, was the self-employed category, which accounted for one-third of the survey sample. The size of this category indicates the importance of informal employment in the peri-urban economy. Only 6 percent of the respondents described their employment as unremunerated family jobs. Business owners constituted just 10 percent of the sample. By combining permanent and temporary employment categories, we can conclude that the peri-urban labor force is integrated into the formal metropolitan economy even though the informal sector is still an important source of local jobs (table 5.5). These findings contrast with much of the literature on Latin American squatter settlements, which are typically depicted as informal economies dominated by microenterprises reliant on unremunerated family labor.

Table 5.5 Economic position of household head, by gender, in five peri-urban neighborhoods of Porto Velho

ECONOMIC POSITION	OVERALL SAMPLE	FEMALE HOUSEHOLD HEAD	MALE HOUSEHOLD HEAD
Business owner	10%	4%	14%
Self-employed	31	29	34
Permanent employee	35	32	37
Temporary employee	7	11	4
Unpaid family labor	6	12	1
Out of work	11	12	9
Totals	100	100	99

Source: Browder, unpublished survey, 1993.
Note: 296 total usable responses; column totals may not sum to 100 due to rounding.

An analysis of the economic positions of Porto Velho's peri-urban labor force reveals gender differences. Among business proprietors, men predominate. Among the temporarily employed, unremunerated family workers, and the unemployed, women are disproportionately represented. Male and female respondents are about evenly represented among the self-employed and permanently employed. These findings indicate that while women are as important as men in their contribution to the formal and informal labor force, females are more likely than males to be economically vulnerable (table 5.5).

In terms of the location of workplaces, the 1993 survey found that most respondents were employed in the peri-urban fringe. (No significant gender differences appeared in the distribution of workplace location.) About 43 percent of the survey sample was employed in its own neighborhood, and 15 percent worked in other peri-urban districts of the city. These figures indicate that the urban periphery is largely self-suffi-

cient in employment generation. Yet one-third of the respondents indicated the city center as their job location, suggesting significant economic
integration of the central business district and the periphery. Rural and
interurban linkages were minimal. Only 6 percent of the respondents
worked in rural areas, and 4 percent indicated working in another city.
Together, these findings reinforce the idea that the peri-urban metropolitan fringe, although geographically close to the rural zone surrounding
the city, is well integrated into the urban economy of Porto Velho.

Finally, the 1993 survey considered the prospects for occupational
mobility among Porto Velho's peri-urban population. Occupational mobility implies more than just a change in jobs. Instead, it indicates a capacity to
move between occupational classes in a manner that normally improves
one's socioeconomic status. Of course, not all changes in occupation result
in economic advancement. Nevertheless, generally the higher the frequency of movement between occupations within a labor force, the greater
the amount of economic opportunity. In these terms, our survey found a
relatively low rate of occupational mobility among Porto Velho's peri-urban
labor force. Overall, two-thirds of the sample indicated no change in occupation since arriving in the *bairro* of current residence. Approximately 30
percent had changed occupation once since moving to the metropolitan
fringe. Three percent had changed occupations more than once. There is little in the official IBGE census data to indicate whether these rates are lower
than elsewhere in the country. The average period of peri-urban residency
in the survey sample was three years, and the probability of changing occupations during this time appeared to be low.

Overall, the population on the peri-urban fringe of Porto Velho is
socially heterogeneous. Stereotypical images of the shantytown dwellers
do not accurately depict this population. The distribution of the labor
force by economic sector in the 1993 survey of 298 households, drawn
from five distinct peri-urban neighborhoods, reveals considerable diversity. Although most peri-urban residents are employed in private service
and commercial sectors of the local economy, many on an informal or
unlicensed basis, some neighborhoods appear to have high concentrations
of middle-income government workers, school teachers, professional
service providers, health care workers, administrators, and students. In
contrast to widely held misconceptions, the peri-urban fringe is not
teeming with miners, who represent less than 4 percent of the labor
force. Moreover, the peripheral population is not composed predominantly of displaced farmers, peasants, or agricultural day laborers. Nor

are these fringe neighborhoods cesspools of the unemployed. Rather, peri-urban neighborhoods are lower-income but active local economies, desperately deficient in physical and social infrastructure but integrated into the socioeconomic rhythms of metropolitan life.

CONCLUSIONS

Before World War II, the evolution of metropolitan centers in Amazonia reflected the vicissitudes of extractive-mercantile economic cycles in natural resources. The trade in rubber latex, the most lucrative commodity, encouraged the formation of a dendritic, riverine urban system, tightly focused on the primate cities of Belém and Manaus and on urban outposts such as Porto Velho during the late nineteenth and early twentieth centuries. Even after the rubber trade waned in the second decade of the twentieth century, these metropolises retained a commanding urban primacy in the region. As the regional economy contracted in midcentury, the depopulated hinterlands continued to depend commercially on fluvial networks leading to the metropolitan entrepôts.

In the postwar period, significant differences in metropolitan growth became apparent, reflecting the heterogeneity of the frontier expansion process in different areas of the Amazon region. Whereas the urban primacy of Manaus has increased because of the program of fiscal incentives in the Zona Franca de Manaus, the other metropolitan centers have not been so favored. As a proportion of the population of the state, the Manaus microregion has grown from 29.0 percent in 1960 to 55.4 percent in 1991. In terms of economic structure, the capital of Amazonas also remains preeminent in the industrial, commercial, and service sectors. The Belém microregion, by contrast, has declined gradually as a proportion of total state population from 31.8 in 1970 to 27.9 in 1991. Likewise, Belém has gradually declined as an economic center relative to the interior of Pará. The most dramatic changes have occurred in Porto Velho, the Brazilian Amazon's third largest city. As a result of the extensive land settlement program of Rondônia, the demographic and economic primacy of Porto Velho has plummeted even while the city's population exploded in absolute numbers. In 1970, 75.7 percent of the state population lived in the Porto Velho microregion, but this figure fell to 26.0 in 1991. The capital has also lost its clear dominance in the industrial, commercial, and service sectors of the state.

The 1993 survey in Porto Velho dispels several misconceptions about the peri-urban population. The findings indicate that most of the residents of the peripheral districts are long-term urban residents who moved within the last three years to their current residence from the same or other peri-urban neighborhoods in the metropolitan area. While the importance of informal and unlicensed employment is evident in the high self-employment rate found in the survey sample, the importance of permanent and legal employment should not be understated. Permanent jobholders represent the single most important class of workers in three of the five *bairros*. Small-scale business owners and unpaid family workers constitute a relatively small share of the peri-urban labor force. These findings contrast with typical depictions of the Latin American squatter settlements as informal economies, characterized mainly by microenterprises and cottage industries dependent on unpaid family labor.

The relative status of female workers on the peri-urban fringe, however, must be described as precarious. While the 1993 survey indicates that women are as important as men in the labor participation rates, both in the formal and informal occupations, women are more likely than men to be found in the unremunerated, temporarily employed, or unemployed categories. In other words, the women of peripheral neighborhoods are economically more vulnerable than their working male counterparts. In terms of government policy, programs targeted to female workers should become a high priority. Neighborhood women themselves should play an active role in any local development plans, identifying common problems and proposing practical solutions with the aid of municipal and nongovernmental resources.

The majority of peri-urban residents work in their own or other peri-urban neighborhoods around Porto Velho, further indicating that the fringe is a dynamic incubator of economic activity and largely self-sufficient in terms of job creation. This finding contrasts with a stereotypical image of peri-urban settlements as parasites on the metropolitan economy. Still, a significant portion of the peri-urban labor force works in the central business district, implying daily work commutes of as much as two hours. Insofar as one-third of all peri-urban residents work downtown, improvements in roads and bus transport that decrease the time needed to travel to work are likely to pay for themselves through increased income generation and tax receipts.

Overall, the intraregional differences reflect the diverse forms of urbanization within Amazonia. In some cases, federal incentives and corporate investments directly guide growth; in other cases, migration and settlement are largely unplanned from the standpoint of government policy, the result of local agency, as in the urbanization of the metropolitan periphery. Indeed, since most of the metropolitan growth in Amazonia now is occurring on the metropolitan fringe, devising participatory local strategies for peri-urban community development should be given a high priority by lead institutions and local growth coalitions within the metropolitan area of Porto Velho. Given the diversity of the peri-urban population, however, uniform policies aimed at all peripheral neighborhoods would not be appropriate. Planners should work with organizations of local residents to ensure that development programs meet the different specific needs of the various communities in urban Amazonia.

Notes

1. As noted earlier, the principal definition of *urban* in Brazil depends, in practice, on a locality's political status: a city necessarily serves as a municipality- *(município)* or county-level seat *(sede)* of government. Generally, a population of at least three thousand is necessary to obtain municipal status. This political criterion complicates a precise measurement of the urban population of Amazonia, but the contemporary proliferation of new municipal seats in Amazonia is in itself a striking indicator of regional urbanization.

2. The rapid growth of Porto Velho during the period of colonization in Rondônia masks the capital's relative economic isolation from the state's dynamic inland interior, a feature commonly associated with disarticulated urbanization, as described in chapter 4.

3. Despite the rapid growth of some Sunbelt cities, most central cities in the United States have declined in population in recent years. The largest city, New York, declined from roughly 7.9 million in 1950 to 7.1 million in 1980. Even the larger metropolitan area of New York, while growing in absolute terms, has experienced a relatively slow growth rate in recent decades. Similar trends have been experienced in Chicago, Philadelphia, San Francisco, and other large cities of North America.

4. The various Brazilian terms for shantytown reflect regional and local characteristics. The term *favela* apparently originated in a squatter settlement that emerged on a particular hill, Morro da Favela, in turn-of-the-century Rio de Janeiro, after urban renewal displaced the downtown population (Godfrey 1991). *Favela* is the most common term for hillside slums in southern Brazil, but it has a very pejorative meaning. The *baixada*, a less offensive term, exists in any low-lying area subject to flooding, as in Rio de Janeiro's Baixada Fluminense. *Mocambo* is a common name for slum complexes in the Northeast. Throughout Brazil, the term *invasão*, or "invasion," refers to any illegal seizure of land, in rural or urban areas, by an organized community.

5. The federal criterion for shantytowns assumes a minimum of fifty-one residences, which have formed during the last ten years on alien property (public or private). Furthermore, these favelas are considered to be densely inhabited, "disorderly" or unregulated in form, and lacking in essential public services such as water, electricity, and sewers (IBGE 1992).

6. The Belém microregion, as defined by the IBGE(1991), includes the municipalities of Ananindeua, Barcarena, Belém, and Benevides.

7. The upper stretch of the Amazon River, east of the confluence with the Rio Negro, is known in Brazil as the Rio Solimães.

8. The Manaus microregion, as defined by the IBGE (1991), includes the municipalities of Autazes, Careiro, Careiro de Várzea, Iranduba, Manacapuru, Manaquiri, and Manaus.

9. For an excellent history of the church in Amazonia see *História da Igreja na Amazonia*, published by the Comissão de Estudos da História da Igreja na América Latina (Hoornaert 1992).

10. The importance of rubber latex extraction to the Guaporé Valley was officially acknowledged in 1873 when the governor of Mato Grosso established a tax collection post in Santo Antonio (Gomes da Silva 1984).

11. Personal communication with Rondônian chronicler and statesman Amizael Gomes da Silva, 19 July 1993.

12. Carlos Macedo Dias, personal communication, 14 July 1993.

13. Amizael Gomes da Silva, personal communication, 19 July 1993.

6

Crossroads of the Dawn: Urbanization of Rondônia's Populist Frontier

❧

Nineteen hundred and seventy-six, I arrived at the crossroads of the dawn and kissed the ground with great pride, where today is the city I saw being born.
From the song "Minha Cidade," by João Baptista Lopes, ca. 1983

"No map in my possession showed a town called Rolim de Moura in the Federal Territory of Rondônia. Yet I was assured in November 1983 by several prominent mahogany exporters in Curitiba, São Paulo, and Rio de Janeiro that, although *meia precária* (primitive), this town did indeed exist.[1] Arriving in Pimenta Bueno, Rondônia, after more than a week of seemingly interminable bus trips, I entrusted myself one final time to another dilapidated, dust-covered, smoke-belching bluebird and its cheerful pilot, an ever-smiling, toothless man sporting a Talking Heads T-shirt and a two-day beard. This would be the last 65-kilometer leg of a 3,000-kilometer journey from Belém, at the mouth of the Amazon River, to the frontier town of Rolim de Moura, *capital mundial de mogno* (world capital of mahogany).

"I first arrived in Rolim de Moura at midday on July 24, 1984. The bus descended into a sprawling bowl of single-story, wood-frame business establishments, buzzing with activity in a continuous whirlwind of dust.

Beyond the choking confusion of downtown, glimpses of a larger land-scape passed in and out of view between great pylons of smoke and dust. A landscape almost randomly strewn with shacks and patches of remnant jungle gradually rose above the pall.

"The municipal prefect, Adegildo Aristides Ferreira, immediately agreed to meet the first known *norteamericano* to set foot in this town in the Brazilian outback. While curious onlookers gathered in the reception area outside his office, Adegildo wasted no time reciting a long list of hardships endured by his fellow citizens, who numbered, he believed, somewhere between 10,000 and 20,000. Rolim de Moura suffered from no running water or plumbing, no electricity, malaria and dysentery of epidemic proportions, terrible communications, daily homicides, prosti-tution, runaway inflation, and choking dust. The prefect blamed these problems, all customary trademarks of frontier urban life in the Amazon, on a '*governo que não vale*' (a worthless government).

"Following our meeting, the prefect handed me over to 'Bezinho' (Baby Bull), the prefect's burly assistant, for a motor tour of the town, thereby sparing me a taxi fare of U.S. $20 per hour (in a country where the legal minimum wage is U.S. $60 per month). Droves of incoming trucks, many laden with mahogany logs or colonists from the South, crisscrossed Avenida 25, a 200-meter-wide swath of dirt that constitutes Rolim de Moura's principal commercial boulevard. Within the span of ten enor-mous city blocks, covered by Bezinho in probably under 30 breathless sec-onds in the *prefeitura*'s only car, a VW Beetle, we broke clear of the smoth-ering cloud of downtown dust to a prominent point on the town's north-ern edge, the cemetery. Here, Bezinho informed me, rested hundreds of pioneer souls, the eerie legacy of violence and disease on the frontier.

"The tour continued to the town's industrial sector, on the western perimeter, and we descended into another choking cloud, this time of mahogany wood smoke produced from the smoldering refuse of the town's 36 lumber mills. (I have never in my life encountered more air pollution than here, in Rolim de Moura, in the midst of the Amazon's rainforest.) 'Rolim is the world's mahogany capital,' Bezinho boasted. This town, like many others I would later visit in Rondônia, was profoundly affected by the international mahogany trade. Most lumber mills I would visit burned more of the precious mahogany than they processed, a function of their antiquated milling machinery and factory management practices.

"As the workday drew to a close, Bezinho suggested that the hour for refreshment was at hand. Although it was only 3:30 P.M., we repaired to his favorite bar, a luncheonette at the rear of the Triangulino supermarket, where a group of congenial local lumbermen already had begun to congregate. Over countless rounds of Brahma beer, I was introduced to the economic titans of this frontier boomtown, and to their particular story, from a life of deforestation and poverty in the South to one of prosperity based on the relentless exploitation of mahogany in the North.

"Well after sunset on my first day in Rolim de Moura, I took a meal at a local *mercenaria*, around the corner from the DeColores Hotel, where Prefeito Adegildo had encouraged me to take a room for the night. I noted that my presence as a stranger (not to mention a foreigner) was not considered unusual, although a general atmosphere of edginess held every stranger in suspicion. The life of the shop was not focused on food, which was exceptionally good for the frontier, but on billiards and *cachaça* (sugar cane brandy). Rolim de Moura, in 1984, is a town of transients, mostly newcomers, who, knowing no one, drift into corner canteens to relive the lives that so many seem to be escaping. Sometimes the past catches up with a few. The next morning I learned from the housekeeper at the DeColores that an unknown man was shot dead at the *mercenaria*'s snackbar only an hour after my departure the night before, one of 380 homicide victims in this small rainforest city in 1984" (John O. Browder, adapted from Log-book, "Rolim de Moura," July 24, 1984).

Ranking among the more obscure events occurring in the year 1976 was the birth of a new town in a remote corner of the Amazon's rainforest. It was a familiar story. Hundreds of dusty frontier towns have appeared across the Amazon landscape over the last twenty years in similar fashion. In July of that year two Brazilian families packed their belongings and ventured 45 kilometers from the Amazon town of Cacoal to establish a claim on a small tract of rainforest land at the junction of two planned roads (Linha 200 and Linha 25) in one of the federal government's hastily conceived colonization projects in the Federal Territory of Rondônia. Without any official support they built a rustic grade school for their children alongside lean-tos covered with plastic sheets. Settlers soon followed to carve out farms from the surrounding jungle. A malaria control post was established. On July 13, 1977, during a brief helicopter reconnaissance flight over the undulating forestlands known as the Chapada dos Parecis, the local director of Brazil's colonization institute, Expedito Rafael, dropped in to officially proclaim this expeditionary settlement at the junction of Linhas 25 and 200 the site of a future city, Rolim de Moura, the "crossroads of the dawn."[2]

The encampment survived for little over a year, serving briefly as a rest stop for pioneers moving further westward along the Linha 25 frontier trail. One of those pioneers, João Baptista Lopes, settled in another encampment 16 kilometers to the West. Favored by a small river (Rio Anta Atirada), the location had attracted some two hundred settlers by June of 1977. Baptista, a teacher, folk music singer, and public servant from Cacoal, drew up the first rough street plan for the pioneer settlement in 1978 and thereafter would be regarded as the founder of Rolim de Moura.

By 1984 the seven-year-old crossroads settlement of Rolim de Moura was teeming with an estimated fifteen thousand inhabitants (SEPLAN 1985). Displaying all the unseemly features of a sprawling Wild West boomtown, Rolim de Moura also became the dominant service center in an emerging system of smaller settlements on this populist frontier. Five years later the boom was over. Although Rolim de Moura's urban population had grown to about thirty thousand and it had become the seat of its own municipality during this brief interval, the vibrant, bustling boomtown days of the early 1980s were over. Shops were closing, houses stood abandoned, lumber mills lay in ruins in the industrial district. Rolim de Moura had atrophied.

In this chapter we examine the patterns of settlement formation on the populist frontier. Based on interviews with 713 urban households we present a portrait of Rolim de Moura and two smaller and more remote settlements (Santa Luzia do Oeste and Alto Alegre) that define the Rolim de Moura–Alto Alegre (Rolim-Alto) settlement corridor in south-central Rondônia. We situate our description in a city-systems framework, for it would be impossible to comprehend fully the dynamics of urban transformation in the Amazon frontier without casting the net of inquiry more broadly to capture the growth patterns of these adjacent settlements that emerged around Rolim de Moura during the 1980s and depended on it.

The establishment of Rolim de Moura was part of the contemporary occupation of the Rondônian frontier, a process that began in the late 1950s. This contemporary movement is a historical culmination of previous episodes of Luso-Brazilian settlement in the Guaporé-Madeira River Valleys, which are reviewed below. Since the early 1970s Rondônia has absorbed over a million migrants, mainly from Brazil's South and Southeast. While many migrants came to Rondônia in search of farmland, which the federal government offered freely, most came to establish shops and services in Rondônia's rapidly growing towns and cities. By 1990, like the rest of the Brazilian Amazon, more than half of Rondônia's population resided in urban settlements. Rondônia, however, is one of the last of the Amazon states to undergo an urban transformation (table 1.3).

PHYSIOGRAPHY OF RONDÔNIA

Rolim de Moura is located in the south-central quadrant of the Brazilian state of Rondônia (fig. 6.1). The state encompasses an area of 243,044 square kilometers (roughly the size of the former West Germany or the U.S. state of Montana) and represents 2.9 percent of Brazil's national territory. Physiographically characteristic of the southern Amazon region, the dominant form of vegetation is classified as open tropical, moist forest, extending over approximately 75 percent of the state. The remaining area, extending southward from the Pacaas Novos Ridge (an ancient sandstone plateau dividing the Rio Madeira basin to the north from the Guaporé Valley watershed to the south) contains a mix of mildly undulating savanna shrublands, natural grasslands, and floodplains. The humid, tropical climate is predominantly *Awi* (Köppen scale) with annual rainfall ranging from 1,800 to 2,200 mm and an annual mean temperature ranging from 21 to 26 degrees centigrade. Approximately 90 percent of the land area of Rondônia is covered by dystrophic latosols with either clayey or concretionary textures. Only 10 percent is classified as *terra roxa* (eutrophic podsols) and is considered suitable for annual or permanent cultivation (World Bank 1981a).

A BRIEF HISTORY OF RONDÔNIA

The Portuguese were the first Europeans to explore the Guaporé Valley in the mid-seventeenth century, establishing Jesuit mission settlements a century later. Yet it was not until the Amazon rubber boom (1850–1920) that the Brazilian settlement of the valley began in earnest. Between 1877 and 1900 some 158,000 migrants were drawn to the rubber estates (*seringais*) established on these remote tributaries of the Amazon (Benchimol 1977; SET 1983). Another 22,000 gandy dancers were

Figure 6.1.
Settlement areas
in Rondônia
Source: Adapted from
Martin Coy (1985).

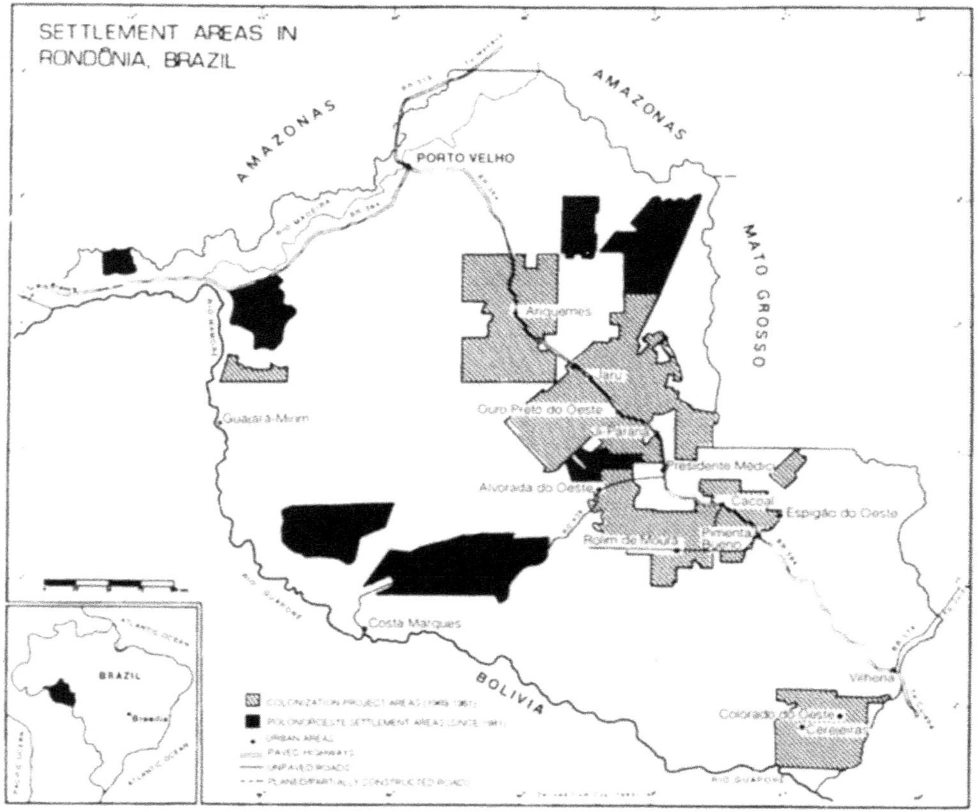

recruited to construct the ill-fated Madeira-Mamoré Railroad between 1907 and 1912, a disastrous British venture to boost local rubber production that, legend claims, "left one corpse for every sleeper in the ground" due to Indian attacks and epidemic illness (Gomes da Silva 1984:49).

The wealth created by the rubber boom gave added urgency to the abiding Brazilian concern to secure its ambiguous borders with Bolivia and Peru. High priority was given to installing an effective communications system between Rio de Janeiro and Acre, a profitable rubber tapping territory that Brazil had previously annexed from Bolivia. In 1906 a Brazilian army officer, Marechal Candido Mariano da Silva Rondon, was commissioned to extend the telegraph line from Cuiabá, the capital city of Mato Grosso, to Rio Branco, the principal town in Acre, a distance exceeding 1,500 kilometers.[3] The telegraph line between Cuiabá and Porto Velho subsequently became the basic route for the interstate highway, BR-364 (*Brasil Rodovia*). Initially opened in 1960 and finally paved in 1984, the BR-364 provided the only all-weather overland link between Rondônia and the Atlantic coast and thus became the central axis from which the settlement of Rondônia spread. Rondon's initial expedition also gave rise to several urban settlements, three of which have survived to today: Pimento Bueno, established in 1908; Jí-Paraná, founded in 1912, which is now Rondônia's second largest city; and Ariquemes, named after the now extinct Arikem Indians (Socorro Pessoa 1988:21–22). In 1956 the Federal Territory of Guaporé was renamed Rondônia in the explorer's honor.

The more efficient production of rubber transplanted by the British to Southeast Asia toward the end of the nineteenth century ended Brazil's virtual world monopoly on the rubber market. With the end of the rubber boom era by 1920, thousands of migrants drifted out of the Guaporé-Madeira region. By 1940 the area's population had declined from its peak of 180,000 in 1900 to 21,000 (SET 1983).

In 1943 the stagnant Guaporé-Madeira region was jolted by renewed demand for rubber sources in the New World, caused by the Japan's seizure of Malaysian rubber plantations during World War II. The region was officially separated from Mato Grosso and designated the federal territory of Guaporé. With new political status, combined with the Washington Accord of 1942, a U.S. wartime measure to promote New World rubber production (SET 1983), the territory's population increased again to 37,000 by 1950 (IBGE 1989). Several short-lived agricultural colonization schemes were established during the immediate postwar period to produce food for the territory's growing number of residents.[4] But Washington's renewed interest in the Amazonian rubber production lasted only for the wartime. Just as it appeared that the Guaporé Valley would again slip into oblivion, another natural resource of much greater commercial value than rubber was discovered.

Migration to Rondônia gained new momentum after 1952 when cassiterite (tin ore or "tin stone") was discovered in large commercial quantities. By 1958, favored with high world prices and easily extracted by hand, popular cassiterite mining brought a new frenzy of extractive activity to the Territory. A "weekend tin rush" followed.[5] With the opening of the BR-364 in 1960, then a compacted soil road between Cuiabá and Porto Velho, a new overland corridor to Rondônia was opened to tin prospectors and settlers. By 1960 the population of the Territory had risen to 70,000, twice its 1950 level. While most of the influx of tin miners took up residence in Porto Velho and Ariquemes, other settlers, mainly farmers, fanned out along the BR-364 to stake land claims. Several private colonization initiatives were launched in the interior, most of which were overwhelmed by settlers from the start. The failure of these corporate colonization projects later added legitimacy to the government's claim that any "rational" settlement of the Amazon had to be government-directed.[6] By 1970 Rondônia's population stood at 111,000.

In 1971 the federal government reached an unpopular decision to prohibit individual tin prospecting in favor of corporate *mineração* groups. In practical terms, the closure of popular tin mines in Rondônia idled thousands of disgruntled miners, a situation the government viewed as potentially destabilizing. The dedication of public lands (*terras devolutas*) to the settlement of smallholders was an easy concession for national policymakers, especially since additional pressures for agrarian reform arising from the modernization of agriculture in the south of Brazil demanded this step anyway.

The most important migratory wave to Rondônia began in the early 1970s. It was linked to land consolidation occurring in the South and Southeast, regions festering with social tension from a decade-long transformation of traditional family coffee farming to large-scale mechanized soybean, wheat, and sugarcane plantations (Mueller 1980; Bakx 1987). Until the early 1960s Brazilian agricultural exports depended on the production of traditional tropical cash crops (cacao, coffee, sugarcane), the real prices of which had been declining worldwide since World War II. In the mid-1960s the military government initiated various rural credit programs to stimulate the "modernization" of Brazil's stagnant agricultural sector by promoting the introduction of nontraditional export crops, especially soybeans, and the mechanization of production. Corporate enterprises linked to agribusiness were favored by these credit programs, although much of Brazil's agricultural heartland in the South and Southeast was held in family-owned coffee farms, from where much of Brazil's food crop production also originated. Between 1967 and 1979 the land area cultivated in soybeans in the southern states of Paraná and Rio Grande do Sul rose from 577,000 to 6.4 million hectares (Romeiro 1987; Millikan 1988). By 1975 Brazil had captured 30 percent of the world soybean market, becoming the second ranking exporter of this crop after the United States by 1976. Meanwhile, the cultivated area devoted to coffee production in Paraná alone fell from 1.8 million to 487,000 hectares between 1963 and 1983 (Quandt 1986:15, in Millikan 1988:48).

The agricultural transformation of Brazil's heartland from mixed-crop (coffee and food crops) to mechanized monocultural soybean production was accompanied by the mass consolidation of small family farms into larger corporate estates. Between 1970 and 1980 in the southern state of Paraná, for instance, 109,000 farms smaller than 50 hectares gave way to 450 new estates larger than 1,000 hectares (Mahar 1989:31). This consolidation also signaled a shift from labor-intensive to more capital-intensive production processes. On smaller farms coffee and food crop production had been based on household labor and sharecroppers (*parceiros*). On larger coffee plantations, production depended on a permanent labor force of tenant farmers (*colonatos*). During the 1970s, Millikan notes, "the state-promoted substitution of coffee and traditional food crops by mechanized soybeans and wheat was associated with a breakdown of traditional tenant and sharecropping arrangements, together with an increasing reliance on a reduced workforce of seasonal wage labor" (*boias frias*) (Millikan 1988:52). As a result net migration of displaced rural workers from Paraná reached 2.5 million during the 1970s (Martine 1988; Mahar 1989:31).

The decision to open Rondônia to the growing masses of landless farmers in the South was encouraged by four other factors. First, anxious to curb the further growth of metropolitan areas, Brazilian officials looked to the frontier zones of the Mato Grosso and Rondônia as labor safety valves to absorb the agrarian population displaced by the modernization of agriculture in the South and Southeast. Interregional land price differentials further encouraged landowners in the South to sell off their holdings there and buy new land in the North. In 1975, for instance, one hectare of land in the Southeast traded for 13 hectares in the Amazon (FGV various years).

Second, the notion that Rondônia could take the demographic pressure off the modernizing South and Southeast was given added legitimacy by initially optimistic assessments of soil suitability for agriculture in Rondônia obtained from a nationwide natural resource inventory (Radam Brazil Project) and a series of smaller edaphic studies completed in the mid-1970s (Falesi 1974; EMBRAPA 1975; FJP 1975). One important study reported that 33 percent of Rondônia's surface area was suitable for permanent crop agriculture under "primitive management systems" and that 51 percent would be of "good suitability under developed management systems" (FJP 1975). More recent surveys showed that only 9.9 percent of Rondônia's surface area consisted of soils of good quality "for annual or perennial crop production with moderate fertilizer inputs after several crop cycles" (World Bank 1981a). However, only 1.5 percent (3,650 square kilometers) is considered naturally fertile with no major limitations for farming (Furley 1986:96). Yet by 1983, 25 percent (60,465 square kilometers) of Rondônian forestland had been officially designated for agricultural settlement (Millikan 1984:90). Backed at the time by favorable "scientific findings," mass peasant settlement of Rondônia could be justified as the "rational" strategy for its development.

A third force operating contemporaneously with the modernization of southern agriculture to drive the settlement of Rondônia was the discovery of mahogany timber (*Swietenia macrophylla*) in the Territory in the early 1970s. This discovery alone would not have had such momentous consequences for Rondônia had it not been for three other factors. First, the effective opening of the BR-364 highway route in the late 1960s gave lumbermen new overland access, although precarious, from the country's wood processing centers in the South to the new mahogany frontier in the southern Amazon. Second, having virtually depleted Paraná pine (*Araucária angustifolia*), a valuable deciduous timber that once covered vast tracts of the southern Brazilian landscape, the South's industrial wood sector was entering a period of sectoral decline during the 1970s. The availability of new mahogany timber provided a fortuitous stimulus to Brazil's stagnant industrial wood sector. Finally, lucrative export incentives offered by Brazil's Central Bank from 1979 to 1985 effectively subsidized the exorbitant costs of mahogany extraction from remote forests in the Amazon and provided the final inducement necessary to attract lumber producers to Rondônia in droves.[7] A brief mahogany boom followed (Browder 1987).

During the 1950s lumber mills in Paraná employed 80 percent of all persons engaged in manufacturing (Foweraker 1981:33). The replanting of Paraná pine never exceeded the pace of its extraction from the wild. Between 1971 and 1983 exports of the araucária wood plummeted by 94 percent, from 995,000 to 55,000 cubic meters. During the same period Brazilian mahogany exports, originating entirely from the Amazon and increasingly from Rondônia, tripled from 19,400 to 61,000 cubic meters, representing over 30 percent of total Brazilian lumber exports (Browder 1986). Encouraged by generous export incentives during this period (Browder 1987), some 400 licensed wood processors, many from Paraná, moved to Rondônia by 1982 (UFRRJ 1985a); their numbers grew to over 1,100 by 1990.

Finally, the occupation of Rondônia was symbolically linked to the failure of the colonization program along the Transamazon Highway in Pará to achieve its "social" objective of resettling 100,000 landless peasants from the arid Northeast to the lush forests of the central Amazon (chapter 3). In 1974, amid mounting political pressure from corporate groups to consolidate land reserved from smallholder settlement along the Transamazon Highway, the government reversed its policy on colonization and auctioned off large blocks of public land in Pará to corporations based in São Paulo (Hecht 1985; Pompermayer 1979, 1982). The policy shift from peasant colonization to large corporate land development— mainly for livestock production—undoubtedly heightened political pressure for an expeditious smallholder settlement solution elsewhere (Mueller 1980; Martine 1990). The consolidation of landholdings by corporate enterprises in Pará could be partially justified by the concession of Rondônia to landless rural workers.

The policy shift of 1974 was part of the First Amazon Development Plan and underscored an ongoing political rift between two federal agencies sharing responsibilities for land development in the Amazon region: the Superintendency for Amazon Development (SUDAM) and the National Institute of Colonization and Agrarian Reform (INCRA). In chapter 3 we discussed the divergent institutional and ideological missions of these two government bureaucracies and how each has given a distinctive character, corporatist and populist, respectively, to those areas of the Amazon frontier it has overseen. The character of each frontier bore the imprimatur of its respective lead institution.

In Pará, INCRA's influence in shaping the social character of frontier expansion was complicated by a host of other institutional players at both the state and federal levels with different political mandates. The colonization of the Transamazon Highway in Pará, sponsored by INCRA, was increasingly viewed as "predatory," "irrational," and antiecological. Pressure applied by corporate groups from the Southeast effectively derailed INCRA's settlement program by 1975. A "developmentalist" ideology, supporting a corporate mode of occupation, prevailed in Pará under the institutional aegis of SUDAM.[8]

Despite losing to SUDAM most of its political foothold in the Transamazon area, INCRA managed to have its way in Rondônia. Playing upon the abiding military fear of foreign encroachment in the Amazon, INCRA succeeded in retaining institutional control of Rondônia by arguing that "our frontier must be quickly populated" (Arruda 1977).[9] With the execution of a federal decree (Decree Law 1164) on April 1, 1971, nearly all of Rondônia was placed under the exclusive federal control of INCRA. The dominant national security ideology became associated with INCRA's colonization mandate to populate Rondônia's agricultural frontier. Agricultural activities were proclaimed a superior form of production to "primitive extractivism," a euphemism for tin mining and rubber extraction dominating the Territory's economy at the time (SUDAM 1968), and therefore a more reliable approach to permanent and effective occupation by Brazilian nationals. Although the program for the settlement of Rondônia that followed in the 1980s maintained the developmentalist pretense made fashionable by the ascendancy of POLAMAZONIA and the First Amazon Development Plan, the underlying character of frontier expansion in Rondônia was decidedly populist.

However, INCRA was overwhelmed from the start. From 1971 to 1975 the agency implemented seven colonization projects, most of which were located along the BR-364 highway (fig. 6.1). Both the Transamazon and Rondônia colonization programs initially drew upon the same wellspring of grandiose and paternalistic ideas of directed settlement. In fact, INCRA tried to provide "comprehensive orientation and control of the settlement process," from land demarcation to lot selection down to the provision of roads, health and educational facilities, and technical assistance, credit, marketing, and storage (Martine 1990:27). Even before the new colonization areas were officially inaugurated, most of the land had been claimed by squatters and land speculators (*grileiros*), forcing INCRA to play catch-up in its land titling role; in the early 1980s that process took on average two years per property claim. As new settlers, like João Baptista Lopes, spontaneously spilled over into unclaimed forest areas adjoining the colonization projects along the BR-364, INCRA was kept hopelessly behind in the race to adequately plan for new settlement areas. Inundated, INCRA hastily enlarged the original projects into the new areas spontaneously penetrated without prior pedological surveys or appraisal of other natural resource and topographic features. The town at the "Crossroads of the Dawn" was an artifact of this chaotic decision process.

The full extent of INCRA's dilemma might be best appreciated by reference to a few revealing figures. Before 1983 most migrants to Rondônia came in search of farmland. Between 1977 and 1983 approximately 44,640 families (some 223,000 people) arrived in Rondônia, suggesting an average of 750 families per month. During the same period INCRA had processed only 19,630 land titles (table 6.1). The influx of migrants accelerated to about 2,500 families per month after the paving of the BR-364 was completed in 1984, but at least half of these migrants arrived in urban centers (SEPLAN 1983, various years). Between 1977 and 1988 approximately 94,240 families staked rural claims on federal lands, Indian and biological reserves, and in national forests. During this period a total of 20,231 families were officially settled in the nine new settlement projects (Millikan 1988:82). Taking into account the original seven settlement projects established between 1970 and 1975, a total of 43,743 families were officially settled by 1988, just 46 percent of those households known to occupy land in Rondônia in that year.

The agricultural development strategy for Rondônia called upon colonists to produce perennial cash crops (coffee, cacao, and rubber) and annual food crops for domestic markets and depended on their ability to quickly adapt their farming practices developed in a subtropical climate to the tropical conditions of the North. This ability was widely thought to be constrained by the lack of efficient transportation and marketing. Anxious to facilitate the marketing of Rondônia's potential cornucopia, the Brazilian government sought a loan from the World Bank to pave the BR-364, and in 1981, after much wrangling, the POLONOROESTE program was launched to "help bring order to the large, spontaneous migratory flow to the Northwest" (World Bank 1981a:iii). Two-thirds of POLONOROESTE's budget of U.S.$1.5 billion was allocated to road construction, new settlements, and rural development activities. About

U.S. $77 million were dedicated to environmental protection, public health, and the defense of Amerindian communities (World Bank 1981a).

Table 6.1 Rondônia: Number of migrants received, 1976–1988; rural land titles distributed, 1970–1984; and estimated total population, 1970–1988 (000s)

YEAR	MIGRANTS	LAND TITLES	POPULATION
1970	-	-	111.1
1976	17.4	2.9	-
1977	6.3	1.6	-
1978	12.7	3.6	-
1979	36.8	1.9	-
1980	49.2	4.6	491.1
1981	60.2	3.7	572.7
1982	58.1	1.4	663.0
1983	92.7	-	767.5
1984[a]	153.3	2.9	943.8[b]
1985	151.6		1,122.8
1986	165.9		1,320.9
1987	103.6		1,460.1
1988	51.9		1,549.8
1990	-		1,588.6
Total	959.7	30.7[c]	1,588.6

Sources: IBGE (1985, 1990); SEPLAN/RO (1985); SET (1983); World Bank (1987, 1990); author's estimates.
[a]*Asphalting of the BR-364 highway was completed in 1984.*
[b]*Author's estimate based on assumed 3.0 percent general population growth rate and 0.5 percent out-migration rate for the years 1984 to 1990.*
[c]*Includes 8,116 titles distributed between 1972 and 1976*

One focal point of the program's settlement component was the implantation of thirty-nine "Urban Rural Service Centers" (NUARs). At full development each NUAR was to provide health, education, research, marketing, and technical training and assistance to an average of 1,900 rural families (Millikan 1984). The World Bank, which predicted an internal rate of return on the first phase of the project of 22 percent, agreed to finance 34 percent of the program's total cost (World Bank 1981a:47). By the end of 1987 much publicized satellite images of the conflagration enveloping Rondônia's tropical forest landscape evoked an international outcry from environmentalists, political leaders, and the press, and POLONOROESTE was condemned as a development disaster and an environmental catastrophe. By 1990 Rondônia's population stood at approx-

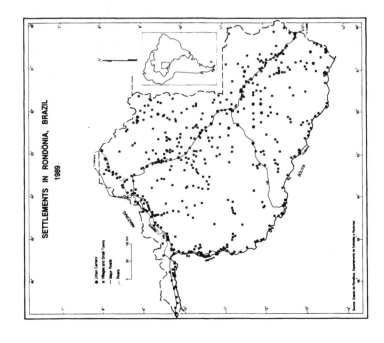

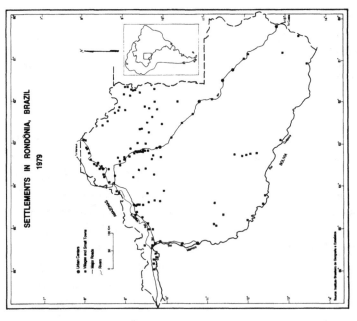

Figure 6.2. Urban settlements in Rondônia, 1979 and 1989

imately 1.6 million, more than half of whom resided in urban places with more than 2,000 inhabitants.[10]

In summary, the 1970s gave a new and essentially populist character to this politically constructed frontier. Rondônia offered the expedient solution to various social dilemmas stemming from the government's closure of the popular tin-mining sector in the territory and from the "modernization" of agriculture in the Southeast, both having added to Brazil's landless peasantry. Dubbed the "greatest land rush since the settling of the American West" (Bridges 1988), the settlement of Rondônia was a chaotic affair from the start. Seemingly a result of both internal and external forces impinging upon the military-authoritarian government, contemporary Rondônia is an artifact of the accumulated effects of changes in various federal policies pursued by the military regime: the failure of the Transamazon experiment of socially oriented colonization, the excesses of Brazil's export promotion policy, the ill-conceived road-building program financed by the World Bank, and the internecine conflicts within the federal bureaucracy.

As northern environmentalists gaped with horror at the carnage consuming Rondônia's tropical forests, few took note of the state's proliferating number of urban settlements, which grew from 70 in 1979 to 286 in 1989 (fig. 6.2) and the possible significance of agrarian urbanization for the tropical forests (see chapter 10). Rondônia's specific evolution as a populist frontier, unlike that of southern Pará, would give the process and initial form of urbanization a distinctive central place character.

CORRIDORS OF FRONTIER URBANIZATION IN SOUTH-CENTRAL RONDÔNIA

As the original seven colonization areas adjoining the BR-364 highway filled in during the late 1970s, several important movements into the interior of Rondônia were spontaneously initiated by colonists and land speculators that eventually provided the major corridors of pioneer expansion into Rondônia's populist frontier during the 1980s. As the population of the interior grew, the seven municipalities Rondônia had in 1980 were subdivided into twenty-three by 1989. By 1989 the new interior *municípios* accounted for 44.2 percent of Rondônia's total population and 82.1% of its incremental population growth since 1980.

Our attention initially focused on the Cacoal–São Miguel do Guapore corridor (Linha 25). Opened in 1976 with the establishment of Rolim de

Moura, this corridor now extends from the BR-364 highway to the frontier town of São Miguel do Guapore, a total distance of 165 kilometers (fig. 6.3). During the 1980s four parallel settlement corridors branched southward from the Linha 25 trunk road: (1) the Triunfo–Primavera corridor, running over 100 kilometers southward through the Chapada dos Parecis toward the *município* of Colorado d'Oeste by 1990; (2) the Nova Estrela–Vila Parecis corridor, penetrating only 55 kilometers into the southern Guaporé frontier; (3) the Rolim de Moura–Santa Luzia–Alto

Figure 6.3. Cacoal—-São Miguel do Guaporé settlement corridor, 1989
Source: Maplan 1989

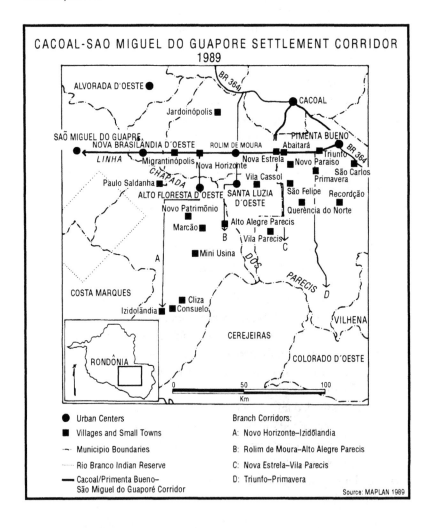

Alegre do Parecis corridor, extending over 70 kilometers southward where it stopped (in 1990) abruptly in the open tropical forests of the município of Alta Floresta; and (4) the Novo Horizonte–Alta Floresta–Izidolandia corridor pushing over 100 kilometers from Linha 25 into the Guaporé River Biological Reserve.[11] In 1990 this emergent network of settlement corridors connected twenty-three urban settlements of various sizes and served an estimated 123,000 people (both urban and rural residents) in a settlement front encompassing approximately 17,000 square kilometers (roughly 7 percent of Rondônia's territorial area).

The Evolution of the Rolim de Moura–Alto Alegre Settlement Subsystem

Rolim de Moura figures prominently in the economic evolution of the Cacoal–São Miguel settlement system. With superior access to larger nodes on the main highway both north and south, Rolim de Moura was afforded a locational advantage for retail trade in the interior that few other inland settlements enjoyed. The growth and eventual contraction of Rolim de Moura's economic influence and market catchment area as a rural service center can be charted in three periods:

Initial settlement, 1976–1980. Shortly after the founding of Rolim de Moura in 1976, INCRA opened feeder roads from Cacoal and neighboring Pimenta Bueno, and this frontier encampment burst into boomtown life. As word of the new town and the available forestland around it spread, droves of migrant homesteaders poured in; a land rush followed.[12] By 1980, within the short span of only four years, Rolim de Moura's population had grown from twelve to an estimated five thousand. Land speculation and agricultural expansion by family homesteaders was the *force motris* behind the initial settlement and expansion of Rolim de Moura. The town, which served mainly as a short-term landing for incoming settlers, provided little for its own consumption and remained within the market catchment area of Cacoal and Pimenta Bueno for most of its consumption supplies.

Rolim de Moura became a recognizable pioneer settlement, as defined in our typology in chapter 4, during this initial settlement period. It was not until the early 1980s that local agricultural production and the infusion of windfall profits from mahogany extraction, both subsidized by government credit programs, generated sufficient surplus capital in the local economy to attract investment in local productive activities, especially in the construction, retail, and service sectors. During the second period of Rolim de Moura's evolution, the town broke its market dependence on Pimenta Bueno and Cacoal for lower-order consumer goods and extended its own market area into the settlement frontier as it expanded west and southward. This expansion closely approximates the sequence schematically outlined in chapter 4 (fig. 4.1C and 4.1D).

The mahogany boom and municipal autonomy, 1981–1985. Between 1980 and 1984 the urban population of Rolim de Moura literally doubled every two years to an estimated 15,000–16,000 in 1984. This period of extraordinary population growth corresponded not only to the rapid spontaneous agricultural settlement of the forestlands around the town, all of which were claimed by 1982, but also to the dramatic expansion in the local industrial wood sector. Encouraged by a nationwide export promotion policy conferring lavish subsidies to qualified corporations, Brazilian trading companies quickly entered any commodity sector that enjoyed a foreign market (Browder 1987). Mahogany, which was progressively overtaking Paraná pine *(Araucária angustifolia)* and Virola *(Virola* sp.) as Brazil's premier export wood, occurred in relative abundance in the native forests of southern Rondônia. This, combined with an urban industrial growth policy that gave free urban land to factory owners, made Rolim de Moura one of the Amazon's major centers for lumber production during this period. The first lumber mill (Natalino e Orli) came to Rolim de Moura on October 1, 1977 (Lopes 1989:114).

By 1980 ten small mills *(picapaus)* operated in the town, producing in total about 12,800 cubic meters of rough lumber and related building construction materials for the growing local market and squares for veneer mills in Paraná (Browder 1986, 1989a). By 1981 several trading companies with preferential access to the export promotion credits moved into Rolim de Moura, contracting with local lumber mills in exclusive service arrangements to produce just mahogany. By 1984 the number of lumber mills established in the town had grown to thirty-six, and annual output exceeded 100,000 cubic meters, nearly all of which was export-bound mahogany lumber (Browder 1986, 1989a). With the lumber industry virtually ceasing production for local construction, urban homesteaders and homebuilders were compelled to import appropriate wood building materials from other towns outside the frenzied mahogany timbershed. Generally the vast mahogany wood wastes produced by the town's lumber mills were used, creating the ironic visual effect of a ramshackle town built of solid mahogany, the substantial wood scraps of the export boom.[13]

The impact of the mahogany boom on the occupation of Rondônia was enormous. The industry, in its relentless pursuit of mahogany, provided new jobs in the territory for those loggers and sawyers idled by the depletion of Paraná pine in the South. The lumber industry became the leading industrial sector in Rondônia, generating over 60 percent of the territory's industrial product by 1980 and directly employing an estimated 30 percent of its urban labor force (Browder 1989a).[14] Agricultural settlers benefited by the income they obtained from the sale of timber rights on their lots.[15] The mahogany boom produced enormous windfall profits for trading companies, not only from export sales but mainly from the recirculation of government-subsidized loan moneys into lucrative high-yielding, short-term money market instruments (see chapter 9). Awash in money from the mahogany trade, small pioneer settlements were suddenly transformed into boomtowns. During this period eight major commercial and public banks established operations in Rolim de Moura, even the Banco do Brasil; one of them even reported its second highest volume of business in the Amazon.[16]

The mahogany boom accelerated the process of frontier expansion with the addition of hundreds of kilometers of logging roads into the mahogany-rich tropical forestlands to the south of the city.[17] Logging trucks, on empty backhauls to the mahogany timbershed, carried consumer goods to pioneers on the fringes of the settlement frontier, thereby reinforcing commercial linkages between growing market towns and their hinterlands. By 1984 loggers, under trading company contracts, had cut out nearly all accessible, mature mahogany trees in an enormous swath of forestland extending some 100 kilometers south of the mills in Rolim de Moura, and some were illegally logging mahogany from the Guaporé Biological Reserve and the Rio Branco Indian Reserve.[18] Settlers sometimes followed the loggers into unclaimed forest areas. Some settlers were recruited and brought to the farthest reaches of logging operations in the reserves on the false promise of securing land that the logging companies illegally controlled with their own militias. In this way many unsuspecting settlers' families were set up as a first line of defense against Indian attacks on remote logging camps (Browder 1986).

In part due to its growing economic influence from the mahogany trade, on August 5, 1983, Rolim de Moura was separated politically from Cacoal and designated its own *município* encompassing (for a short time) the entire settlement frontier south to Costa Marques *município* in the Guaporé Valley.[19] Although the town could boast little in the way of amenities (see section below on public health and utilities), its designation as an autonomous *município* accorded it a status comparable to that of the more developed cities on the BR-364 highway and gave it access to considerable financial and institutional resources from the state for public works, service, and patronage.

In April 1985 the mahogany boom crashed. Foreign markets, glutted with subsidized Brazilian mahogany, collapsed. Letters of credit from U.S. mahogany importers were canceled. Thousands of cubic meters of cut mahogany roundwood were left to rot in the mill yards of Rolim de Moura alone,[20] where 40.6 percent of the urban labor force had been employed (Browder 1989a; UFRRJ 1985). The sudden crash in the mahogany market in April 1985 sent local lumber producers into a tailspin as mahogany prices plummeted by 50 percent in two weeks. In the United States consumers paid less for imported mahogany veneer than for domestic birch (Browder 1986). The large trading companies in São Paulo that had taken control of the local lumber production by 1982, pulled out (Browder 1986, 1987, 1989a). Most lumber producers stayed on, for a while at least, producing sawed wood for local and national markets. By 1990, however, the lumber industry in Rolim de Moura had contracted from thirty-six to twenty mills. The big mills had closed. Increased state government regulation of the industry, combined with the depletion of mahogany and other valuable timbers from Rondônia's forests, made continued production inefficient for the larger mills.

During the early 1980s, however, the penetration of the mahogany timbershed and the expansion of the agricultural frontier followed identical paths southward from Rolim de Moura, serving to expand its market area for trade goods and strengthen its local economic base (fig. 4.1D). By 1982 all 2,100 INCRA lots in the Rolim de Moura colonization area were legally claimed, the agrarian frontier around the town closed. Typically, incoming migrant families lodged briefly in the town as their adult male members scouted south in search of unclaimed forestlands. Concurrently, mahogany logging had virtually ceased within a forty-kilometer radius of Rolim de Moura, and by 1984 loggers hauled mahogany an average of 94 kilometers north from the environs of Cliza, Consuelo, Izidolandia, and the Guaporé Biological Reserve to Rolim de Moura's mills (Browder 1986:175). The demographic and economic southward movement provided conditions for the emergence of many other settlements, including Santa Luzia, founded as "Bambu" in 1979 and then separated from Rolim de Moura *município* in 1985, and Alto Alegre de Parecis, founded in 1986–87 in the Alta Floresta *município* (table 6.2), all of which were served for a time by Rolim de Moura's expanding trade goods, finance, and service sectors.

Santa Luzia d'Oeste's formative history was very different from that of Rolim de Moura. Unlike Rolim de Moura, which began as a populist movement, Santa Luzia's origins as a frontier settlement are closely linked to the personal history of a single enterprising pioneer family. The Cassols were lumbermen, politicians, and public servants from Maravilha, in the southern Brazilian state of Santa Catarina. In 1975 Reditário Cassol responded to an INCRA solicitation and received 2,000 hectares in a land grant from the federal government. His family initially settled in Vilhena on the border of Mato Grosso in 1977. In that year Reditário trekked to the site of his claim just 20 kilometers south of the Crossroads of the Dawn, only to find his land claim teeming with intruders. Upon learning that INCRA had tried to subdivide his property in response to pressures from land grabbers, Cassol traveled to Brasília in 1978 to secure a definitive land title from the INCRA president. Rebuffed in the capital, he returned to Rondônia and petitioned the governor of the Territory for intervention. None was forthcoming. In frustration, Cassol returned to his temporary home in Vilhena, gathered his sons and their firearms, and set out to their contested land grant to forcibly evict the interlopers. During a 1990 interview, his son César Cassol, then prefect of Santa Luzia, pulled off his shirt to point out the gunshot scar on his back. "This is how we secured our property rights in Rondônia," he exclaimed. In 1982 INCRA finally granted the Cassols a definitive land title to their original 2,000-hectare claim.

The Cassols were successful entrepreneurs, buying and selling sand in the construction industry and applying the profits to the mahogany trade. With further profits they invested in their own furniture factory and lumber mill, expanded their cattle herd to 2,000 heads on 900 hectares of pasture (becoming one of the largest ranches in the state), and most recently diversified into cash crops, expanding their original holdings to 4,000 hectares. Next to their plush country-club-like estate with private electricity generation and swimming pool is the neatly groomed company town, Vila Cassol, consisting of some forty single-family dwellings, each painted robin-egg blue. Since the beginnings of Santa Luzia, the Cassols have been the town's patrons. Family members, in succession, have served as prefect since the town and surrounding forestland were dismembered from Rolim de Moura and designated an autonomous *município* in 1985. The family patriarch, Reditário Cassol, was elected state senator and then federal deputy, assuring his family's influence in the political life of the state. With a strong family commitment to public service and their own wealth already secure, the Cassols enjoyed an unusual reputation for honest leadership and fairness in their administration of the *município*. This reputation is borne out by a series of recent public works projects in the town (hospital and stadium construction and street paving) with state funds.

During its early years between 1979 and 1984, however, Santa Luzia was little more than a satellite of Rolim de Moura; its principal function in the emerging subregional economy was as a crossroads rest stop and service station for colonists and loggers. Until the late 1980s, Santa Luzia depended heavily on Rolim de Moura for its most basic goods and services. In a 1988 socioeconomic profile of Santa Luzia, the State Planning Secretariat concluded that "commercial activities in Santa Luzia d'Oeste depend heavily on imports from Rolim de Moura," and the "provision of services is virtually nonexistent, limited to just one bank office, some restaurants, dormitories and bars. The reduced number of professionals indicates the great dependence of [Santa Luzia] on the nearby polarizing *município* of Rolim de Moura" (SEPLAN 1988a:16–17). The lumber industry found in Santa Luzia little advantage over Rolim de Moura, just 20 kilometers away, and due to deficient energy infrastructure only eight mills operated there in 1985 (SEPLAN 1988a:16), decreasing to seven in 1990. After the closure of Rolim de Moura's settlement area in 1982, subsequent colonists moved on to unoccupied areas beyond Santa Luzia. By 1985 Santa Luzia's urban population totaled 3,960 in an area of 16 square kilometers. Another 9,240 persons (roughly 1,800 families) occupied rural homesteads in the new *município* (SEPLAN 1988a:8).

Post-mahogany market area contraction, 1986–1993. The years following the mahogany crash of 1985 brought festering recession to Rolim de Moura. Between 1984 and 1990 the number of industrial wood processing companies, once the mainstay of the urban economy, plummeted by 45 percent. The number of banks declined by 37.5 percent. Commercial service and retail establishments on the main business avenue (Avenida 25 de Agosto) fell by 22.1 percent and 24.2 percent, respectively.[21] Truck traffic conveying wood products fell from 78 percent to 14 percent of total road commerce (table 9.4). Agriculture also suffered soil degradation from continuous cultivation without fertilizers, and yields of most major cash crops began to decline in the *município* during this period.[22] In 1990, 69 percent of the farmers interviewed in a rural sector survey of the Rolim de Moura–Alto Alegre corridor reported no net income from agriculture (chapter 9). But despite the local recession, the population of Rolim de Moura continued to grow from about 15,000–20,000 in 1985 to 30,000 in 1990, after which it stabilized.[23]

The economic downturn in Rolim de Moura corresponded to a modest upturn in the economy of Santa Luzia. Privileged by a location closer to the active settlement front, many basic service businesses and commercial shops moved to Santa Luzia, as local demand for tools, dyes, fuels, machine services, and cereal processing shifted southward in the Rolim de Moura–Alto Alegre settlement corridor. With enlightened political leadership that was also effective in securing state resources for urban infrastructure development, Santa Luzia became a more attractive location for shopowners, cereal processing companies, and service providers. With the growth of the local business community, the market area once dominated by Rolim de Moura became fragmented by the towns of Santa Luzia and Alta Floresta (fig. 4.2E). By 1987 all available rural properties in the Santa Luzia *município* were occupied; the agricultural frontier there closed, but the migratory flow pressed southward toward the tall woodlands of Alta Floresta from where news had come of yet another new pioneer settlement, Alto Alegre, at the very fringe of the agrarian frontier.

Virtually nothing is known from secondary sources about the origins of the village of Alto Alegre; our 1990 visit was the first comprehensive household survey of this community. The village evidently was founded in 1986–87. It lies within the *município* of Alta Floresta and has absorbed the spillover of land-seeking migrants turned away from Santa Luzia. By 1990 the village population stood at an estimated 800. There were seven lumber mills, one restaurant, one dormitory, and a dozen or so rustic but surprisingly well-provisioned shops. The village received regular bus service from both Alta Floresta and Santa Luzia. In July of 1990 the active settlement front of the Rolim de Moura–Alto Alegre corridor extended southward only 15 kilometers beyond Alto Alegre, and then the road abruptly ended at the edge of the forest (plate 10.2). Rolim de Moura's influence as a market center was clearly diminished in this frontier outpost, some 55 kilometers beyond Santa Luzia (chapter 9). In 1990 Alto Alegre existed in the same dependent relationship to Santa Luzia that the latter had had to Rolim de Moura five years earlier. The future of Alto Alegre as an emerging urban settlement and whether it will evolve beyond a ramshackle pioneer settlement into its own *município* service center (fig. 4.1F) will likely depend on how far and how fast the populist frontier expands and fills in to form a new market hinterland for this town to serve.

In summary, the evolution of the Rolim de Moura–Alto Alegre settlement corridor into a subregional settlement system can be explained by three factors. First, the continuous outward population growth from the primary settlement in the system (Rolim de Moura) due to the influx of new settlers. This growth, especially during the formative period (1977–1983) extended Rolim de Moura's effective market area and enhanced opportunities for capital formation in the local service and commercial sectors of the urban economy serving its growing hinterland.

Second, the Brazilian government-subsidized mahogany boom (1980–1985) provided a major stimulus to the expansion of the financial, service, and commercial sectors by generating enormous windfall profits and a new demand for basic industrial goods and services for the wood processing sector. This stimulus to the growth of the economic base of the city bolstered the primacy of Rolim de Moura in the settlement system emerging from it and propelled the expansion of the settlement frontier southward, deep into the forestlands of Alta Floresta and toward the floodplains of the Guaporé Valley.

Third, the most recent phase (after 1985) is characterized by the progressive contraction of the local urban economy in Rolim de Moura and the fragmentation of its original market hinterland, aggravated by the sudden collapse in the mahogany boom. While the flurry of activity associated with an extractive cycle provided the catalyst for transition from one stage to the next, the underlying force in the evolution of the Rolim de Moura–Alto Alegre settlement system was the shifting market envelop associated with the geographic expansion of the settlement frontier. In general terms, then, the marketing principle that undergirds central place conceptualizations of settlement patterns (chapter 2) appears relevant to the explanation of the emergence of the Rolim de Moura–Alto Alegre settlement system in the populist frontier. Nonetheless, the unique histories of these towns, shaped by local forces, condition the importance of the marketing principle of spatial organization.

In the next section, we test several hypotheses regarding populist frontier settlements at different stages of integration into the region's urban network, based on Rondônia household survey data collected in 1990. Generally, we found that the different settlements in the Rolim de Moura–Alto Alegre settlement system displayed different key characteristics associated with their age, location, and economic functions relative to each other. These differences, we suggest, can be largely explained by the marketing principle of central place theory, but notable exceptions exist. We address the following questions: What are the important physical, economic, and social characteristics that distinguish lower-order settlements in the agrarian frontier? What types of economic activities are prevalent in frontier towns? What functions do these settlements provide within the urban systems of which they are a part? And what do their distinctive attributes reveal about the nature of urbanization in different geographic frontier settings?

GENERAL CHARACTERISTICS OF ROLIM DE MOURA, SANTA LUZIA, AND ALTO ALEGRE

Rolim de Moura, the oldest settlement (founded in 1977) in the Rolim de Moura–Alto Alegre urban system, is the closest to the national market (40 kilometers from the BR-364 interstate highway) and was the largest of the three settlements in 1990 (estimated urban population: 30,000). Alto Alegre, the youngest of the three settlements (founded in 1986–87), is the furthest from the national market access link (109 kilometers from the BR-364) and the smallest (population: 800). Santa Luzia falls between the two in terms of age, distance, and size (table 6.2). This roughly symmetrical pattern of distance, age, and population is consistent with hypothesized central place tendencies. Implicit in the marketing principle is the notion of a distance gradient in the market integration of the frontier. More remote (and typically younger and smaller) settlements will display characteristics (e.g., fewer urban functions, lower levels of development) that are significantly different from those of older and larger settlements located closer to the market core. In our definition of a populist frontier (chapter 4), the marketing principle provides the initial logic behind spatial organization. This symmetrical pattern was not found in the Xingu corridor, which suggests a very different form of urbanization on the contested corporatist frontier there (chapter 7, table 7.2).

Our 1990 survey included 713 household interviews representing 6.7 percent, 16.0 percent, and 51.3 percent of the total urban populations of Rolim de Moura, Santa Luzia, and Alto Alegre, respectively. Average household sizes ranged from 4.624 to 4.773 members, smaller than was

Table 6.2 General characteristics of the Rolim de Moura–Alto Alegre corridor settlements, July 1990

	ROLIM DE MOURA	SANTA LUZIA	ALTO ALEGRE
Year founded	1977	1979	1986–87
Distance to BR-364 (km)[a]	40	60	109
Approximate 1990 urban population[b]	30,000	6,000	800
Number of urban dwellings surveyed[c]	419	208	86
Estimated % of total population surveyed	6.7	16.0	51.3
Household size (persons/HH)	4.773	4.625	4.663
Gender (% of population)			
Female	48.8%	47.3%	49.1%
Male	51.2	52.6	50.9
% Female-headed households	6.5%	3.4%	4.8%

[a] The BR-364 is the asphalted all-weather federal highway extending from Rio Branco, capital city of Acre, through Rondônia to Cuiabá, the capital city of the state of Mato Grosso, and from there to all major national markets.
[b] Based on demographic data provided by the malaria control organization SUCAM (Superintêndencia de Campanha Contra a Malária).
[c] Excludes households headed by persons who arrived in the survey area on or after January 1, 1990 (or within the preceding six months of the survey date).

found in the Xingu corridor. The population was disproportionately male in gender distribution (50.9 percent to 52.6 percent), the reverse of our findings in Pará. Female-headed households, ranging from 3.4 percent (Santa Luzia) to 6.5 percent (Rolim de Moura) of all households, were significantly less common than in the Xingu corridor settlements. The differences in these general characteristics between the Rondônia and Pará survey samples (comparable Pará data is presented in chapter 7) suggest fundamental differences in the populations and histories shaping these two frontier urbanization experiences.

Physical Characteristics and Public Health Problems

During the 1980s the number of new frontier settlements in Rondônia increased dramatically (fig. 6.2). In addition, the size of those providing market-center functions to an expanding agricultural hinterland, like Rolim de Moura, also grew at such a rapid rate that even considerable efforts at urban planning were typically overwhelmed from the start. Three sets of images of Rolim de Moura illustrate the pace of physical growth of this urban area (fig. 6.4).

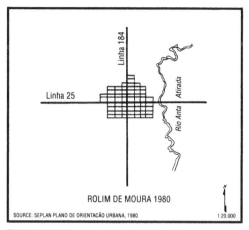

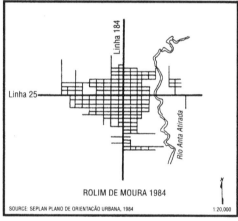

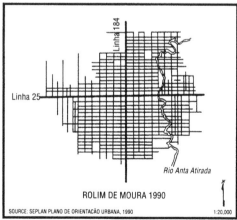

Figure 6.4. The spatial evolution of Rolim de Moura, 1980, 1984, and 1990

Source: 1980: Seplan Plano de Orientação Urbanaa; 1984: Seplan Diagnóstico de Rolim de Moura; 1990:Spot Satellite Image

In an expanding populist frontier, guided initially by the spatial economic logic of the marketing principle, we would expect the physical development of frontier towns to vary in direct relation to a distance gradient from the settlement center to its frontier. Therefore, more centrally located (and typically older and larger) settlements are likely to enjoy a higher degree of physical development than more remote (younger and smaller) settlements. This assumption may be measured by the degree to which households must provide their own physical infrastructure and basic services.

Hypothesis 1: The rates of informal self-provisioning of housing, running water, sewage disposal, and energy supply increase with distance from the principal center and decrease with age of the settlement in the frontier urban settlement system.

Table 6.3 Indicators of urban physical infrastructure and public health, Rolim de Moura–Alto Alegre corridor, July 1990 (percentage of households)

INDICATOR	ROLIM DE MOURA	SANTA LUZIA	ALTO ALEGRE	WEIGHTED MEAN
Self-built housing[a]	18.2%	24.0%	37.5%	22.2%
Informal water supply[b]	44.2	52.4	75.0	50.3
Informal sewage disposal[c]	67.0	79.8*	86.4*	73.0
Individual power supply[d]	18.2	40.9*	37.5*	27.1
Gastro intestinal disorders[e]	10.5*	12.5*	8.0	10.8
Hepatitis in household[f]	2.4*	2.4*	0*	2.1
Malaria in household[g]	41.2*	36.9	41.0*	

[a]*Self-built housing by current occupant excludes those houses purchased already constructed, or those built by contractors by order of the owner.*
[b]*Informal water provision includes wells without pumps, local creeks, and open springs. Formal water provision indicates deeper wells with pumps and piped water systems.*
[c]*Informal sewage disposal reflects lack of a septic tank or sewer line and reliance on outhouses or the outdoors.*
[d]*Individual power supply means lack of electricity (from any source) or self-provision of energy by means of individual gas or kerosene stoves or lamps.*
[e]*Gastrointestinal disorders refer to diarrhea and dehydration affecting anyone dwelling in the house during the preceding twelve months.*
[f]*Percentage of households in which someone dwelling therein contracted hepatitis during the preceding twelve months.*
[g]*Percentage of population residing in all households surveyed that contracted malaria during the preceding twelve months. Or percentage of household heads contracting malaria during the twelve-month period from July 1, 1989 to June 30, 1990.*
All difference of proportions are significant at .05 level of confidence, except for those pairs noted by an asterisk.

The survey data (table 6.3.) support Hypothesis 1 and indicate that the rates of household self-provision on the populist frontier increase with distance from the central node in the settlement system and decrease with the age of each settlement. For instance, self-built housing was reported by only 18.2 percent of the households surveyed in Rolim de Moura but by 37.5 percent in Alto Alegre. Similarly 44.2 percent of Rolim de Moura's households provide their own water supply (either from a private well or hand-carried water from the Anta Atirada River), while 75 percent of Alto Alegre's residents are dependent on individual water sources. In each of the settlements no less than two-thirds of all households provide for their own sewage disposal. Meanwhile the rate of informal household energy provision (i.e., diesel generator) is twice as high in the outlying settlements as it is in the primary urban node.

As rates of informal water supply, sewage disposal, and individual power supply decline, so too, we expect, would certain public health ailments. For example, diarrhea and other gastrointestinal disorders as well as any number of varieties of hepatitis are commonly contracted from the consumption of contaminated food and water, which would more likely occur in those settlements with both high population densities and high rates of informal self-provisioning of sanitation, power, and water services. In contrast to these largely urban illnesses, the most widespread disease afflicting the human rural population in Rondônia is malaria (Sawyer and Sawyer 1987). We would expect, therefore, the incidence of malaria to be highest in the smallest, most remote settlements, where a higher proportion of the population would be exposed to the anopheles mosquito.

The survey data, however, do not clearly support these intuitive suppositions about public health, as there are no significant differences in the incidence of these ailments among the separate urban populations surveyed. The below-average rate of both gastrointestinal disorders and hepatitis in Alto Alegre is probably related to the continued availability of uncontaminated natural water sources (wells and nearby streams) in this new, sparsely populated settlement. Such resources are no longer available in the older, more congested, and polluted towns. The lack of significant differences in the rate of malaria among the three settlements perhaps attests to the blurred functional distinction between urban and rural sectors and the continuous movement of significant proportions of the population of all three settlements in and out of rural areas, an issue we explore in greater detail in chapter 8.

These data also suggest that public investment in health and sanitation facilities in higher-order settlements has not kept up with demand, a supposition that is evident from secondary data sources describing the provision of such public services (SEPLAN 1988a, b) Although the difference is not statistically significant, a larger proportion of households in Santa Luzia are connected to a public water supply network and sewage disposal system than in Rolim de Moura.[24] Moreover, despite a population that is five times larger than that of Santa Luzia, Rolim de Moura's basic health network contained only seventeen medical facilities, versus fourteen in Santa Luzia (SEPLAN 1988b).[25] With no commensurate increase in health and sanitation facilities, continued population growth only exacerbates the threats to public health.

There are significant differences in physical infrastructure and public health between the Rondônia and Pará study sites (see chapter 7). The rates of gastrointestinal disorders, hepatitis, and malaria are significantly higher in Pará (table 7.3) than in Rondônia.

Sectoral Distribution of the Urban Labor Force

The structure of employment of a community reflects its unique social and economic history, comparative locational advantages, and resource endowments. These undeniably important determinants of labor employment are acknowledged here, but also must be contextualized in the logic of the populist agrarian frontier. Six observations concerning employment structure are noteworthy in this regard.

First, since a frontier is given much of its character by a dominant economic sector, we would expect the populist agrarian frontier to be heavily reliant on agricultural and related activities. The distribution of the labor force by sector in the Rolim de Moura–Alto Alegre corridor reveals some interesting nuances in the expected pattern.

> *Hypothesis 2:* Employment rates in the agricultural sector progressively increase along the distance gradient of the agrarian frontier such that outlying pioneer settlements have higher rates of agricultural employment than more centrally placed urban nodes.

As predicted, agricultural employment rates progressively increase outward along the settlement corridor, from 10.3 percent in Rolim de Moura to 25.9 percent in Santa Luzia to 30.5 percent in Alto Alegre, a tendency that requires little additional explanation (table 6.4). As towns grow in population and structural complexity, the proportion of the urban population employed in farming tends to decrease. Moreover, the number of persons engaged in farming is likely to decline over time due to land consolidation near the older settlements. This proportional relationship between agricultural employment and distance from the primary node of the settlement corridor is not evident in the Xingu corridor, reflecting the different economic orientations of the contested corporatist frontier.

Table 6.4 Sectoral distribution of the labor force, Rolim de Moura–Alto Alegre corridor, July 1990 (percentages)[a]

SECTOR	ROLIM DE MOURA	SANTA LUZIA	ALTO ALEGRE	WEIGHTED MEAN
Extraction	*3.9*	*2.6*	*4.2*	*3.5*
Mineral	1.8	0.7	0.6	
Timber	2.1	1.9	3.6	
Non-wood forest	0.0	0.0	0.0	
Agriculture	*10.3*	*25.9*	*30.5*	*17.1*
Industry	*18.3*	*13.0*	*16.2*	*16.5*
Wood processing	9.6	4.7	7.8	
Construction[b]	3.9	5.9	2.4	
Agro-industry[c]	4.8	2.4	6.0	
Metallurgy	0.1	0.0	0.0	
Commerce	*13.7*	*13.4*	*15.0*	*13.8*
Transport	*6.2*	*4.0*	*3.0*	*5.2*
Services	*30.8*	*21.0*	*21.0*	*26.8*
Finance	0.9	0.7	0.0	
Industrial, auto, mechanical	7.3	3.8	3.0	
Food, lodging, entertainment	2.8	3.1	5.4	
Building repair	1.2	0.7	0.0	
Unskilled[d]	16.6	12.0	12.0	
Skilled[e]	0.7	0.2	0.0	
Professional	1.3	0.5	0.6	
Public	*10.8*	*17.7*	*9.0*	*12.6*
Education	5.4	5.4	6.0	
Government	5.4	12.3	3.0	
Other and Inactive	*6.0*	*2.4*	*1.1*	*4.4*
TOTAL (Percent)	100.0	100.0	100.0	
Persons Employed	867	424	167	

[a]*Refers to primary occupation of economically active population in the RM-AA corridor study sites. The economically active population includes all persons aged ten years and over listing a money-earning occupation or employment, either formal (licensed) or informal.*

[b]*Includes industrial ceramics, brick-making, and cement.*

[c]*Includes food processing.*

[d]*"Unskilled" personal services.*

[e]*"Skilled" personal services.*

A second economic characteristic that we would postulate about the populist frontier is the greater importance of extractive activities in younger, more remote settlements, than in older, more centrally placed cities whose immediate hinterlands are likely to have been mined of natural resources in earlier periods.

Hypothesis 3: The extractive sector will employ higher rates of workers in outlying, newer settlements than in centrally placed older towns.

What is most telling from the survey data (table 6.4) is the minuscule importance of extraction in the local economy and the corresponding lack of statistically significant differences among settlements in the Rolim de Moura–Alto Alegre settlement corridor. Extraction accounted for only 3.5 percent of total employment in the corridor overall, ranging from 2.6 percent in Santa Luzia to 4.2 percent in Alto Alegre. Santa Luzia deviates from the predicted pattern presumably because its close proximity to Rolim de Moura affords it no real locational advantages in terms of access to resources. In the populist agrarian frontier the rate of labor participation in the extractive sector is virtually negligible.[26] By contrast, in the Xingu corridor extractive activities (mainly mining) claimed more than three times the share of the labor force (table 7.4), although even there the sector ranked fifth (after service, commerce, industry, and agriculture) in terms of labor force participation.

Third, neoclassical economic growth theory has long maintained a positive correlation between urbanization and industrialization (Kuznets 1966; Higgins 1968) although exceptions known as overurbanization or pseudo-urbanization are widely acknowledged (Dogan and Kasarda 1988). Tendencies toward production specialization and agglomeration economies are benchmarks of the urbanization process in industrialized countries. At the subregional level, the appearance of such tendencies in frontier regions of the global South may be more problematic. Nevertheless, we are prompted by the prevailing wisdom of spatial economics to hypothesize the obvious:

Hypothesis 4: Industrial employment rates in the urban system of the populist frontier are higher in the centrally placed, primary settlements than in the peripheral pioneer settlements.

Industry is only the third most important generator of employment in the frontier settlements of the Rolim de Moura–Alto Alegre corridor, overall employing 16.5 percent of the labor force. In this regard, the Rondônia and Pará tendencies converge. While the data to some extent support Hypothesis 4, two interesting nuances require explanation. First, we must note the high industrial employment rate in the outlying pioneer settlement of Alto Alegre, owing principally to the presence of eight lumber mills (a large number for a town of only 800 inhabitants). Second, we note the relative paucity of industrial sector employment in Santa Luzia, except for the construction subsector, which exceeded that of Rolim de Moura (in economic recession) and Alto Alegre (where self-provisioned housing prevails). In 1990 Santa Luzia was awash in dry-season construction contracts for road paving, stadium construction, and other civic works. Therefore, the town's labor force was enjoying a construction boom during the period in which our survey teams interviewed the local population.

Apart from these local idiosyncrasies, the paucity of manufacturing industries in Rolim de Moura is noteworthy. Rolim de Moura is not uncharacteristic of interior settlements in Rondônia. Even where robust local markets for various nondurable and intermediate consumer goods (e.g., beverages, packaged foods, automobile tires, packaging products) are manifest, value-adding manufacture industries are scarcely found. Industrial capital has chosen not to invest in the populist frontier despite the growing urban labor force and consumer market, choosing instead to export consumer goods from the industrial South (2,000–3,000 kilometers away). We will revisit this strategy of corporate capital in chapter 9, as it bears upon larger questions of surplus extraction, global circulation of capital, and the institutional role of the state in shaping frontier urbanization.

Fourth, the importance of the service sector stands out, especially in relation to the primary and secondary sectors of the local economy. Intuitively, services gravitate toward locations of higher and more concentrated population. This theorem would apply to both skilled professional services and unskilled personal services.

Hypothesis 5: Service sector employment rates are positively related to the population size of specific settlements.

Again, with some qualification, the hypothesis is confirmed by the employment structure data (table 6.4). Overall, the service sector employment rate in Rolim de Moura is higher than that in the other two smaller settlements, where the rates are identical. Disaggregating the service sector, we find other probable affirmations of central place tendencies: service sector employment rates dissipate over distance gradients

from the dominant node for financial services, industrial services, building services, professional services, and skilled personal services (all
absent in Alto Alegre).

Fifth, we expect public sector employment to be higher in the
older, more established settlements where long-term political ties
have been given more opportunity to mature into patronage jobs for
favored local citizens.

> *Hypothesis 6:* Government employment rates follow a distance
> gradient and are therefore higher in centrally located, older, pri-
> mary settlements in the settlement system of the populist frontier.

In this case, we must reject Hypothesis 6 as false. Santa Luzia com-
mands the highest rate of public sector employment among the three
study sites, due, we must suppose, to the greater effectiveness of the
município's patron family, the Cassols, in securing resources from the
state for jobs and public works in the *município*. There are no statistically significant differences in public education employment rates
among the settlements. This important deviation from the hypothesized pattern anticipated by the administrative principle of central
place spatial organization underscores the importance of local history
and the often decisive role of local politics in determining the economic structure of settlements.

Another telling detail is the much higher rate of labor force participation in the public sector of the populist frontier (12.6 percent)
than in the contested corporatist space of the Xingu corridor (2.6
percent). In the postauthoritarian period the corporate sector has little use for local government, which is now likely to side with populist forces resisting subordination and is therefore usually no ally of
capital. The history of municipal adjudication of urban land disputes
in Xinguara is illustrative of this antagonism between corporatist
interest and municipal government (chapter 7). The populist frontier
is more likely to benefit by local government, and public sector
employment rates in the Rolim de Moura–Alto Alegre corridor bear
out this supposition.

Quality of Urban Life in the Populist Agrarian Frontier

Although central place theory does not explicitly treat matters of public
investment and quality of life, implicit in the notion of network hierarchy

is the assumption that, given economic and demographic growth, over time various indicators of the quality of life in frontier settlements will improve (e.g., public health, material amenities) and will give rise to an increased subjective sense of satisfaction and contentment. Hence, older and larger *município* service centers are likely to be more hospitable environments than pioneer settlements. Having examined public health problems above, we divide this question into two hypotheses that pertain to material well-being and attitudes of contentment:

Table 6.5 Material possessions, intentions to stay, and major urban problems, by percentage of households, Rolim de Moura–Alto Alegre corridor, July 1990

	ROLIM DE MOURA	SANTA LUZIA	ALTO ALEGRE
Possessions:			
Radio	68.7%	59.1%	45.5%
Clock	76.3	72.1	68.2
Bicycle	57.7	47.6	43.2
Sewing machine	51.0	51.0	44.3
Gas lamp	55.3	51.0	42.0
Pressure cooker	93.1	89.4	88.6
Water filter	79.7	73.1	77.3
Running water	43.3	33.7	21.6
Shower	48.8	36.1	19.3
Indoor toilet	47.1	30.8	17.0
Gas stove	90.2	71.6	81.8
Refrigerator	55.0	46.6	34.1
Automobile	21.8	17.8	13.6
Television	56.5	39.9	28.4
Telephone	13.4	4.8	1.1
Median number of possessions from list	9	7	6
Household head intends to stay in town	*73.4%*	*87.0%*	*86.4%*
Most often cited problem in town (% household heads)	*Dust, poor air quality (24.2%)*	*Lack public services[a] (31.3%)*	*Lack of public services (43.2%)*
Second-most cited urban problem	*Lack of electricity (14.4%)*	*Lack of private services[b] (6.3%)*	*Lack of medical assistance (20.5%)*
Third-most cited urban problem	*Lack of work, insufficient income (12.4%)*	*Lack of work, insufficient income (5.3%)*	*Lack of work, insufficient income (5.7%)*

[a]*Public services are telephone, running water, and electricity services.*

[b]*Private services include supermarkets, churches, gasoline stations, banks, private transport services.*

Hypothesis 7: Residents of the larger center in a regional urban system enjoy more material possessions than residents of the lower-order settlements.

Hypothesis 8: Residents of the larger center in a regional urban system are more content with their current situation than those of the lower-order settlements.

We accept Hypothesis 7. We asked respondents whether they possessed fifteen material amenities we had listed. The median number of possessions claimed per household followed the hypothesized correlation to the distance gradient of the populist frontier (table 6.5). The median number of possessions per household in the larger center (Rolim de Moura) was nine (from a list of fifteen), compared to seven and six in Santa Luzia and Alto Alegre, respectively.

Additionally, household informants were asked to indicate from this list the number of material amenities they owned before arriving at their current residence, thus giving us some indication of material accumulation tendencies over time. In the cases of Rolim de Moura and Santa Luzia there were significant increases in the number of material amenities acquired by households after establishing current residency. In Rolim de Moura, for example, the percentage of households possessing ten to fifteen of the items listed increased from 26.3 percent to 35.4 percent. In Santa Luzia, the group possessing ten to fifteen items increased from 14.9 percent to 21.6 percent. Only in Alto Alegre was there a slight (insignificant) reduction in households possessing this amount of material amenities, from 15.9 percent to 12.5 percent.[27] Based on these material indicators, we would expect that Rolim de Moura's residents perceive their quality of life as somewhat better than do those of the other two settlements. This, however, was not true.

We reject Hypothesis 8. With the lack of employment opportunities being most frequently cited as the source of discontent, 22.5 percent of Rolim de Moura's household heads interviewed said that life had worsened since their arrival (and 26.6 percent indicated an intention to leave town), compared to only 13.9 percent and 8.0 percent in Santa Luzia and Alto Alegre, respectively (table 6.5).

While the residents of Rolim de Moura have acquired more material amenities than the populations of the lower-order settlements in the Rolim de Moura–Alto Alegre settlement corridor, the economic downturn in the economy during the post-mahogany period has left much of the population unemployed, disgruntled, and restless. While daily life is harder in the peripheral towns, residents there are more content with their prospects. Yet, when asked to identify the "most serious problem" in their community, the most frequently cited problem in Santa Luzia and Alto Alegre was "lack of public services", while in Rolim de Moura it was "poor air quality." Lack of other vital services (e.g., electricity, medical assistance) ranked second among the concerns expressed by residents. In all three samples, "lack of work/insufficient income" was the third most frequently cited community concern (table 6.5).

CONCLUSIONS

The short history of Rolim de Moura and its hinterland illustrates the complexity of the urbanization phenomenon in a populist agrarian frontier. Although the marketing principle of central place theory provides a useful conceptual starting point for understanding the processes of urban network expansion and hinterland segmentation, many idiosyncrasies eventually emerge. These idiosyncrasies are linked to distinctive local histories and to the differential impacts of national policy and market forces, all of which affect the social orientation, rate of expansion, and urbanization of the frontier.

During the mahogany fever of the late 1970s and early 1980s, Rolim de Moura enjoyed direct links to global markets. Lumber traders, with direct satellite links to U.S. hardwood importers and ample government-subsidized capital, controlled local industrial wood production to meet the requirements of the international mahogany trade dominated by corporate interests in Brazil's industrial core, the United States, and the United Kingdom. The mahogany boom years brought ephemeral prosperity to this once modest pioneer settlement, briefly turning it into an industrial rainforest city and extending its sphere of influence virtually to the Bolivian border. Santa Luzia, by contrast, garnered only the financial breadcrumbs of the mahogany trade as a truck stop between the mahogany timbershed and the lumber mills in Rolim de Moura. With the collapse of the mahogany boom in 1985 and the continued agrarian settlement of the forestland surrounding this former pioneer settlement, Santa Luzia was quickly elevated to a *município* service center. However, this was also due to the growing political influence and ambitions of the town's patron family, the Cassols. In response, Rolim de Moura's tenuous grip on the active frontier began to loosen, and its market catchment area contracted. Rolim de Moura became an urban relict, as postulated in chapter 4. Alto Alegre emerged spontaneously in the post-mahogany years, and in 1990 it was a bustling pioneer service center just 15 kilometers from the perimeter of the settlement frontier. Central place tendencies are evident in the early evolution and market segmentation of the Rolim de Moura–Alto Alegre settlement system. Yet, the different histories, characteristics, and patterns of growth of these three settlements demand an explanation beyond the marketing principle, one that indicates their differential articulation to the global economy.

Despite the absence of a monogenic master principle responsible for spatial organization, the physical, economic, and social characteristics of frontier settlements in the populist frontier do reveal significant evidence of a distance gradient. As one moves from the hub of Rolim de Moura to the fringe of Alto Alegre in this settlement corridor, informal self-provisioning of vital services (e.g., dwelling construction, water, sewage disposal, energy) increases; employment in agriculture increases; the service sector decreases in importance, and there are fewer material amenities in the households. Several other characteristics, however, are idiosyncratic to a distance gradient and are not explained by the marketing principle: the roughly comparable incidence of disease, comparably low rates of employment in extractive and industrial sectors, the higher rate of dissatisfaction with the quality of life in the more advanced centers of the urban system, and the asymmetrical distribution of public sector employment.

Multiple forces have shaped the patterns of populist urbanization observed in our Rondônia research. This conclusion leads us back to a pluralistic approach to understanding urbanization of the populist frontier. Local patterns of urbanization reflect the composite effects of differential articulation of settlements to influential factors operating at local, regional, and national levels—influences that arise both simultaneously and sequentially in historical successions, creating, in the aggregate, a disarticulated mosaic of spaces. Yet, in fundamentally patterned ways the populist frontier of Rondônia stands in marked contrast to the contested corporatist frontier of southern Pará where a different mosaic emerges.

Notes

1. At the time author John Browder was working on his doctoral dissertation, a study of the Brazilian mahogany trade that ultimately led him to this timber town of Rolim de Moura in the interior of Rondônia (Browder 1986).

2. The original encampment at the intersection of two planned roads, Linha (Line) 200 and Linha 25, was abandoned in favor of a more hospitable site located 16 kilometers to the

west at Linha 184 and Linha 25 where Rolim de Moura stands today. The town takes its name from Antonio Rolim de Moura, the Portuguese Count of Azambuze, who was the first governor of the province of Mato Grosso from 1762 to 1775. This town, the subject of our study, is not to be confused with the small gold-mining settlement of the same name founded two centuries earlier on the Guaporé River in Rondônia.

3. Candido Rondon will be best remembered, at least among North Americans, for his brief association with Theodore Roosevelt. In late 1913 Roosevelt and Rondon launched a nearly ill-fated expedition to collect exotic wildlife for the American Museum of Natural History. Their voyage down the treacherous waters of the unknown "River of Doubt" in the Madeira Valley, nearly left both men and their crews dead. In honor of Roosevelt's heroism, the Brazilian government christened the waterway the Rio Roosevelt.

4. The Colonia Agricola Presidente Dutra was established near Guajará Mirim in 1945 and was followed by similar projects (*Candeias, Paulo Leal, Periquitos, Iata,* and *Areia Branca*) near Porto Velho, of which only *Iata* and *Paulo Leal* survived (Gomes da Silva 1984).

5. Anecdotal accounts report over one hundred air flights to Porto Velho per weekday and two hundred on weekends chartered by "recreational miners" during the 1960s (SET 1983).

6. A fitting example is provided by the Calama S.A. colonization enterprise located near Jí-Paraná (then Vila Rondon). This private colonization project intended to sell about 1,500 plots to migrant families in the 1960s. Inundated with new settlers, land speculators, and miners the project lapsed into chaos because of conflicting land claims, deficiencies in basic infrastructure and support services, and the disorganized growth of urban areas (Mueller 1980; Millikan 1988). The failure of corporate colonization initiatives in Rondônia, such as Calama, gave INCRA additional evidence to argue for a "rational," government-directed colonization program that would culminate in POLONOROESTE.

7. By mid-1984 the cost of mahogany timber extraction was U.S. $132 per cubic meter (lumber equivalent), representing 82 percent of the total FOB unit production cost of mahogany lumber (Browder 1986:283).

8. In a 1973 speech Interior Minister Reis Velloso, acting under pressure from corporate groups represented by the São Paulo–based Association of Amazon Entrepreneurs, presaged the policy shift formalized in the First Amazon Development Plan (1975–79) that would effectively undermine INCRA's influence in the Transamazon area: "Until now, the Transamazon has emphasized colonization, but the necessity of avoiding predatory occupation with consequent deforestation, and of promoting ecological equilibrium, leads us to invite large enterprises to assume the tasks of developing the region" (translation: Hecht 1985:673).

9. In a 1978 speech at the University of Brasília, INCRA's director, Helio Palma de Arruda, reiterated the pervasive national security fear in the military establishment associated with the Amazon's extensive undefended (and in several places undefined) international borders, fears that INCRA would feed upon to support its own institutional survival: "The Amazon, by offering the world conditions for meeting its food deficits and absorbing its surplus population, raises the problem of securing our frontiers with neighboring countries. We need to populate the region to secure our defense and sovereignty" (Arruda 1977).

10. We note a significant discrepancy between our estimate of 1,588,000, based on a projection from 1984 population data using a 2.5 percent net growth rate, and 1,132,692 reported by the 1992 census.

11. Izidolandia is named after a former mayor of Alta Floresta who in 1988 organized a rudimentary lumber settlement adjoining the boundaries of the Guaporé Biological Reserve, one of the few areas in Rondônia where mahogany timber has not been exhausted. It is not known how many residents the settlement supported in 1990; probably fewer than one hundred.

12. Rolim de Moura's days as an expeditionary resource settlement were short-lived, probably spanning no more than eighteen months from mid-1976 to the end of 1977. Yet it would be inaccurate to suggest that the town's origins were oriented toward resource extraction, which would more accurately characterize the second period in its evolution as a populist frontier. In its earliest days the town served as a gathering place for land-grabbers and speculators, farmers, and petty merchants.

13. Between 1980 and 1984 mahogany log recovery rates steadily declined from 64 percent to 39 percent due to the progressive depletion of mature mahogany trees from the wild and the increasing harvest of juveniles (Browder 1989a:8).

14. Statewide industrial output in Rondônia grew from U.S.$15.7 million in 1975 to U.S.$73 million in 1980. The contribution of the industrial wood sector (all wood processing activities except fuelwood and cellulose production) in Rondônia rose from 28.6 percent to 61.2 percent of these amounts during this same period (Browder 1989a). In 1980 Rolim de Moura's ten mills accounted for an estimated 2.2 percent of total statewide industrial output.

15. Before 1984 an estimated 53 percent of colonists surveyed in the Jí-Paraná colonization area derived cash income, in-kind goods, or other useful services from loggers in exchange for stumpage rights (Browder 1989a).

16. These eight banks operating in Rolim de Moura in 1984 were: Bamerindus, Itáu, Bradesco, Banco do Brasil, Beron, Banco Mercantile do São Paulo, Finanza, and Comind. By 1985 only the first five banks were still in operation. In 1984 one major bank, Bamerindus, reported its Rolim de Moura cash turnover as the second highest in the Amazon.

17. In 1982 alone, during the height of the mahogany boom in Rondônia, an estimated 190,000 hectares of forestlands were opened by loggers, a small fraction of which were subsequently settled permanently by farmers (Browder 1986).

18. Quite apart from both export subsidies and revenues from mahogany lumber, the local trade in mahogany logs proved to be an extremely lucrative and socially exploitative business in itself. Between 1981 and 1984 the rate of increase in the mill-gate price of mahogany logs in Rolim de Moura exceeded that of Brazilian treasury bonds (ORTN)—which was used during this period as an indicator of general price inflation—by a factor of 2.3 (Browder 1987:295). See chapter 9 for a discussion of the adverse social impacts of the government-subsidized mahogany trade.

19. On May 20, 1986, the município of Alta Floresta d'Oeste was separated from the southwest flank of the Rolim de Moura município. The town of Alto Floresta has a story unlike that of Rolim de Moura. It was founded in 1981 by Sr. Isidoro Stédilli, who in that year obtained title to a large land grant from the governor of the territory of Rondônia to form a spontaneous colonization nucleus (SEPLAN 1988c).

20. One mill operator (Nelson Botelho of the Brasforest Company) reported acquiring a stock of 2,000 cubic meters of mahogany logs in the harvest of 1984. The following year these logs remained in rotting stacks in the company's abandoned Rolim de Moura millyard.

21. This reduction of business on main street was offset by a larger increase in both service and retail establishments on neighborhood streets during this period, representing a 17 percent increase overall.

22. From 1984 to 1987 average yields of rice in Rolim de Moura dropped from 1,550 to 1,429 kg/hectare (-7.8 percent), corn from 1,600 to 1,550 kg/hectare (-3.1 percent), beans from 600 to 509 kg/hectare (-15.2 percent), and coffee from 709 to 650 kg/hectare (-8.3 percent) in 1989 (SEPLAN 1983, various years; SEPLAN 1988d; SEPLAN 1990).

23. The 1992 demographic census reports 28,272 urban residents in Rolim de Moura in 1991 (IBGE 1991).

24. While public infrastructure and health facilities were abysmally poor in both Rolim de Moura and Santa Luzia and utterly nonexistent in Alto Alegre, it is noteworthy that Santa Luzia enjoyed the highest rate of households reporting access to municipal water and sewage disposal, an outcome of its more successful political strategy of leveraging state resources for local development projects and evidence of the importance of local agency in explaining diverse patterns of frontier urbanization.

25. This 1988 government socioeconomic report on Rolim de Moura noted that "there is a general deficiency in basic sanitation such that more than one-half of the population is vulnerable to illnesses," especially those linked to inadequate sewage disposal (SEPLAN 1988b:33–34).

26. It is important to point out that many other urban settlements in the populist frontier, those we would classify by our model as expeditionary resource settlements, are almost entirely dominated by employment in mineral extraction (e.g., Bom Jesus and a myriad of *garimpo* settlements along the tributaries of the Madeira and Guaporé Rivers). The essential feature of these expeditionary settlements is that they do not autonomously evolve into subregional urban systems but become fixed over time as bivouacs, regimented by world resource markets and utterly disarticulated from the urban systems in which they are situated.

27. The data on previous material possessions are not reported in table 6.5 due to space limitations and are reported only in summary form here.

7

Instant Cities of Southern Pará: Urbanization of a Contested Corporatist Frontier

%..%

In 20 months the town of Xinguara emerged, from virtually nothing, and became the third most important population center in the municipality of Conceição do Araguaia, the model "Wild West" of Amazonia. Eight thousand inhabitants installed themselves precariously while they awaited the realization of the dream that brought them from various parts of the country: a piece of land.
Lúcio Flávio Pinto (1978)

The Tucumã-Ourilândia complex showed more clearly than most urban places how the contradictions between rich and poor and between planned and spontaneous settlements generated patterns of population distribution that even the best-laid plans had neither intended nor anticipated. Linked to one another through a kind of symbiotic rivalry and interacting to produce a textured and dynamic urban system, the side-by-side towns of Tucumã and Ourilândia stood as a kind of metaphor of contemporary frontier change in Amazonia.
Marianne Schmink and Charles Wood (1992:208–209)

One of Brazil's most bitterly contested frontiers of contemporary settlement lies in eastern Amazonia, along the margins of highways recently constructed between the Araguaia and Xingu rivers (fig. 7.1). Historically, migratory pressures have emanated from the more heavily populated Northeast and Center-West of Brazil and swept westward across the states of Maranhão, Goiás, Tocantins, and into the *terra firme* uplands of southern Pará. Contemporary interregional migrations have accompanied the expansion of corporate cattle ranching, agriculture, logging, and mining operations. Subsequent disputes among cattle ranchers, peasant farmers, land speculators, miners, native peoples, and others have pervaded the various expansion fronts. As boomtowns sprang up in the settlement corridors initiated by new pioneer roads, pitched battles over land and resources in the surrounding countryside became a regular feature of urban life. Against the backdrop of a regional land rush, such new or reinvigorated centers as Marabá, Conceição do Araguaia, Redenção, Xinguara, Tucumã, and Ourilândia gained notoriety throughout Brazil for their rampant violence and frontier justice (Godfrey 1982, 1990; Schmink and Wood 1992).[1]

These boomtowns of eastern Amazonia serve as modern counterparts of what historian Gunther Barth (1988), in his studies of the nineteenth-century North American West, calls "instant cities."[2] The instantaneous growth of such urban centers has become commonplace in the receding forest fringe of southern Pará, where the in-migration associated with resource exploitation has created burgeoning boomtowns virtually overnight. Within a decade of their founding many instant cities became locally administered municipal seats (*sedes de município*), the primary criterion for urban status in Brazil. Although some urban nuclei have been carefully planned as centerpieces for corporate development projects, most fledgling settlements have seemed to observers rather like cities of peasants or rural shantytowns, even after becoming "urban" by virtue of municipal status.[3] In contrast to the neatly planned company towns, the spontaneous settlements have experienced overwhelming poverty, pronounced rural folkways, and rudimentary physical infrastructure. As journalist Lúcio Flávio Pinto (1978) noted above, most original residents of Xinguara arrived from other regions in hopes of obtaining land to farm, but the paralysis of planned colonization projects forced migrants to pile up in town, where they worked intermittently in local lumber mills, forest clearing, agriculture, construction, commerce, street vending, and so forth. With the accumulation of a mobile labor force in such spontaneous settlements, the region has witnessed a blurring of distinctions between the city and the countryside. The very categories of urban and rural have become problematic in the contemporary settlement frontiers of southern Pará, which are marked by modern communications, rapid transportation links, and consequent migratory flux.

Regional development policy in southern Pará contrasted with that of Rondônia, where the state's alliance with the peasantry sustained a populist agrarian frontier. Government-sanctioned corporate interests played more substantial roles in shaping the social space of southern Pará, a region dominated by such large-scale development projects as the Carajás iron-ore complex and the Tucuruí hydroelectric operations

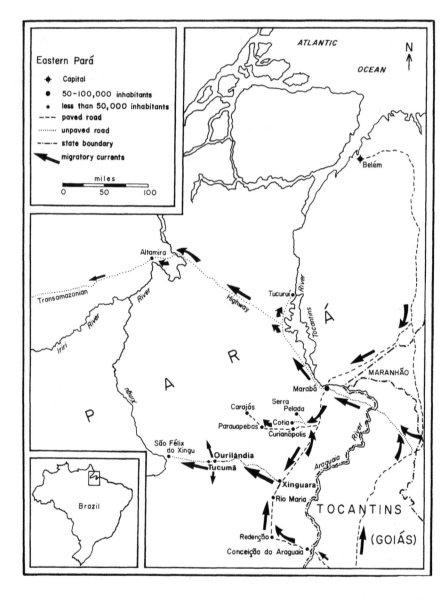

Figure 7.1. Contemporary migration patterns in southern Pará
Source: Brian Godfrey 1990, with permission of the American Geographic Society

(described below). Yet, outside the corporate enclaves of these "big projects," small farmers, migrant miners, native peoples, and other popular sectors have succeeded in asserting their own claims in many areas.

Moreover, state and federal governments have not been monolithic in their support of corporate interests: state factions occasionally have supported the small-scale migrant, as in the Transamazon Highway colonization scheme and other broad-based settlement programs. Rather than a rapid conversion to capitalism legitimized by the state, southern Pará has experienced an uneven transition involving complex linkages among diverse economic sectors, volatile patterns of resource extraction, a dramatic growth of informal activities, and widespread occupational mobility among migrants.[4]

Any generalizations about contemporary settlement and development in southern Pará are necessarily perilous for, as Schmink and Wood (1992:13) rightly point out, the region exhibits "not a single process of linear change but instead a diversity of contested frontiers with highly varied outcomes." Yet in recognizing the complexities and idiosyncrasies of evolution on the local level, we should not eschew the value of a regional analysis of settlement structure. We argue in this work that the asymmetrical patterns of urbanization in southern Pará have been driven by the prevalent development policy of the state in concert with private capital, coupled with the dynamic influx of popular sectors of the population. The clash of social classes has led to the formation of contested corporatist frontiers. The large-scale, centrally planned development projects have relied on the cheap labor, which has participated in a more directed fashion than has been the case in a fully populist frontier. For example, migrant labor is necessary in the early phases of project construction, and later many workers and their families remain in the region, sometimes employed in subcontracted work and informal activities linked to the development project. In these cases, the hybrid urban structure of a dual city emerges, juxtaposing a company town and an informal settlement outside the restricted corporate area, as discussed in our presentation of the corporatist frontier model (chapter 4). In addition, migrants often engage in small-scale resource extraction, such as timber or gold, often at odds with centralized corporate activities. Overall, the vast majority of migrants reside in spontaneous settlements in the region, not in the company towns themselves. Nevertheless, the corporate projects provide the dominant political-economic context of contemporary regional development and mediate the pressures for social change in eastern Amazonia.

In the contested corporatist frontiers of southern Pará, in contrast to the populist frontier of Rondônia, we cannot necessarily begin with the marketing principle and central place theory as appropriate conceptual points of departure. Even in southern Pará, however, migrants should not be reduced analytically to mere pawns of capital. Massive in-migration to Amazonia also stems from exogenous "push" factors, such as land concentration, unemployment, and poverty in the sending areas. As Sawyer (1984:190) notes:

> In sum, it does not seem appropriate to view frontier settlement as the result of some preconceived plan or even an implicit process to mobilize labor power for future capitalist activities. Brazil's surplus population needed few if any special incentives to migrate to the Great Frontier. Most settlement can be considered "spontaneous."

Given the strong popular impulses behind interregional migration and subsequent settlement in the region, it would be too simplistic to explain the recent evolution of the eastern Amazon region solely in terms of an alliance of the state and private capital. Residents of frontier regions have repeatedly asserted their own interests, often in alliance with local governments and sometimes to the detriment of the government and large companies. For example, small-scale gold prospectors, the *garimpeiros*, defied the governmental authorities and the state-owned mining company, Companhia Vale do Rio Doce, at the Serra Pelada and elsewhere in southern Pará (see chapter 10). This chapter recounts how the small farmers and *garimpeiros* of Ourilândia do Norte battled against exclusion from the nearby corporate colonization project at Tucumã, forcing the demise of a large-scale private project sponsored in the mid-1980s by the Andrade Gutierrez Company, one of Brazil's largest corporate entities. Whatever the fate of such local battles, however, the alliance between state and private capital initially opens the corporatist frontier to settlement through massive infrastructure investments, which create initial conditions of occupation not found in populist frontiers.

The geographic juxtaposition of large-scale corporate projects and small-scale popular sectors results in a spatially fragmented and socially contested rural-urban continuum in southern Pará. This part of eastern Amazonia exhibits the volatile and irregular dynamics of a corporatist frontier challenged by populist impulses. The result is a region divided by social polarization, uneven infrastructural development, and spatial asymmetry in its urban structure. In short, contemporary frontier urbanization embodies the spatial consequences of a long and continuing history of land conflict. The settlement systems emerging in such contested corporatist frontiers often diverge from the spatial logics of such grand models as central place theory and extractive mercantile models of frontier expansion, as seen in the case of southern Pará. This chapter briefly reviews the physical geography and the history of the region before turning to the contemporary evolution of the Xinguara–Ourilândia do Norte–Tucumã settlement corridor. Finally, we analyze the results of our extensive 1990 survey of 903 urban households in these three towns.

PHYSIOGRAPHY OF SOUTHERN PARÁ

Xinguara and the nearby towns studied in this chapter lie in the southeastern quadrant of the state of Pará (fig. 7.1). This huge and diverse state resists facile generalizations. Located in the eastern part of Amazonia, Pará is the second largest state in Brazil, smaller only than the neighboring state of Amazonas to the west. Pará encompasses an area of 1,227,530 square kilometers, more than 14 percent of Brazil's total national territory, an area larger than every American state except Alaska, twice as large as France, and about the same size as Peru. Given the state's considerable territorial extensions and its equatorial location in the lower Amazon River Basin, Pará is marked by significant internal variations in precipitation, land forms, soils, vegetation, human settlement, and many other factors. Put in the simplest terms, the state of Pará exhibits a fundamental cultural and geographic dichotomy. River transport has linked *várzea* and *igapó* lowland environments along the major rivers to the Belém côre region; adjacent *terra firme* upland areas in the Bragança agricultural region were once linked to the capital by railroad lines and more recently have been accessible by highways. The state's vast interior, however, consists mostly of *terra firme* uplands separated from Belém by considerable distance, rough terrain, and increasing regionalist sentiment.

Southern Pará, for example, is virtually a separate geographic province. Regional isolation from the capital at Belém is accentuated by southern Pará's vast territorial extension, distinctive physical geography, singular settlement history, and strong economic linkages to other regions of Brazil. The eighteen municipalities of this region, according to jurisdictions recognized in the Brazilian census of 1991, jointly extend over an area of nearly 240,000 square kilometers, about a fifth of Pará's total territory and roughly equal to the entire Brazilian state of Rondônia or the U.S. state of Minnesota. Although southern Pará remains marginal to the center of power in Belém, the region now presents the most dynamic profile of sustained population growth anywhere in the state. In the last three decades, the population of this region of southeastern Pará has increased nearly twentyfold. Census statistics indicate that the regional population exploded from approximately 41,000 (40.2 percent urban) in 1960 to nearly 775,000 (50.7 percent urban) in 1991. Despite this impressive demographic growth, in 1991 the population density of southern Pará was still only about 3.2 people per square kilometer (IBGE 1977, 1991).

Southern Pará is dominated in geological terms by the Xingu Complex, a massive Precambrian formation of highly weathered materials, which exhibits substantial variations in physical relief. Mineral riches abound. Near the Carajás mountains are found the world's largest known deposits of high-grade iron ore, along with the third largest reserve of bauxite and large quantities of manganese, nickel, copper, gold, cassiterite, and other minerals. Some of Brazil's largest recent gold strikes have been discovered in this region, such as at the Serra Pelada. Most of the soils in the region are considered red-yellow podzols, a moderately fertile type in the Brazilian system of classification. Yet substantial areas of highly fertile "pink earth" (*terra roxa*) soils can be found in the São Félix do Xingu region (Becker 1990:64–65; Governo do Estado do Pará 1976; Prefeitura Municipal de Tucumã 1990; RADAM 1974).

The climate of southern Pará is markedly tropical and variably humid. The annual medium temperature is a warm 24 degrees centigrade (about 75 degrees Fahrenheit). Rainfall averages about 2,000 mm a year, concentrated in a wet season from roughly November to May, though significant local variations occur in precipitation. The region's climate is classified as tropical wet and dry in the Köppen system, ranging from a tropical seasonal (*Aw*) regime in the east along the Araguaia River to tropical wet (*Am*) in the west along the Xingu River.[5]

The natural vegetation in southern Pará is largely transitional tropical forest, ranging from patches of low-lying *cerrado* scrublands in the east to lush tropical rainforest in the west. The easterly areas of *cerrado* shrubs and grasslands extending from Brazil's Center-West region are marked largely by poor soils, seasonal rains, and inadequate drainage. Cattle ranching is a long-standing vocation of the relatively impoverished *cerrado* grasslands. Yet most of southeastern Pará's natural vegetative cover consists of open tropical forests, adapted to a dry season, which becomes gradually less pronounced in the areas of greater precipitation to the west, near the Xingu River, an area dominated by lush tropical rainforests (RADAM 1974; CONSAG 1978).

The heterogeneity of the settlement system in southern Pará stems in part from the sheer serendipity of the discoveries of natural resources. Due credit should be given to the environment itself in explaining regional evolution: mineral strikes, mahogany trees, fertile soils, and other natural riches are not ubiquitous, but rather display specific spatial distributions in Amazonia. Yet the spatial heterogeneity of contemporary frontier expansion in southern Pará reflects far more than the irregular distribution of natural resources. The contemporary regional geography also stems directly from the institutional forces promoting uneven development around selected poles of modernization.

A BRIEF HISTORY OF SOUTHERN PARÁ

Although attempts to explore and settle southern Pará began in the colonial period, the area remained marginal to the Portuguese mercantile system. After national independence, the accounts of foreign explorers, missionaries, and travel writers received wide attention abroad and piqued the interest of the Brazilian authorities.[6] However, access was difficult: roads were nonexistent, and river transportation on the Amazon tributaries, the Xingu and Tocantins-Araguaia rivers, was particularly trying during the dry season. Large-scale Brazilian settlement began only in the late nineteenth century, when the rubber boom and other extractive resource cycles encouraged in-migration from adjacent coastal areas, particularly from the Northeast. The region's first permanent urban settlements in modern times originated during this period. Populations clustered along the major watercourses at entrepôts formed for the extraction and transport of rubber, Brazil nuts, minerals, and animal skins—and displaced native Indian groups in the process (Velho 1972; Godfrey 1982). Until after World War II—when cattle ranching, agriculture, and mining encouraged movement into the *terra firme* uplands—the expansion and retraction of extractive activities near the rivers in large part determined the demographic fluctuations of the principal towns in southern Pará: Marabá, Conceição do Araguaia, and São Félix do Xingu (fig. 7.1).

Marabá was founded in 1898 as a rubber trading post at the confluence of the Tocantins and Itacaiunas rivers. By the early years of the twentieth century, a small village emerged on the site, which became a municipal seat in 1913. Rubber tappers, mainly migrants from the Northeast, entered the local forests in the summer and extracted latex. In the win-ter, when the rivers were high, the tappers floated the accumulated latex downstream to Marabá. The most common type of rubber tree found in the area was known as *caucho* (*Castilloa ulei*), which generally required felling to extract the latex, as opposed to *seringa* (*Hevea brasiliensis*), which could be tapped repeatedly. With the exhaustion of local *caucho* trees and the fall of rubber prices after the First World War, the production of latex in the area fell dramatically throughout the region. Fortunately for Marabá's economy, other extractive activities followed in the wake of the declining rubber trade. Increasing production of Brazil nuts (*castanha do Pará*) during the 1920s prevented a total collapse of the local economy. In the late 1930s the discovery of diamonds in the rivers near Marabá initiated another extractive cycle. After World War II cattle ranching and agriculture gradually emerged as the dominant economic activities in the Marabá region. The opening of new roads encouraged land speculation and conversion of large Brazil nut operations into cattle ranches (Velho 1972; Godfrey 1979, 1990).

This general pattern of economic cycles and landscape changes was replicated in other river towns, such as Conceição do Araguaia and São Félix do Xingu, which also emerged as rubber trading posts around the turn of the century. Unlike Marabá, however, these other two towns possessed fewer alternative resources to replace rubber when the commodity prices declined early in the twentieth century. Both the Lower Araguaia and the Xingu region therefore entered a prolonged period of economic stagnation and demographic decline after World War I. The municipal population of Conceição do Araguaia, for example, plummeted from about 15,000 in 1910 to 6,000 in 1950. After their heyday during the rubber boom, these towns fell for decades into localized patterns of subsistence production with weak interregional linkages (Godfrey 1979, 1990). In short, before the pivotal postwar development of transportation networks in the *terra firme* uplands, the urban system of southern Pará consisted essentially of a series of small riverine settlements, such as Marabá and Conceição do Araguaia, linked by river trade to the region's metropolitan center, Belém.

The dramatic recent increases in the regional population date from the opening of the Belém-Brasília Highway in 1960 and of the Transamazon Highway in the early 1970s. Expansion of the contemporary settlement frontier in southern Pará occurred as new transportation corridors—primarily penetration roads, but also landing strips and even railroads—allowed migrants to enter previously isolated areas of the *terra firme*

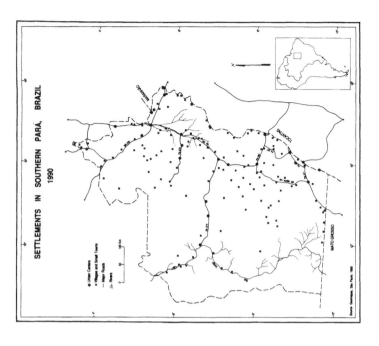

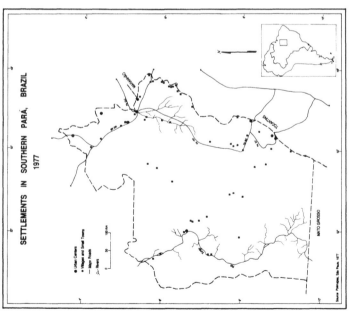

Figure 7.2. Settlements in southern Pará, Brazil, 1977 and 1990

Source: Brian J. Godfrey 1992, with permission of the American Geographic Society

212 Instant Cities of Southern Pará

uplands. Besides the dramatic increase in the total population of southern Pará (table 7.1), the proliferation of new urban centers at key junctures in the regional highway network is a notable aspect of regional change: only eight urban settlements existed in 1977, but there were twenty-one cities in this area of southern Pará by 1990 (fig. 7.2).[7]

Table 7.1 Population growth and urbanization in southern Pará, by municipality, 1970 and 1991

MUNICIPALITY, YEAR CREATED	1970 POPULATION	/% URBAN	1991 POPULATION	/% URBAN
Brejo Grande do Araguaia, 1988	—		11,892	26.8%
Conceição do Araguaia, 1909	29,830	23.8%	54,900	54.4%
Curianópolis, 1988	—		38,590	38.9%
Itupiranga, 1947	5,368	26.7%	72,521	12.7%
Jacundá, 1961	2,229	24.6%	42,890	51.4%
Marabá, 1913	24,798	59.6%	122,231	83.7%
Ourilândia do Norte, 1988	—		28,742	37.9%
Parauapebas, 1988	—		53,312	51.5%
Redenção, 1982	—		55,895	80.3%
Rio Maria, 1982	—		26,522	54.6%
Santa Maria das Barreiras, 1988	—		7,230	11.2%
Santana do Araguaia, 1961	9,651	18.0%	15,920	53.4%
São Félix do Xingu, 1961	2,397	38.7%	24,834	32.9%
São Geraldo do Araguaia, 1988	—		38,522	22.8%
São João do Araguaia, 1961	15,348	11.1%	19,787	6.7%
Tucumã, 1988			31,393	9.7%
Tucuruí, 1947	10,091	57.6%	81,635	56.4%
Xinguara, 1982			48,007	57.0%
TOTALS	99,712	34.1%	774,823	50.7

Source: Instituto Brasileiro de Geografia e Estatística, various years.

Overall, recent development projects and interregional migrations have steadily increased the ties between southern Pará and the Brazilian economic core regions to the southeast. Although access to and from the northern capital at Belém also has improved, national integration has effectively weakened southern Pará's dependence on the traditionally dominant capital region. Moreover, economic integration and national security on Brazil's northern periphery, important goals of the military's road-building campaign, have replaced the parochial self-interests of regional elites with the aspirations of Brazil's postwar business elites, based in the Southeast core region. Yet this regional development strategy, while improving national access to Amazonia, has contributed to speculative pressures, disputes over land ownership, and even regional separatism. To understand the recent history of the urban system in southern Pará, we must now turn to the region's history of land conflicts.

Southern Pará became one of Brazil's most violent arenas of land conflict during the military-authoritarian period. At the time a Brazilian national magazine once called southern Pará "probably the tensest region in the country, at odds with land disputes almost daily" (*Veja* 1978:61–62). More recently, the Human Rights Watch (1991:72) singled out the region for special mention in a report on rural violence in Brazil:

> Indeed, the violence against rural activists registered in the state of Pará is concentrated in the southern part of the state: of the 17 murders in 1988, nine were in southern Pará; of the 10 in 1989, seven were here; and six of the seven killings in the first half of 1990 were also in southern Pará.

More detailed statistics gathered by the Institute of Socioeconomic Development of Pará (IDESP 1990a) confirm that contemporary land disputes have centered on the southern part of the state. Of the 303 documented land conflicts between 1964 and 1988, 212 or 70 percent of the state's total occurred in the municipalities located between Marabá and Conceição do Araguaia. In addition, 63 percent of Pará's total of 529 agrarian murders during this twenty-five-year period occurred in these municipalities of southern Pará. More than one hundred of these rural deaths occurred in Xinguara, the state's most violent municipality in terms of agrarian murders during this period. Of all the municipalities in the region, the following five ranked highest in agrarian-related murders committed between 1964 and 1988: Xinguara, 101 deaths; Marabá, 73; Conceição do Araguaia, 54; Santana do Araguaia, 30; Redenção, 22.[8]

These contemporary land conflicts in southern Pará have deep roots in the past. Historically, secure legal title to land has been notoriously rare in a region long dominated by extractive economies. Until recently, large landholders extracted rubber, Brazil nuts, and other forest products based on perpetual leases (*aforamentos*) obtained from the state of Pará. These quasi-legal leases referred more to the rights of forest extraction than to the occupation of clearly demarcated tracts. Traditional populations clustered along the rivers and left vast inland areas officially classified as "unoccupied lands of the state" (*terras devolutas do estado*) even if already inhabited by indigenous peoples, *caboclos*, or other rural residents. Another element of confusion stemmed from the long-established Amazon tradition of gaining title to land through habitual occupation, cultivation, and general improvement of unclaimed tracts; by proving de facto possession (*posse*), squatters (*posseiros*) have hoped to gain legal land titles (Laraia and da Matta 1967; Velho 1972). Land tenure in southern Pará became even more confused after World War II, when successive state political administrations mounted contradictory, often corrupt land-titling programs (Hecht and Cockburn 1989).

The Brazilian military became actively concerned about land disputes in southern Pará in the late 1960s, when a small group of partisans of the Communist Party of Brazil, inspired by Maoist tactics of rural revolution, fled suppression in southern cities and settled along the lower Araguaia River. The rebels settled inconspicuously in small towns located in the region between Marabá and Conceição do Araguaia, where they mounted a quiet campaign of political mobilization. Although the guerrilla activity was small in scale and probably represented no real threat to the Brazilian state, the country's military regime mounted a large-scale counterinsurgency program in 1972, when the National Security Council assumed jurisdiction over colonization in the area and summoned the army to intervene. From its base near Marabá, the Twenty-third Jungle Infantry Brigade mounted an aggressive program of counterinsurgency and rural stabilization in the early 1970s. After expelling suspected guerrillas in 1974, the army's special forces attempted to bring calm to the area by administering a local program of agrarian reform: when hostilities ended, 100-hectare lots were distributed among hundreds of families of former informants, army guides, and other collaborators. The lots were located on the margins of "operational roads" constructed to make way for troop passage (Schmink and Wood 1992:72–74; Hecht and Cockburn 1989:108).

The military's campaign of rural pacification in the lower Araguaia region, along with the state-directed agricultural colonization program along the Transamazon Highway, marked a populist inclination in the state's land policies in southern Pará during the early 1970s. Elements of this military-style populism continued into the 1980s with the creation of security-conscious land agencies.[9] Yet, the authoritarian regime's commitment to agrarian populism gradually waned, as the task of resolving the region's many land conflicts loomed ever larger. Despite localized success in the military's efforts at rural pacification along the Araguaia River, other areas of southern Pará were racked by land disputes as in-migration proceeded along new roads extending inland toward the Xingu River. In addition, priorities in regional development shifted away from small-farmer colonization to large-scale corporate investments in the late 1970s.

Disillusioned with the prospects for small-farmer settlement in government-directed colonization projects like the Transamazon Highway, Brazil's military regime shifted the emphasis of Amazon regional development programs to large-scale corporate investments in the late 1970s (see chapter 3). In southern Pará the continuing proliferation of settlements reflects a massive governmental investment in transportation infrastructure and resource development, launched under the POLAMA-ZONIA program (1975–1979). The Carajás pole, featured as a preeminent area for "agromineral" activities, garnered nearly a quarter of the total fiscal incentives of this program (SUDAM 1976). Several large-scale projects in the Carajás pole by state-owned companies have led to continued in-migration and sustained (if uneven) economic growth: urbanization has occurred near development projects like the Tucuruí hydroelectric plant, the industrial park at Marabá, and the mechanized iron-ore extraction at Carajás. These state-sanctioned corporate projects, which have led the way in regional growth, have been planned around strategic strongholds, company towns with controlled access. Nearby, popular sectors have resisted government control over regional development by constructing their own strongholds. By alternating support for populist and corporatist interests, Amazon regional development policies have exacerbated conflicts associated with ambiguous land tenure.[10] In southern Pará these fluctuating policies left a legacy of competing federal and state agencies, ambiguous and conflicting land claims between rival social groups, and contrasting settlement forms (Schmink 1982; Schmink and Wood 1992:76–78).

Although not all violence in southern Pará has been associated with social conflicts over land-based resources, the region has witnessed a continuing interclass tension in which the "law of the strongest" (*a lei do mais forte*) has generally prevailed in land tenure. The settlement process often features agricultural migrants, who enter southern Pará along newly constructed roads in search of their own farmland. These migrants often squat on unoccupied (but seldom unclaimed) lands in hopes of obtaining legal title after a period of occupation and cultivation. At times they succeed in winning land titles, especially when helped by government expropriations of rural areas marked for colonization projects. More often, the putative landowners—cattle ranchers, speculators, and more powerful corporate interests—resist such efforts by peasant farmers to occupy claimed lands. Frequently after clearing tracts of forest and planting crops, these migrant *posseiros* are forcibly expelled, sometimes legally backed by judicial orders of eviction, sometimes simply through brute force. Several means are employed to evict the *posseiros*. The police can be summoned, but this is costly and time-consuming. Cattle ranchers and land grabbers more commonly resort to hired gunmen or *pistoleiros*. Often the small farmers, realizing the difficulty of establishing permanent residence in such circumstances, make improvements in hopes of obtaining a larger indemnity upon eviction. As agriculturalists continually press on into newly opened lands, they in effect serve as a moving wedge, pushing back native Indian groups and tropical forest. Behind this pioneer front, landholdings become consolidated into larger holdings, abandoned crop fields are converted into cattle pastures, and the population clusters in tenement towns. Frontier urbanization therefore plays an integral part in the process of land settlement and class conflict (Browder and Godfrey 1990; Godfrey 1979, 1988).

Conflicts over land are not limited to the rural sector. Urban land conflicts are on the rise throughout Amazonia. Nowhere is urban land more hotly contested than in and around the megadevelopment projects that form the core of the corporatist frontier. One example of a corporatist development is the massive iron-ore project at the Serra dos Carajás in southern Pará, which is owned and operated by the state-owned Vale do Rio Doce (CVRD) Company. The Carajás complex, first discovered in the mid-1960s, was planned as a large-scale, debt-financed "agromineral" growth pole project in the late 1970s, garnering nearly a quarter of the total fiscal incentives of the POLAMAZONIA program (SUDAM 1976). Export of iron ore began with the inauguration in 1985 of the 890-kilometer railroad to the port of Itaquí, near São Luis, Maranhão, running on electricity generated from the Tucuruí hydroelectric plant on the Tocantins River (Becker 1990:, 65–73). The CVRD built a company town at Carajás, protected from unwanted incursions by strict security and a heavily guarded entrance (plate 7.1).

Outside the fortified gates of the Carajás project, the town of Parauapebas quickly sprang up (plate 7.2). Peopled by migrants to the region, this spontaneous settlement grew by annual rates of over 20 percent during the late 1980s to reach a population of roughly 27,000 by 1989—five times the population of the company town of Carajás.[11] The labor force of Parauapebas provided services for the mechanized mineral

Plate 7.1. Heavily guarded gate at Carajás, Pará, at the entry to the iron-ore export complex, 1989 *Photo: Brian J. Godfrey*

extraction of the state-controlled company, particularly activities sub-contracted from the CVRD, showing how the corporatist frontier spawns adjacent informal economies. Facing high rents and hemmed in by large landholdings on all sides of the burgeoning settlement, recent migrants organized land invasions of adjacent properties. Municipal authorities, anxious for electoral support, have quickly recognized these squatter invasions, though the provision of promised urban services has been uneven. Roberts (1991, 1992) discusses how urban squatter invasions have shaped the spatial structure of a boomtown on a corporatist frontier of Amazonia:

> Leveraged from the grass roots, the ad hoc city planning in this Amazonian town has fascinating ramifications. For one, although rural-born migrants to Amazonian towns are rapidly becoming urbanized, their urban resocialization occurs in a center built mostly by the poor for their own survival. The interactions among squatters, landowners, the state, and the politicians, in turn, are shaping not only local society but also the layout of the streets and neighborhoods. The streets themselves reflect a logic different from that of the urban planners and engineers. (Roberts 1992:456)

Plate 7.2. View of the central commercial district in Parauapebas, Pará, 1989 *Photo: Brian J. Godfrey*

Another example of the hybrid nature of the corporatist frontier, again uneasily juxtaposing a planned company town and an adjacent popular settlement, is the huge Tucuruí hydroelectric project of southern Pará. Tucuruí is part of Brazil's ambitious program to substitute domestic hydroelectric power—estimated at 100,000 megawatts—for imported petroleum, which consumes about 23 percent of Brazil's merchandise imports each year (Caufield 1983; World Bank 1992:247). Planned since 1973, the Tucuruí dam began operating in 1984 and, presently generating about 4,000 megawatts, is expected to become Brazil's largest hydroelectric power producer (plate 7.3). The Tucuruí facility was a pharaonic undertaking, costing U.S.$4 billion and requiring thousands of construction workers. The hydroelectric dam inundated 2,430 square kilometers of land (300 kilometers south of the port city of Belém), including six older cities, thirteen diamond mines, 21.5 million cubic meters of wood, an Indian reserve, and 120 kilometers of the Transamazon Highway (Scherl and Netto 1979). To house workers employed by the project, ELETRONORTE, Brazil's parastatal power utility enterprise, constructed a core settlement (plate 7.3). The official government literature, issued from the Office of the Presidency of the Republic, spoke of a bold new model of urbanism:

> The new Tucuruí is today a model city of urbanism, infrastructure, public services, and comfort for its inhabitants. It was constructed by ELETRONORTE to house the managers and workers of the hydro-electric construction project. The city is an example of the genuinely Brazilian technology used for the benefit of the community that lives there. (Presidência da República, n.d., pp. 14–15).

Yet the nearby river settlement of Tucuruí predated the company town by decades. It had emerged as a small village that served as a Brazil-nut transshipment center on the Tocantins River by 1947 when the railroad then serving as the logistical backbone of the extractive trade was abandoned. Predictably, publicity of the hydroelectric project, beginning in the 1970s, sparked massive in-migration to Tucuruí. The municipal population exploded from 10,091 in 1970 to 61,140 in 1980, an increase of 505 percent; by 1991 the population had reached 81,635, a further

Plate 7.3. Aerial view of Tucuruí, Pará, 1989. The planned company town is in the foreground; the old river town, now engulfed by recent in-migration and informal house construction, lies in the background. In the distance the hydroelectric dam can be seen upriver from the town. *Photo: Brian J. Godfrey*

increase of 33 percent over 1980 (table 7.1). The results of this rapid urbanization diverged dramatically from the official record:

> Reactions to the project are felt most intensely in the city of Tucuruí. For three years the city has suffered from an enormous invasion of construction workers; the population has increased from 13,000 to 35,000 inhabitants, without the most nominal adjustments of infrastructure. The city has no INPS (National Institute of Social Welfare); the official government hospital with its 15 beds has neither surgeon nor anesthesiologist. Only 35 percent of the residents have running water, and none of its schools offer classes beyond the 5th grade because they lack teachers due to low salaries. . . . Today, old freight cars serve as living quarters for dozens of families who have no means to pay the exorbitant rents resulting from rampant real estate speculation that has seized the city. No room is found for less than $75 per month and in spite of the emergence of numerous makeshift hotels, no accommodations are available for the hundreds of families that regularly arrive there. The situation worsens during the rainy season when part of the city disappears beneath the river's rising flood waters. (Scherl and Netto 1979:24–27).

These two accounts of corporatist-led urbanization in southern Pará suggest a hybrid spatial structure, polarized between company towns and spontaneous settlements. This urban dualism reflects a continuing regional conflict between large-scale corporate interests and small-scale populist forces, often mediated by would-be local political bosses, as illustrated below by the case of Chapeu de Couro in Xinguara. Most land disputes have been resolved in favor of the powerful corporate entities, but sometimes popular sectors have wrested controls of local areas of production from them. Although these megaprojects most dramatically illustrate the clashing social dynamics of urbanization on the corporatist frontier, virtually every urban center in southern Pará bespeaks a social history of land disputes, class conflict, and institutional rivalry.

In a larger territorial sense, the regional separatist movement represents the greatest land dispute looming in southern Pará. The region's vast area, sense of isolation from the state capital, and rich natural resources have led many regional loyalists to suggest the formation of a separate state. The distinctive cultural origins of the region's population also provide a driving force for regional separation. Residents of southern Pará are overwhelmingly migrants, mainly from the Northeast and Center-West of

Brazil, and so they differ noticeably in speech, dress, food, and other manners from the riverine *paraense* style to the north. Additionally, the prevailing economic ties of southern Pará are to adjacent regions, particularly to the states of Tocantins, Goiás, and Minas Gerais, rather than to Belém and the rest of Pará. Most of all, residents mention the region's poor physical infrastructure and social services, which are blamed on the inattention of state authorities. Just as northern Goiás recently formed the new state of Tocantins, so goes a common argument in the region, southern Pará should secede and take control of regional resources away from the grip of entrenched special interests in the distant capital of Belém. A popular name for the proposed new state would be Carajás, after the mineral-rich mountain range often pointed to as the economic linchpin of the region.

Federal deputy Giovanni Queiroz introduced Legislative Decree Number 159/92 with the intent of creating a new state of Carajás, but the measure still awaits approval in Brasília. The bill envisions a plebiscite in thirty-two municipalities extant in southern Pará in 1993. If ratified by Congress and the voters, the new state would encompass 281,636 square kilometers, ranking eleventh in size among the twenty-six Brazilian states. The state of Carajás would even be a decent-sized country in its own right, larger in territory than Ecuador, the United Kingdom, or the U.S. state of Colorado. A successful separatist movement would have important ramifications for the urban system of southern Pará. A powerful new Amazonian state would lay claim to a significant share of federal revenues for its own development; it would designate its own capital city, and undoubtedly it would enfranchise many more new municipalities than if southern Pará retained its status within the current state. All these eventualities would greatly affect the evolution of urban systems in eastern Amazonia.

EVOLUTION OF THE XINGU SETTLEMENT CORRIDOR

Contemporary settlement of the PA-279 highway in southern Pará, stretching between the towns of Xinguara and São Félix do Xingu, provides a case study of massive in-migration, often contradictory state policies, and intense conflicts among diverse social groups over land and valuable resources. This region may be denominated the Xingu settlement corridor (fig. 7.3). It includes the four recently created municipalities of Xinguara, Ourilândia do Norte, Tucumã, and São Félix do Xingu. In 1991 the jurisdictions of these four adjacent *municípios* jointly encompassed over 120,000 square kilometers, an area larger than Cuba, Portugal, or

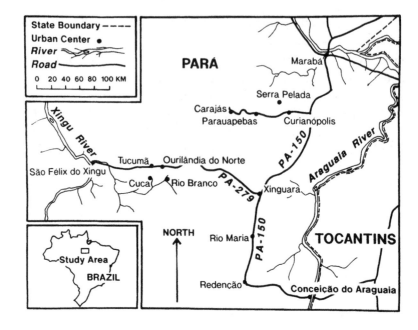

Figure 7.3. Xingu settlement corridor, southern Pará
Source: Brian J. Godfrey 1992, with permission of the American Geographic Society

the U.S. state of Pennsylvania. The 1991 resident population of these four municipalities exceeded 133,000 (IBGE 1991).

The main attractions drawing migrants to the Xingu settlement corridor have been land acquisition and the extraction of timber, gold, and other natural resources. The population of this area began to grow rapidly in the late 1970s, when the PA-279 initially was projected to São Félix do Xingu. As news of the imminent road spread throughout the region, migrants poured into present-day Xinguara, and land conflicts between small farmers and cattle ranchers proliferated. Timber extraction, based on the region's prized mahogany trees, quickly became a leading economic sector, though local stands of this once abundant resource were as widely depleted during the 1980s as they were in Rondônia. By the late 1980s gold mining revitalized the corridor's extractive economy. The social conflict that erupted at the entrance gates to the corporatist colonization project of Tucumã, led to the formation of a neighboring populist "instant city" in Ourilândia. Though not as massive in scale as the Carajás or Tucuruí megaprojects, settlements along the Xingu corridor vividly illustrate the social dynamics of a contested corporatist frontier. In addition, the shifting ripple effects of resource-extractive cycles are apparent in the demographic and commercial evolution of the principal towns in the Xingu road corridor: Xinguara, Ourilândia do Norte, and Tucumã.

Initial Settlement at Xinguara, 1976–1980

Xinguara was born of social conflict over land. Initially called Xingu Junction (Entroncamento do Xingu) and later officially renamed Xinguara as the meeting point between the Xingu and Araguaia rivers, the town emerged in the midst of the late-1970's land rush in southern Pará. When a small-farmer colonization program announced in 1976 by the state of Pará floundered, cattle ranchers seized land occupied by peasant *posseiros*. The level of tension mounted as land disputes proliferated and local shoot-outs became commonplace. The local authorities were unable to control the rising tide of violence, so the population armed itself with guns. Xinguara, along with nearby towns like Redenção, quickly became notorious throughout Brazil for its Wild West atmosphere on the Amazon frontier. During the late 1970s Xinguara fell under the sway of local bosses, generally powerful merchants, ranchers, and land speculators, who surrounded themselves with hired gunmen. These charismatic figures used the threat of violence and the promise of patronage to establish their authority (Godfrey 1982; Kelly and London 1983; Schmink and Wood 1992).

The profile of a particularly notorious character in the fledgling town personifies the human drama of Amazon frontier life. Probably Xinguara's most intimidating public figure during the late 1970s was José Ferreira da Silva, better known as "Chapeu de Couro" (Leather Hat), because of his use of the attire characteristic of his native northeastern backlands *sertão*. Chapeu de Couro fit the familiar image of the Brazilian rural boss: sullen, burly, rough in speech and manner, and prone to violence, feared as a gunslinger throughout the region, he also had a generous streak for friends and allies.[12] Local residents maintained that he had killed scores of people in land disputes. Yet in our 1978 interview, held on the porch of the modest hotel he then owned, Chapeu de Couro defended his record as a public-spirited citizen. He claimed to have killed only one man, in self-defense, and in recent years to have begun attending church. He opened a primary school for local residents. Chapeu de Couro called himself one of Xinguara's first residents, though later municipal publications did not acknowledge him among the original founders.[13] Yet he was widely identified with the political machinations surrounding the town's founding, and in that sense his personal story illustrates the diverse social and institutional forces at work in southern Pará's recent settlement history.

Chapeu de Couro began his work in the region as a labor contractor (*empreiteiro*) and, many claimed, a hired gunman for local ranches attempting to evict *posseiros* from their land claims. From 1969 until the mid-1970s he worked at several ranches in southern Pará and regularly traveled to northeastern states to recruit workers. In 1976, as Xinguara emerged as the administrative headquarters for a state colonization project, Chapeu de Couro became the right-hand man of the administrator of the local state agricultural agency (ITERPA). Widely feared for his brutal methods, he took charge of demarcating and awarding urban lots to new arrivals in town—and, many claimed, he profited handsomely in the process. When ITERPA turned the administration of the town over to the municipality of Conceição do Araguaia in March of 1977, Chapeu de Couro built the town's second hotel, which became known by his own nickname.

After its opening in August of 1977, the Hotel Chapeu de Couro quickly became a haven for Xinguara's sizable population of rural workers, who filled the available beds on weekends and between jobs on local ranches. At the time the other hotel in town, the Hotel Comaxim, attracted an entirely different, more prosperous clientele: businesspeople, government functionaries, truckers, ranchers, and others. Chapeu de Couro maintained close ties with the local ranchers and labor contractors, who regularly sought out short-term workers at his hotel in Xinguara until the early 1980s when he moved to the more active new frontier at Tucumã. Although in his later years Chapeu de Couro reportedly became a prosperous gold trader and a prominent citizen in the Xingu region, he apparently met a violent end in 1986 when he was killed on a trip to Redenção by an unknown assassin. His personal saga reflects the generalized violence so endemic to the history of southern Pará.

Against this backdrop of social turmoil, the town mushroomed during the brief span of a few months in 1976–1977. Seemingly overnight, Xinguara metastasized from a small cluster of houses to a burgeoning population of 8,000 in 1978 at the juncture of a new road, the PA-279, under construction westward to the Xingu River. The growing local population and the increasing vehicular movement through the crossroads town attracted scores of small businesses. Timber, the most lucrative resource near Xinguara, provided an initial basis for industrial employment. By 1978 ten lumber mills operated on the edges of Xinguara, employing an average of about 320 full-time workers, nearly all male and by conservative estimates constituting at least 7.1 percent of the total labor force of the town.[14] Of course, many more workers were involved in the extraction of mahogany trees from the surrounding forests (Godfrey 1979, 1990). The mills, located at readily accessible roadside locations around Xinguara, consisted of little more than little open sheds for sawing the piles of timber stacked up in the adjacent yards.

As the population of Xinguara increased through in-migration during the late 1970s, pressure for urban lots mounted in the growing town. The urban area typically expanded through land invasions organized by squatters, and through subsequent municipal subdivisions and the titling of invaded areas. As in Parauapebas, most new residents in Xinguara settled on the urban periphery, encroaching on small farms, cattle ranches, and lumber mills. After such invasions, local authorities inevitably bowed to popular political pressures and titled the new urban lots, after settling claims with the previous landowners. In fact, the municipality has repeatedly expropriated larger areas than the initial squatter settlement for new urban subdivisions, since new lots could be sold at a profit to the municipality.[15] In only a few instances, such as the northerly subdivisions of Itamaraty I and II, did the local prefect plan in advance for urban expansion without the immediate impetus of land invasions.[16]

Despite irregular fits and starts in construction due to uncertain state funding and changing contractors, gradually the PA-279 pushed westward from Xinguara toward the small former rubber depot town of São Felix do Xingu on the Xingu River, approximately 265 kilometers away. Small farmers, land speculators, and prospectors pressed the road ever forward. By 1978 the road reached to about 80 kilometers west of Xinguara, where a new settlement called Agua Azul was founded. Initially landless migrants, frustrated over the paralysis of the state colonization project, poured in to the new boomtown in a land rush at the end of the road. Although some predicted Agua Azul would soon rival Xinguara in importance, once valuable timber was exhausted locally, the area surrounding the settlement was quickly deforested and consolidated into large cattle ranches. Agua Azul began to lose population by 1980, when discoveries of gold, cassiterite, and other mineral deposits in the Xingu region prompted migrants to proceed further westward along precarious local roads and into isolated landing strips. In 1981 the PA-279 finally passed the midway point in the trajectory between Xinguara and São Félix do Xingu; at that time the former boomtown of Agua Azul had declined in importance because migrants had moved on to newer areas (Godfrey 1979:118; Schmink and Wood 1992:157). The abortive urbanization of Agua Azul again demonstrated a hollow frontier. Agua Azul's collapse points to the lack of central-place functions in the disarticulated urban system of southern Pará.

In short, Xinguara's strategic location in the regional transportation system ensured an influx of migrants during the late 1970s. Intermittent expansion of the agricultural, ranching, timber, and gold-mining sectors allowed the town to emerge as a market center offering a variety of services. Xinguara's urban area expanded under the pressure of squatter invasions and the subsequent demarcation and titling of lots near the crossroads location (plate 7.4). The vital road connections permitted continued expansion to new frontiers along the PA-179 to the west of Xinguara, but few viable local towns emerged as secondary settlements; such initially promising settlements as Agua Azul did not thrive for long. As natural resources were exhausted and rural lands became concentrated in large holdings, many settlers moved on from Xinguara to newer frontier areas. In the early 1980s the new towns of Tucumã (a corporate planned project) and Ourilândia do Norte (a spontaneous settlement) grew rapidly to the west on the Xingu settlement corridor. These towns, along with Xinguara, soon became the seats of independent municipalities. A disarticulated urban system emerged early on in the Xingu corridor: a few large towns became independent municipal seats, but very few secondary settlements accompanied them. Overall, weak local central place functions characterized the Xingu corridor from the initial period of frontier settlement.

The Contested Corporatist Frontier, Tucumã and Ourilândia, 1981–1989

As elsewhere in Amazonia, Xinguara's aspiring elites lobbied effectively for municipal independence. The town, previously under the jurisdiction of Conceição do Araguaia, was declared the seat of an independent municipality in 1982. Municipal autonomy reflected the city's demographic growth and increasing socioeconomic importance as a market center in southern Pará. By the mid-1980s the town's population stood at 15,000. The north-south PA-150 was paved by this point, while the still unpaved PA-279 finally reached westerly São Félix do Xingu in 1983. Xinguara was electrified in 1986 after the construction of a local substa-

Plate 7.4. Aerial view of Xinguara in 1988. Approximately 20,000 residents are clustered at the intersection of the PA-150 and the PA-279 highways. *Photo: Brian J. Godfrey*

tion from the Tucuruí hydroelectric plant; soon after that the implantation of a telephone post and satellite dishes for television reception increased communication ties with the rest of Brazil. Gradually Xinguara, along with southern Pará, became integrated into a larger system of market exchanges, dominated by the distant metropolitan areas of the Brazilian Southeast (Godfrey 1990).

By the mid-1980s Xinguara was an established market center. Many lumber mills previously located in Xinguara had moved westward to Tucumã and other new frontier areas in the Xingu area.[17] Rural land in the municipality became increasingly concentrated into larger holdings. In 1985 registered latifundia and rural firms, which accounted for only 39 percent of the titled holdings, controlled 85 percent of the rural land in the municipality of Xinguara (Godfrey 1990). The local landscape had been transformed from tropical forest to open fields of pasture, farming plots, and generally a *campo sujo* (dirty field) of scrub vegetation dominated by the presence of rapidly proliferating, fire-resistant *babaçu* palm trees. Despite the exhaustion of local timber and mineral resources near Xinguara, the expansion of resource extraction westward along the Xingu corridor in the mid-1980s revived the stagnant regional frontier. The discovery of gold, along with valuable mahogany and fertile farmland, encouraged a new wave of regional in-migration in the mid-1980s and led to the emergence of the new towns on the PA-279 between Xinguara and São Félix do Xingu. The history of Tucumã and Ourilândia do Norte, neighboring cities only about ten kilometers apart on the PA-279, puts into sharp relief the conflict between corporatist and populist forces in Amazonia. The evolving physical designs and socioeconomic functions of these two towns reflect interacting corporatist and populist nodes in the same contested area. Xinguara remained marginal to the resulting struggle between Tucumã and Ourilândia, reflecting a regionally disarticulated urban system.

Tucumã originated as the headquarters town for a vast private colonization scheme. In 1978 INCRA opened bidding on the Gleba Carapanã, a 400,000-hectare area in the municipality of São Félix do Xingu known for its valuable timber and fertile *terra roxa* soils, which were considered excellent for agriculture. The giant Brazilian construction and engineering firm, Construtora Andrade Gutierrez (CONSAG), with annual profits of approximately U.S.$350 million at the time, was selected to implement the Tucumã project in the area in January of 1979. CONSAG, contractor for the PA-279 highway, had planned the project for several years. The selection of CONSAG gave rise to charges of corporate favoritism, leading to heated hearings in the Brazilian Senate before final approval in 1980 cleared the way for project implementation. In 1981 CONSAG began construction of the planned town of Tucumã, located about 160 kilometers west of Xinguara on the new road. Road construction stopped at the Tucumã project, due both to state budget problems and CONSAG's desire to prevent invasions of its settlement area. CONSAG considered roads within the Tucumã project to be private, inaccessible to outsiders, which left the municipal seat of São Félix do Xingu isolated, cut off from road access (Schmink and Wood 1992:158–163).

The Tucumã project became the dominant feature in the area's settlement in the early 1980s. Tucumã's urban nucleus, the headquarters of the colonization project, witnessed the layout of an elaborate, baroque-style plan, located off the main highway, featuring wide avenues, functional land-use specialization in various different sectors, and several separate residential districts. Despite a promising beginning with careful technical planning and corporate backing, the Tucumã project proved unrealistic in social terms. In the early 1980s CONSAG sold some 3,000 agricultural lots to small-farmer families, largely from the Brazilian South. Project administrators reasoned that the presence of hardworking farming families, largely of European descent, would improve the prospects for successful agriculture. The rigorous requirements CONSAG set up for settler selection were intended to favor aspiring small farmers from the southern states, especially from São Paulo, Paraná, and Santa Catarina, considered more entrepreneurial and technically skilled than the migrants from the impoverished Brazilian Northeast. The exclusion by CONSAG of the northeastern migrants, predominant in southern Pará, created growing social unrest in the region (IDESP 1990c).

The Tucumã project was bitterly resented by growing contingents of local migrants, mainly from the Brazilian Northeast, camped outside the project gates. These migrants were numerous, generally impoverished, and accustomed to squatting on unoccupied lands. They regarded CONSAG as an exclusionary elite, denying people's rightful access to promising lands both for agriculture and mining. Indeed, miners formed the vanguard of the migrant hoards. Already an illicit gold-mining camp at Vila do Cuca (Village of Cuca), named by tradition after the reputed founder, had been detected in the early 1980s on lands reserved for the Tucumã project. Ominously, CONSAG was unable to evict the intractable miners at Cuca. News of both the land settlement project and the gold strikes further intensified in-migration from adjacent regions. In 1982 these migrants began massing outside the fortified gates of the Projeto Tucumã. This restless, burgeoning settlement initially was called Guaritaí, after the name of the main gate to the large-scale private colonization project. Rechristened Ourilândia do Norte (Goldland of the North), the town soon exceeded the nucleus at Tucumã in population. By 1984 Ourilândia had a population of some 10,000, as opposed to some 3,000 in Tucumã, and tens of thousands of additional miners worked in the surrounding areas.

To CONSAG officials, the burgeoning population at Ourilândia do Norte appeared to be a chaotic squatter settlement, teeming with peasants and miners, poised to invade the Tucumã project. As in the case of Xinguara, at Ourilândia do Norte a rectilinear grid-iron plan simply enveloped the existing highways, turning such roads into the principal urban thoroughfares. As the town of Ourilândia mushroomed outside the gates of the Tucumã project, which prevented entrance to the restricted corporate lands, popular hostility toward CONSAG rose to the breaking point. Tensions between the private company and the hostile local population led to a series of large-scale invasions of the Projeto Tucumã in late 1984 and early 1985. The land invasions, carefully planned as mass movements, involved thousands of migrants, who simply overwhelmed the outnumbered company guards. Unable to dislodge the squatters, CONSAG soon abandoned the project and began negotiating with the federal government for indemnification. With the demise of the official settlement project, land tenure in the area remained confused, and spontaneous invasions continued to be a problem (Butler 1985; Schmink and Wood 1992:207; IDESP 1990b).

After CONSAG negotiated an agreeable financial compensation for project lands with the federal government in 1988, the state government of Pará was obligated to recognize the population growth in the Tucumã–Ourilândia do Norte axis with a new municipal administration. Local pressure for municipal autonomy had been building for years in the area. São Félix do Xingu was one of Brazil's largest and least densely inhabited municipalities. Although the PA-279 had reached the municipal seat by this point, the road remained in difficult condition, isolating the town of São Félix do Xingu from the larger and wealthier settlements to the east, where the benefits of natural resource extraction accumulated. Both Tucumã and Ourilândia do Norte lobbied fiercely for the privileged status of the new municipal seat. Trying not to offend either town, the state government ultimately compromised and created two new and separate *municípios*, whose administrative seats lay less than 10 kilometers apart on the PA-279. Thus, the rivalry between the two interdependent towns with radically different social origins was perpetuated administratively into the future.

Spatial Fragmentation, Market-Area Consolidation, and Urban Stabilization, 1990–1993

By 1990 the three urban centers along the Xingu corridor had established themselves as independent municipal seats, small to medium-sized towns, and service centers for the local and even regional markets. According to population estimates of local agencies, our 1990 field surveys, and the 1991 Brazilian census, the urban populations no longer grew rapidly in any of the three settlements. Xinguara, the largest of the towns, officially had a 1991 population of 27,292.[18] Tucumã, favored by its origins as a company town with better physical infrastructure and banking and commercial services, grew to a population of 12,455. The town of Ourilândia do Norte, once larger than Tucumã, saw its 1991 population stabilize at 10,893 (table 7.2). The former mining camp of Cuca, now a local service center in its own right for other gold mines in the municipality of Tucumã, reached a population of more than 2,250.[19]

Xinguara, the largest town and the regional market center for the Xingu corridor, shed the rough-and-tumble image of a wild frontier outpost by the 1990s. This "instant city" between the rivers increasingly took on both the size and the appearance of a medium-sized, modestly appointed town of the Brazilian Center-West, such as might be found in northern Goiás or Tocantins, the states of origin of many local residents. Public authorities landscaped central public areas, and Xinguara residents benefited from the addition of street lights and electricity generated by the Tucuruí dam.[20] Despite Xinguara's improved but still uneven urban infrastructure, the town's economic future is assured by its strategic location in the regional road network. A detailed analysis of bus routes and communications linkages in 1988 indicated Xinguara's growing ties to the metropolitan areas of the Brazilian economic heartland in the Center-West and Southeast, along with significant but secondary ties to Belém, the state capital (Godfrey 1990). Further studies in 1990 confirmed these regional patterns of commercial interaction (chapter 9).

By the early 1990s Tucumã and Ourilândia do Norte were struggling toward economic consolidation of their respective municipalities. In economic terms, these two towns still relied heavily on the exploitation of gold and other natural resources as well as on diverse urban services for workers in those primary sectors. The municipal government of Tucumã estimated in 1990 that 16,000 miners locally produced a monthly average of 150 kilograms of gold, though the illicit nature of the mining activities made any estimate subject to speculation (Prefeitura Municipal de Tucumã 1990). Tucumã's economy also depended on the activities of some the municipality's nineteen sawmills, many of them large operations ringing the town. Ourilândia do Norte showed less reliance on timber, but true to its name depended more on the activities of goldminers, as seen later in our surveys of the local labor force. Future growth in both towns depended largely on new gold strikes, continued timber exploitation, and other natural resources. Yet by the early 1990s the resource-extractive economy of the region appeared stagnant. The accessible timber had rapidly diminished, and the primitive state of local roads hindered transport to more remote areas. Gold mining was on a clear decline: the easy alluvial strikes had been worked for some time and reaching the deeper veins of ore required a significant capital investment.

As commercial centers, the two interdependent towns grew in tandem to serve local interests. Tucumã remained the headquarters for more established companies and capitalized regional businesses, such as banks, communications, air transport firms, and so on. Our 1990 research team counted 203 total businesses in Tucumã, including two banks, four gold-trading stores, several large supermarkets, and stores selling construction and hardware materials, agricultural equipment, and other specialized commodities. In Ourilândia do Norte, we counted 253 total businesses, mainly smaller, less capitalized businesses, such as the many lodging houses, bars, and houses of prostitution catering to miners. The town had four gold-buying posts and only one commercial bank. Ourilândia do Norte remained more of a labor reserve for extractive industries and a center for a burgeoning informal, often illicit economy catering to transplanted migrants, mobile miners, small-scale merchants, and other local residents (see chapter 9).

Between 1977 and 1991 the pattern of frontier urbanization in the Xingu corridor manifested both populist and corporatist tendencies. While rural land disputes in southern Pará typically have been resolved in favor of corporatist interests and large cattle ranchers, urban land contestations have more frequently been adjudicated in favor of landless migrants. The expansion of the settlement frontier westward along the Xingu corridor entailed the construction of a corporatist center (e.g., the Tucumã project), the restricted access to which resulted in the establishment of a populist mining settlement (e.g., Ourilândia). This pattern of urban system change roughly corresponds to the sequence outlined in chapter 4 (fig. 4.3).

As in the preceding chapter on Rondônia's populist urban frontier, we now consider the physical, economic, and social characteristics and functions of lower-order settlements in the contested corporatist frontier of the Xingu corridor. Specifically, we analyze the results of an extensive survey of 903 households in the three cities of the Xingu settlement corridor: Xinguara, Ourilândia do Norte, and Tucumã. As in the previous chapter on Rondônia, the results of the extensive household surveys in southern Pará allow us to test several hypotheses on the physical and socioeconomic characteristics of frontier urbanization.[21]

GENERAL CHARACTERISTICS OF XINGUARA, OURILÂNDIA, AND TUCUMÃ

The Xingu corridor contrasts in fundamental ways with the urbanization pattern evident in the Rolim de Moura–Alto Alegre corridor. First, the spatial distribution of settlements by size and age in the former does not follow a continuous distance gradient from the core to the periphery of the urban subsystem, as is apparent in the Rondônia study. The most obvious evidence of irregularity is that the settlement ranked third in terms of size and age (Ourilândia) is not positioned in third place along the corridor, but rather in second place, between Xinguara and Tucumã (table 7.2). This idiosyncrasy may seem trivial given the short distance (10 kilometers) separating Ourilândia from Tucumã; but it is a distortion that must be attributed to the historical relationship between the outlying two settlements (described above) and not just to the pure economic logic of the marketing principle underlying central place theory.

Table 7.2 General characteristics of the Xingu corridor settlements, July 1990

	XINGUARA	OURILÂNDIA DO NORTE	TUCUMÃ
Year founded	1976	1982	1981
Distance to PA-150 highway (km)[a]	0 km.	150 km.	160 km.
1991 urban population[b]	27,292	10,893	12,455
Number of urban dwellings surveyed[c]	410	173	320
Estimated % of total population surveyed	7.5	6.9	10.0
Household size (persons/HH)	5.014	5.473	5.119
Gender (% of population)			
Male	49.1%	48.1%	52.6%
Female	50.9%	51.9%	47.4%
% female-headed households	14.2%	15.1%	7.1%

[a]The PA-150 is the asphalted all-weather state highway extending north to Belém, capital of Pará, and south to Conceição do Araguaia and which in turn links into the Belém-Brasília federal highway and national markets in the Southeast.
[b]IBGE, Sinopse do censo demográfico 1991, Rio de Janeiro, 1991.
[c]Excludes households headed by persons who arrived in the survey area on or after January 1, 1990 (or within the six months preceding the survey date).

Second, demographic differences are evident. While the Pará study sites showed similar average household sizes for the three settlements, ranging from 5.473 persons per household in Ourilândia and 5.119 in

Tucumã to 5.014 in Xinguara, these are significantly larger than those in the Rondônia study area (ranging from 4.6 to 4.7 persons per household). Like the Rondônia sites, however, all three of the towns in southern Pará had larger average household sizes than the national average of 4.033 household members (IBGE 1989:119).

Third, in terms of the composition of the population by gender, some significant differences among the three towns emerge in the 1990 survey. Tucumã had the lowest proportion of female residents (47.4%) compared to Xinguara (50.9%) and Ourilândia (51.9%). In addition, Tucumã had the lowest proportion of female-headed households (7.1%) compared to Xinguara (14.2%) and Ourilândia (15.1%). These characteristics contrast sharply with the Rolim de Moura–Alto Alegre settlement corridor where the population is predominantly male, and female-headed households are considerably less common. Tucumã's low proportion of females probably reflects the settlement's origins as a planned company town, populated by a high proportion of bureaucrats, professionals, and middle-class migrants from the South (Butler 1985). The other two towns presumably include more families of footloose migrants, often males, who frequently leave their households behind to work in mining, timber extraction, cattle ranches, and other activities at the more active pioneer fronts. In addition, both Xinguara and Ourilândia are spontaneous market towns with high levels of such informal economic activities as prostitution, pedestrian vendors, and cottage industries, in which female participation is marked. Such sectors are not as well represented in the more formal, planned economy of Tucumã.

These data on household composition in the Xingu corridor reflect the gendered nature of migration and employment linkages on the Amazon frontier. Women, children, and other household members are anchored more firmly in urban centers, providing essential family income through multiple activities, both formal and informal. Male migrants, on the other hand, often proceed into the active resource zones, beckoned by more lucrative activities in mining and timber extraction (Godfrey 1990:112). Frequently men avoid urban households for extended periods, and sometimes they never return at all, leaving female-headed households without external means of support. For example, too often the income from male goldminers in remote camps does not filter back to urban households, which are by default headed by females (Godfrey 1992).

These differences in spatial organization and demography between the Pará and Rondônia survey samples suggest fundamental differences between the populations of the corporatist and populist urban frontiers.

Physical Characteristics and Public Health Problems

As in Rondônia and other areas of rapid urbanization in Amazonia, the rapid growth of urban centers in southern Pará generally has overwhelmed efforts at urban planning, leading to severe problems of physical infrastructure, social services, and public health. In the Xingu corridor, authorities initially attempted to lay out regular ground plans adapted to the prerequisite roadside locations. In the spontaneous towns of Xinguara and Ourilândia, the settlements simply enveloped the highway at strategic sites and proceeded to grow in the form of a regular gridiron plan. In Tucumã, a more elaborate sectoral plan was designed to separate functional land-use districts. Except possibly in the early history of the planned company town of Tucumã, the provision of services has lagged behind the growth of the settlement. Over time, however, the towns have tended to obtain improvements in physical amenities. We found in the Rondônia study area that the older and more centrally located towns tend to have a higher degree of physical development than the newer, more remote settlements. We tested the same hypothesis about the provision of physical amenities in the Pará study sites:

> *Hypothesis 1:* The rates of informal self-provisioning of housing, running water, sewage disposal, and energy supply increase with distance from the principal center and decrease with age of the settlement in the frontier urban settlement system.

The findings of the household surveys do not conclusively support Hypothesis 1, as they do in the Rondônia surveys. While table 7.3 indicates a higher rate of self-built housing and individual power supply in the outlying settlement (Tucumã) than in the central settlement (Xinguara), differences in rates of informal water supply and sewage disposal are not significant between the two. Overall, the highest rates of self-provisioning of physical infrastructure are in Ourilândia, the second settlement in the corridor. The multiple asymmetries apparent in the Pará field data suggest the presence of forces other than those associated with central place dynamics.

Table 7.3 Indicators of urban physical infrastructure and public health, Xingu corridor, July 1990 (percentage of households)

INDICATOR	XINGUARA	OURILÂNDIA	TUCUMÃ	WEIGHTED MEAN
Self-built housing[a]	26.0%	31.2%*	33.6%*	29.8%
Informal water supply[b]	71.7*	94.8	69.4*	75.1
Informal sewage disposal[c]	85.5*	94.8	86.1*	87.2
Individual power supply[d]	6.4	49.6*	44.1*	20.9
Gastrointestinal disorders[e]	24.2	22.1	18.9	21.9
Hepatitis in household[f]	4.0	4.1	3.4	3.8
Malaria in household[g]	48.5*	56.5	45.7*	49.0

[a]*Self-built housing by current occupant excludes those houses purchased already constructed, or those built by contractors by order of the owner.*

[b]*Informal water provision includes wells without pumps, local creeks, and open springs. Formal water provision indicates deeper wells with pumps and piped water systems.*

[c]*Informal sewage disposal reflects lack of a septic tank or sewer line and reliance on outhouses or the outdoors.*

[d]*Individual power supply means lack of electricity (from any source) or self-provision of energy by means of individual gas or kerosene stoves or lamps.*

[e]*Gastrointestinal disorders refer to diarrhea and dehydration affecting anyone dwelling in the house during the preceding twelve months.*

[f]*Percentage of households in which someone dwelling therein contracted hepatitis during the preceding twelve months.*

[g]*Percentage of the household heads who had previously contracted malaria.*

All difference of proportions are significant at .05 level of confidence, except for those pairs noted by an asterisk.

Other locational criteria and history reveal important factors contributing to these distortions. For example, the most noticeable difference is in energy provision: Xinguara, located on the power grid from the hydroelectric plant at Tucuruí, has very few households without electricity (6%), while the more recent settlements show nearly half the population without electric light—and even those households with electricity depend entirely on local generators for their power. Also, we must consider Tucumã's history as a planned urban settlement in the early 1980s. By 1990 little in the way of physical infrastructure development remains to distinguish it from the spontaneously created gold mining town of Ourilândia located next to it, except for water supply and sewage disposal. This blending of physical characteristics corresponds to the final phase of urban consolidation anticipated in our model of the corporatist frontier (fig. 4.3E and 4.3F).

Interestingly, the rates of infrastructure self-provisioning are higher in Pará's corporatist frontier spaces than in Rondônia's populist frontier. More than three-quarters of the residents interviewed in the three towns in the Xingu corridor provided for their own water supply from easily contaminated surface sources—shallow wells, local creeks, and open springs—without the benefit of piped water supply systems or pumps to deeper wells. By contrast, 50% of the Rolim-Alto corridor relied upon informal water sources. Rates of reliance upon individual power supply were lower in the Xingu corridor only because of Xinguara's connection to the Tucuruí electrical grid. Even higher proportions of the townsfolk, in both Pará and Rondônia, relied on informal sewage disposal systems, lacking use of septic tanks or sewer lines.

Not surprisingly, public health problems were more severe in the Pará than in the Rondônia samples. One-fifth of the Xingu corridor residents indicated that someone in their household had suffered from gastrointestinal disorders during the previous year, significant numbers had contracted hepatitis, and about half of all heads of household previously had malaria (table 7.3). Like the Rondônia experience, the incidence of these illnesses is not always significantly different among the settlements in the Xingu corridor, though several local anomalies do stand out. The rates of gastrointestinal disorders and hepatitis were lowest in Tucumã, due in large part to the superior infrastructure and better hygiene of the former company town. Also, malaria appeared to be most prevalent in Ourilândia, where 56 percent of the household heads had contracted the disease, often while working as goldminers in the malaria-infested local *garimpos*. Overall, the incidence of these illnesses in the Xingu corridor is much higher than in the Rolim de Moura–Alto Alegre corridor (table 6.3). Numerous possible local explanations exist for these differences, such as the role of gold mining in the Xingu corridor. In addition, institutional explanations should be considered. Corporatist groups and their institutional sponsors are not especially interested in the long-term, permanent settlement of the corporatist frontier by large independent populations; hence one finds less investment in social overhead as might be found in the populist frontier. The corporatist frontier, deficient in life-support infrastructure, except in corporatist bedroom communities, displays higher rates of debilitating preventable illnesses.

Sectoral Distribution of the Urban Labor Force

In the preceding chapter on Rondônia we hypothesized that the structure of employment in a community would reflect its unique economic history, its special social orientation, and institutional character.[22] We revisit those hypotheses, in order, below in the context of the contested corporatist frontier model.

Hypothesis 2: Employment rates in the agricultural sector progressively increase along the distance gradient of the agrarian frontier such that outlying pioneer settlements have higher rates of agricultural employment than more centrally placed urban nodes.

Hypothesis 3: The extractive sector will employ higher rates of workers in outlying, newer settlements than in centrally placed, older towns.

Hypothesis 4: Industrial employment rates in the urban system of the corporatist frontier are higher in the centrally placed primary settlements than in the peripheral pioneer settlements.

Hypothesis 5: Service sector employment rates are positively related to the population size of specific settlements.

Hypothesis 6: Government employment rates follow a distance gradient and therefore are higher in centrally located, older primary settlements in the corporatist frontier settlement system.

In the Rolim de Moura–Alto Alegre corridor the social character of the urban communities studied was decidedly agrarian in nature, and the distribution of the labor force clearly indicated a positive distance gradient from core to periphery in agricultural employment. In contrast, the findings from the Pará surveys clearly do not support Hypothesis 2 (table 7.4). The structure of economic opportunity in the Xingu corridor is fundamentally different from that in the Rolim de Moura–Alto Alegre corridor. There are no significant differences in agricultural employment rates among the three Xingu corridor cities. Overall, we find a more diverse, less monogenic pattern of urban employment distribution. Here more localized patterns of differentiated articulation arise: In the case of Tucumã the employment orientation is to agriculture; in Ourilândia to mineral extraction; and in Xinguara to a more diverse mix, but generally to higher rates of employment in commerce. These variations warrant some further interpretation.

Clear differences between the towns stand out in terms of extractive activities and, to lesser degrees, in terms of commercial and industrial activities. With regard to extractive sector employment, the survey data do not clearly confirm Hypothesis 3, the notion that outlying secondary frontier settlements have higher employment rates in resource-extractive activities than do primary nodes in frontier urban systems. The extractive sector absorbs nearly one-fifth of the total labor force in Ourilândia do Norte, reflecting the importance of the local gold-mining fields in the

Table 7.4 Sectoral distribution of labor force, Xingu corridor, July 1990 (percentages)[a]

	XINGUARA	OURILÂNDIA	TUCUMÃ	WEIGHTED MEAN
Resource extraction	*10.6*	*19.8*	*10.2*	*12.3*
Mineral extraction	9.0	19.2	7.9	
Timber extraction	1.6	0.6	2.3	
Non-wood products	0.0	0.0	0.0	
Agriculture	*16.0*	*12.6*	*16.4*	*15.4*
Industry	*17.1*	*12.0*	*19.6*	*16.8*
Wood processing	5.7	2.0	9.8	
Ceramics, bricks, cement, etc.	0.4	0.0	0.0	
Textiles	2.2	1.7	2.4	
Agro-industry & food processing	3.3	4.0	0.8	
Metallurgy	0.0	0.0	0.2	
Construction	5.5	4.3	6.4	
Commerce	*20.5*	*15.8*	*16.8*	*18.4*
Transport	*3.6*	*4.6*	*3.0*	*3.6*
Services	*29.4*	*31.9*	*31.6*	*30.6*
Finance	0.6	0.6	1.1	
Auto/mechanical	4.5	2.0	4.1	
Food, lodging, & entertainment	3.3	5.2	2.6	
Building repair	0.7	0.0	1.3	
Unskilled	15.9	21.2	16.0	
Skilled	1.5	1.7	9.4	
Professional	0.7	0.6	1.3	
Administrative	2.3	0.6	4.1	
Public	*2.5*	*3.4*	*2.4*	*2.6*
Education	1.8	2.0	0.8	
Government	0.7	1.4	1.7	
Other and inactive	*0.6*	*0.0*	*0.0*	*0.3*
TOTAL (percent)	100.0	100.1	100.0	
# of Persons Employed	830	349	531	

[a]*Refers to primary occupation of economically active population in the Xingu corridor study sites. The economically active population includes all those aged ten years and over listing a money-earning occupation or employment, either formal (licensed) or informal.*

city ranked second. Yet, in Tucumã, the most remote settlement, extraction employs the lowest rate of workers. Nonetheless, all three cities in the Xingu corridor display much higher employment rates in extraction than do those in the Rolim de Moura–Alto Alegre corridor. Indeed, among heads of household in Pará the prevalence of extractive activities is even more pronounced: in Xinguara roughly 10 percent of the household heads were involved in mining in 1990, in Tucumã 11 percent, and in Ourilândia do Norte a full 24 percent of the heads of household declared gold mining as their primary occupation (Godfrey 1992).[23] The asymmetrical distribution of extractive sector employment rates indicates the geographically selective presence of populist economic forces in the Xingu corridor. The heads of household, even more than the general labor force, tend to work in the most lucrative economic sectors, such as gold mining. The fact that miners are concentrated in Ourilândia reflects the historical struggle between corporatist Tucumã and the populist pressures around it. That gold mining in the Xingu corridor is differentially articulated to popular classes stands in poignant contrast to the timber extraction and wood processing subsectors that are differentially articulated to corporatist interests in the Brazilian Southeast (see chapter 9).

Industrial employment rates show few significant interurban variations. Contrary to Hypothesis 4, industrial employment rates are slightly higher in the third (most remote) settlement in the Xingu corridor (Tucumã) than in the core settlement of the Xingu urban system (Xinguara). In terms of employment generation, the wood-processing sector is a relatively more important industrial employer in Tucumã, the most active zone of timber extraction, where timber mills employ nearly 10 percent of the workforce. The size of the workforce engaged in wood processing in Xinguara (5.7 percent) reflects in part the implantation of plywood and laminate factories in town in recent years, industries not found in the interior of the corridor (Godfrey 1990).

The commercial sector is comparatively more important in Xinguara, representing over one-fifth of the labor force, and in part it reflects the city's function as a regional transportation hub. There was no significant difference in commercial sector employment between Ourilândia (15.8%) and Tucumã (16.8%). In contrast, commercial sector employment was slightly less important in Rolim de Moura than in Xinguara, due in part to the former's more insular location (40 kilometers) from the closest interstate highway (BR-364).

In terms of employment generation the service sector is of roughly equal importance to all three settlements in the Xingu corridor, ranging from 29.4% (Xinguara) to 31.9% (Ourilândia). Given the substantial differences in the population of these settlements, these findings lead us to reject Hypothesis 5. In the Rolim de Moura–Alto Alegre corridor, in contrast, the survey data generally supported this hypothesis, further indicating the basic differences in economic structure between the two corridors. Services are significantly more important in the outlying settlements of the Xingu corridor than they are in those of the Rolim de Moura–Alto Alegre corridor. Particularly noteworthy are the higher rates of participation by unskilled service providers in the former, suggesting the presence of a much larger informal sector labor force in the *garimpo* mining sector of the Xingu than in the agrarian Rolim de Moura–Alto Alegre area.

Finally, as in the Rondônia case study, we reject Hypothesis 6 for southern Pará. Public sector employment rates show no significant differences among the urban centers of the Xingu corridor. The comparability of public sector employment may reflect the fact that by 1990 all three settlements were municipal capitals, each having roughly comparable political claims on government jobs. However, the rates of public sector employment are significantly lower in both the education and the government employment subsectors than in the Rolim-Alto corridor, perhaps reflecting the lower level of importance given to public services in southern Pará where the population, of both populist and corporatist social orientations, is more unstable (see chapter 8).

In sum, the southern Pará and Rondônia study areas are different in four fundamental respects. First, extractive sector employment is much more important in the Xingu region (12.3 percent) than in the Rolim de Moura area (3.5 percent), due chiefly to the importance of gold mining in Pará. Second, commercial sector employment is also significantly higher in the Xingu corridor (18.4 percent) than in the Rolim de Moura–Alto Alegre area (13.8 percent). This difference is most evident in the disparities between the primary cities (Xinguara and Rolim de Moura) rather than among the secondary periphery towns. Due to Xinguara's strategic commercial location as a regional trade hub adjoining a major state highway (PA-150), 20.5 percent of the labor force works in commerce; in Rolim de Moura, a town handicapped by virtue of its more insular location, only 13.7 percent works in commerce. Third, unskilled service workers represent a higher proportion of the labor force engaged in the service sector of the Xingu economy (16.9%) than in Rolim de Moura–Alto Alegre (13.6 percent). This difference is most evident in Ourilândia, where 21.2 percent of the economically active population is unskilled, reflecting the informal mining economy. In further contrast to the Rondônia study, where the primary urban node in the Rolim de Moura—Alto Alegre corridor had the highest rate of skilled and professional labor employment, we found the opposite in the Xingu corridor: the highest rates of skilled, professional, and administrative employment were found in Tucumã, the most remote of the three settlements surveyed, which had a prior history of corporate sponsorship. Finally, we note the significant differences in public sector employment owing, we surmise, to the different demands populist and corporatist jurisdictions make upon local and state budgets. Government patronage employment is a vital aspect of the institutional orientation of Rondônia's populist frontier. The contested corporatist frontiers of southern Pará, more oriented to extraregional economic interests in the Southeast, remain less dependent on local and state political patronage for employment.

The divergent distributions of urban employment in the Xingu and Rolim de Moura–Alto Alegre corridors reflect two different patterns of urbanization. In Rondônia the relatively symmetrical spatial distribution of employment in the Rolim de Moura urban system more closely corresponds to that predicted by the marketing principle of central place theory. In southern Pará the spatial structure of employment activity in the Xingu corridor is asymmetrical, representing a more diverse social character of contesting interests, generally associated with either centralized corporate capital or autonomous local labor.

Quality of Urban Life in the Contested Corporatist Frontier

We hypothesized that residents of larger, higher-order settlements in a regional system possess greater material wealth and exhibit a greater subjective sense of satisfaction and contentment with their living situation than residents of the smaller, lower-order service centers. In the context of the towns of the Xingu corridor, the older and larger center would be more hospitable than the more recent pioneer settlements. These assumptions may be tested on two interrelated hypotheses:

Hypothesis 7: Residents of the larger center in a regional urban system enjoy more material possessions than residents of the lower-order settlements.

Hypothesis 8: Residents of the larger center in the regional urban system are more content with their current situation than those of the lower-order settlements.

We have already established that the distance-age gradient characterizing the spatial distribution of certain key household attributes found in the Rondônia survey data does not apply to southern Pará, since the settlement array in the latter is asymmetrical with respect to distance and age. Not surprisingly, table 7.5 indicates that the assumptions about quality of life implicit in Hypothesis 7 are only partially valid in the Xingu corridor. In the index of fifteen common material possessions, Tucumã ranks overall as highly as Xinguara, each with a median of seven possessions. This **u**-shaped curve in the distribution of amenities and possessions over distance can be explained. Xinguara's households enjoy possessions (appliances) that are more energy-intensive (e.g., refrigerators, televisions, telephones), reflecting the town's greater integration into the broader regional infrastructure, such as the Tucuruí power grid and the state's all-weather highway network. Tucumã residents are more likely to possess their own radios, clocks, gas lamps, pressure cookers, running water, indoor toilets, and even automobiles—amenities and possessions that are less energy intensive than those used in Xinguara, but more diverse and complex than those enjoyed by residents of Ourilândia— due to the city's corporatist heritage. Residents of Ourilândia do Norte generally have fewer possessions and household amenities than do residents of either of the other two cities. Only a few residents of this town enjoy in-door toilets or showers, for example, indicating higher rates of urban poverty. The dualism between corporatist and populist frontier social spaces is abundantly evident in the differences in material possessions and household amenities between residents of Tucumã and Ourilândia.

In terms of the general levels of satisfaction with urban life in the different towns, the survey results generally support Hypothesis 8. Over three-quarters of the household heads interviewed in Xinguara (75.9 percent) indicated an intention to remain living in the town, while some

Table 7.5 Material possessions, intentions to stay, and major urban problems, by percentage of households, Xingu corridor, July 1990

	XINGUARA	OURILÂNDIA	TUCUMÃ
Possessions:			
Radio	54.1%	55.5%	60.2%
Clock	63.6	64.2	68.8
Bicycle	47.7	30.6	33.6
Sewing machine	42.0	38.7	38.3
Gas lamp	20.4	55.5	52.2
Pressure cooker	86.7	89.6	93.8
Water filter	76.9	86.7	86.1
Running water	17.4	5.2	21.9
Shower	22.6	5.8	21.9
In-door toilet	22.1	7.5	28.1
Gas stove	91.2	90.8	91.7
Refrigerator	51.4	27.7	34.6
Automobile	10.3	5.2	11.1
Television	53.3	30.6	42.3
Telephone	9.1	0.0	7.4
Median number of possessions from list (x / 15)	7	6	7
Household head intends to stay in town	75.9	66.5	72.2
Most often cited problem in town (% households)	Lack of work, insufficient income (16.0%)	Lack of electricity or lights (25.4%)	Lack of electricity or lights (29.6%)
Second-most cited urban problem	Lack of public services[a] (15.7%)	Lack of work, insufficient income (20.2%)	Lack of public services[a] (15.7%)
Third-most cited urban problem	Other community issues[b] (15.5%)	Lack of public services[a] (14.5%)	Lack of medical assistance (11.1%)

[a]*Public services include schools, mail delivery, street and road maintenance, trash collection, and so forth.*
[b]*Other community issues include the perception of social life, local entertainment, whether a place is "boring."*

what smaller proportions expressed a desire to do so in Tucumã (72.2 percent) and Ourilândia do Norte (66.5 percent). Two of the top three complaints in the newer settlements related to the deficiencies of urban infrastructure: lack of electricity or lights was cited most often in both Tucumã (25.4 percent of households) and Ourilândia do Norte (29.6 percent); lack of public services such as schools, mail delivery, street and road maintenance, trash collection, and so forth is a frequent complaint in all three towns. Tucumã residents also strongly felt a need for more medical assistance, a problem cited by 11.1 percent of the urban households. The lack of work and insufficient income rated highly in Xinguara (15.7 percent) and Ourilândia do Norte (20.2 percent). Many households in Xinguara (15.5 percent) also complained about the lack of local entertainment and felt bored with their community's social life. Issues of entertainment and boredom are a luxury of the oldest settlement in the Xingu corridor, Xinguara, where greater access to external influences via improved transportation and communications highlights the deficiencies of local social life.

CONCLUSIONS

As opposed to the generally progressive temporal and spatial transition seen in the case of Rondônia's urbanizing frontier, where populist agrarian policies prevail, southern Pará displays a more fragmented, heterogeneous urban transformation. The complex dynamics of settlement along the Xingu corridor do not easily submit to master principles of explanation. From our perspective, the complexities of regional urbanization stem largely from institutionally mediated and locally variable conflicts among corporatist and populist forces.

Xinguara emerged as a regional commercial hub and crossroads town, from which the settlement of the Xingu corridor could be launched. Since the late 1970s the town's strategic location in the regional road system has encouraged an influx of small farmers, cattle ranchers, and loggers. Although authorities touted a philosophy of agrarian populism in initial state-proposed, small-farmer colonization projects near Xinguara, the subsequent consolidation of landholdings, a feature of corporatist frontier policy, caused the local population to pile up in the town or move to more active areas of resource extraction farther on in the Xingu corridor. Local municipal government policy encouraging urban land invasions reinforced this pattern of rural land consolidation. Meanwhile, new road connections subsequently permitted the expansion of resource frontiers and allowed a hinterland to form around the growing urban centers, which attracted new services to municipal seats. Nevertheless, interests based in Xinguara did not dominate events in the new towns to the west, Tucumã and Ourilândia, suggesting the weak regional central place functions.

The collision of competing corporatist and populist forces is most apparent in the juxtaposition of the former company town of Tucumã and the spontaneous settlement of Ourilândia do Norte. Land-hungry migrants and gold prospectors excluded from a large-scale corporate colonization project at Tucumã, owned by the Andrade Gutierrez Company, congregated in Ourilândia. The discovery of gold in and around the Tucumã corporate project area led to the settlement of satellite mining camps and eventually to the demise of the company's hold in the area. Even so, the initial urban planning and infrastructure development of Tucumã subsequently favored the settlement's demographic growth and commercial evolution. Ourilândia do Norte appears to be the poorest of the three towns in terms of both urban amenities and general satisfaction of residents.

In the 1990 survey Xinguara generally rates highest on those indicators reflecting market relations, while Ourilândia and Tucumã remain somewhat less dependent on commercial functions and more reliant on primary activities and resource extraction. After an initial spurt of growth related to the implementation of agriculture and ranching, frontier settlements in southern Pará have risen and fallen in relation to the expansion of resource-extractive activities and commercial functions in the regional urban system. The different boom-and-bust cycles cumulatively leave an impact in terms of improved infrastructure and heightened social stratification, but continued expansion depends largely on the exploitation of new resources and the growth of the related service sector. Only a few fortunate towns like Xinguara, favored by strategic locations in the regional transportation network, can expect to develop as commercial centers of more than purely local importance. Even so, relatively weak regional central place functions prevent Xinguara and the other towns from emerging as higher-order cities: continued economic vitality depends largely on the exploitation of new natural resources and on extraregional market forces emanating from the Brazilian core. The long-term prospects for sustainable frontier urbanization, based on a regionally integrated and self-reliant city-system, remain uncertain in southern Pará.

Notes

1. Frontier settlement in southern Pará was the subject of Godfrey's (1979) master's thesis, based on field research carried out during lengthy stays in the region during 1976 and 1978. In recent years, the author has returned to update data on the region in 1988, 1990, and 1993. For a full description of the research protocol used in the extensive 1990 survey of 903 urban households, the analytical focus of this chapter, see the technical appendix.

2. By calling the contemporary boomtowns of Amazonia "instant cities," we do not mean to imply that the region replicates the North American West's historic patterns of develop-

ment. In both cases, however, the discovery of valuable resources sparked massive in-migration and transformed rustic bivouacs into dynamic urban centers.

3. As noted in earlier chapters, the official definition of *urban* in Brazil depends on a locality's political-administrative status.

4. In his critique of contemporary Amazon frontier studies, Cleary (1993) notes that the political-economy approach once assumed a steady expansion of capitalism on the Amazon frontier (see Cardoso and Muller 1977; Velho 1976; Foweraker 1981). Recent work on southern Pará and other regions has emphasized the ways in which the transition to capitalism has been incomplete, juxtaposing speculative land concentration and noncapitalist relations of production (see Hecht and Cockburn 1989; Sawyer 1984: Schmink and Wood 1992).

5. According to local estimates, the average annual rainfall in southern Pará ranges from 2334 mm at São Félix do Xingu to 1426 mm at Marabá (CONSAG [Construtora Andrade Gutierrez] 1978:4. 44; Prefeitura Municipal de Tucumã 1990).

6. Among the nineteenth-century and early twentieth-century foreign observers who recorded their journeys on the Tocantins-Araguaia and Xingu rivers were Henry Coudreau (1897 and 1977), Karl von den Steinen (1940), Adalberto da Prússia (1977), and Emilia Snethlage (1910).

7. It is difficult to document with precision at the regional level the increasing number of cities, because in practice the main criterion for an urban center is the presence of a municipality- (*município*) or county-level seat (*sede*) of government. In effect, as the population of a place increases in Brazil, this locality exerts political pressure on state authorities to gain municipal status. The resulting political fragmentation has been particularly notable in the state of Pará, which now has more than a hundred local municipalities, and complicates regional demographic analysis.

8. Outright murder is only the most extreme manifestation of rural violence. From 1964 to 1988, 110 death threats were reported in southern Pará, 75 percent of all those reported in the entire state; 2,678 families were expelled from their lots, 70 percent of the state's total; and 305 people were imprisoned in cases of land disputes, 66 percent of those in Pará. It is widely known that many cattle ranches in Amazonia resort to forced labor by rural workers, known as *peões* or peons, to clear forest and create pastures. During the twenty-five-year period, 1,142 documented cases of "slave labor" were reported in southern Pará, 72 percent of the cases of forced labor reported in the entire state (IDESP 1990a). These figures lend credence to persistent press reports of slave-labor conditions in *fazendas* in the region. One hellish case of a *fazenda* near Xinguara illustrates a common complaint of rural workers in the region. In 1984 escaped workers reported to authorities that dozens of *peões* were lured to the Fazenda Santa Cristina with promises of an air-conditioned bus trip, which turned out to be a four-day incarceration in a refrigerated meat truck. On arriving at the remote *fazenda*, abusive gunmen held the rural workers for months against their will, threatening with death anyone who resisted the torturous fourteen-hour days of clearing forest for pasture formation (*O Liberal* 1984; Provincia do Pará 1984).

9. The most notable of these security-conscious land agencies was the Executive Group of Araguaia-Tocantins Lands (GETAT), established by the military regime in 1980 to replace the National Institute of Colonization and Agrarian Reform (INCRA). Later, in 1987, GETAT was abolished. For a good account of the institutional history of land agencies in southern Pará, see Schmink and Wood (1992).

10. Concern with attracting large-scale corporate investment, featured in Operation Amazonia of the 1960s, gave way to the government-directed colonization by small farmers along the Transamazon Highway in the Program of National Integration in the early 1970s. The implementation of the POLAMAZONIA program (1975–1979) returned to massive devel-

opment projects. Poles 1 (Xingu-Araguaia) and 3 (Araguaia-Tocantins) targeted massive tracts of land in southern Pará for mineral development, timber extraction, and cattle ranching.

11. The municipality of Parauapebas, created in 1988 by the state of Pará, had a total population of 53,312 (51 percent urban) in 1991 (IBGE 1991).

12. Based on an interview with Chapeu de Couro, carried out by the author in Xinguara in mid-August 1978. Also see Schmink and Wood (1992:168–169) for references to this colorful personality in early Xinguara's history.

13. Glossy municipal pamphlets, filled with colored photographs, later published to publicize the attractions of life in Xinguara—and to celebrate the achievements of elected officials—listed the following as the first town residents in 1976: Raimundo Henrique de Miranda, José Henrique de Miranda, and Geraldo Martins de Andrade.

14. The conservative 7.1 estimate for labor participation in local lumber mills is based on the assumption that of the approximately 8,000 residents in Xinguara in late 1978, roughly 56 percent (4,480 people) were economically active and over ten years of age. (The 1991 survey in Xinguara found 56 percent of the surveyed population to be employed and over ten years old.) Of course, if considering only the *adult males* employed in 1978, the proportion working in the local lumber mills probably would have been much higher than 7.1 percent.

15. The legitimization of urban land invasions in Xinguara is not readily documented. However, a few well-known cases exist. For example, the Fazenda Marajoara, a large private ranch near Xinguara's urban perimeter, suffered squatter invasions in both the late 1970s and later in the early 1990s. At both times the local municipal government legalized the situation of the squatters by demarcating a larger area than originally invaded, providing land titles to new residents for small fees, compensating the *fazendeiros*, and keeping the profit for the municipality—or, it has been rumored, lining the pockets of politicians! The two resulting urban subdivisions, known as Marajoara I and II, allowed a significant expansion of the northeastern urban perimeter of Xinguara. Such *loteamentos* have occurred periodically since the late 1970s.

16. After gaining independent municipal status for his city in 1982, the newly elected prefect, Itamar Rodriques Mendonça, initiated the two new subdivisions, which were named in his honor: Itamaraty I and Itamaraty II.

17. In 1994 there were twenty-six lumber mills identified in the Xingu corridor. In a survey of 88 percent of these, Xinguara still remained the dominant center with eleven mills, while there was one in Agua Azul, four in Ourilândia do Norte, six in Tucumã, and one in São Felix do Xingu (Mousasticoshvily 1994).

18. Probably the most reliable source of demographic data is the malaria-control organization, known as SUCAM (Superintendéncia da Campanha Contra a Malária), which regularly counts all the buildings and estimates the population of settlements. In the first half of 1990 SUCAM counted 4,541 buildings and estimated 26,680 residents in Xinguara. In late June of 1990, our own research teams conducted field surveys and estimated 4,880 buildings in the town, thereby corroborating the general SUCAM figures. The 1991 census, which later put Xinguara at 27,292, confirmed the validity of these previous surveys (IBGE 1991). The population figures used in devising the sampling procedures for the household surveys in Xinguara, Tucumã, and Ourilândia do Norte were all based on such sources of demographic data.

19. The population figures for Cuca are based on SUCAM demographic data for the first half of 1990.

20. Since the founding of Xinguara the wide expanse of PA-279 running east-west through the town, known as Avenida Xingu, featured a barren central strip with no urban graces. In 1991–1992 the outgoing municipal prefect, anxious to improve the town's Wild West image, transformed Avenida Xingu into an attractively landscaped boulevard, replete with traffic cir-

cles, parking lots, and public plazas with benches, commercial kiosks, and lush gardens. Yet away from this new municipal "postcard" view, Xinguara still displayed unpaved streets, filthy drainage ditches instead of enclosed sewers, and a general lack of public services. Residents also complained about underground wells, the only source of water for urban households, drying up during the summer months.

21. This survey, carried out during June and July of 1990, interviewed 903 households in the three southern Pará towns, not including households that had arrived during the previous six months (or after January 1, 1990). Urban households were defined to include those individuals normally dwelling in a common living unit. As in Rondônia, the interviews in Pará were conducted according to a careful protocol to ensure a representative sample of households and comparability of results. After a pretest and revision of the questionnaire, each settlement was divided into various residential sectors, based on estimates of the number of houses and population densities, to ensure an even distribution of interviews. Within each sector, blocks were randomly selected for inclusion in the surveys. The households interviewed represented a total of 4,247 people or 8.4 percent of the approximate 1990 urban population of 50,000 in the three settlements.

22. The economically active population, all persons over age ten (according to the Brazilian census definition), does not include those solely engaged in unpaid household activities, nor does it include students, or other nonpaying activities. According to these criteria, all three settlements averaged just over two workers per household. The total economically active population ranged from a low of 52.0 percent of the entire population of Ourilândia do Norte to 56.1 percent in Xinguara and 58.6 percent in Tucumã. The weighted average for all three settlements is 55.9 percent, slightly lower than the comparable nationwide figure of 57.1 percent (IBGE 1989:123).

23. Figures for labor participation of heads of household were based on a separate analysis of the 1990 survey questionnaires.

8

Migration, Social Mobility, and Income Generation in Urban Amazonia

✧.✧

Historically, Brazil has always been a country on the move, especially since World War II.[1] In 1950, 5.2 million Brazilians, representing merely 10.3 percent of the nation's total population, resided outside the municipalities of their birth. By 1980, 40 million Brazilians, constituting 33.6 percent of the total population, had moved at least once from their municipalities of birth (CEM 1986:20). Not surprisingly, today the population of the Amazon is predominantly a migrant population. Frontier urbanization must be situated analytically in a national context of extensive interregional migration. In this chapter we examine the patterns of labor migration both to the Amazon and within the region, based on our 1990 surveys in Pará and Rondônia, in relation to the following questions: What processes of labor migration characterize Amazonia's urban population? To what extent does migration lead to occupational change and socioeconomic advancement? How much income do urban households in Amazonia produce, save, or remit? We would expect some differentiation in these phenomena between populist and corporatist frontiers.

PATTERNS OF INTERREGIONAL MIGRATION IN BRAZIL: 1940—-PRESENT

Three major interregional migratory flows have characterized population redistribution since the 1940s. First, the most massive continuous movement has been from the impoverished Northeast to other regions of Brazil, mainly to the metropolitan core regions of Rio de Janeiro and São Paulo in the Southeast (Martine 1992). Between 1950 and 1980 no less than two-thirds of all interregional migrants originating in the Northeast moved to the Southeast (Sampaio and Rocha 1989:12—14). During the decade of the 1960s, 58 percent of the urban population growth of the Southeast was attributed to internal migrations, mainly from the Northeast (IBGE 1979b:48).[2] The second important migratory flow, mainly consisting of farmers from São Paulo, Minas Gerais, and Santa Catarina, was to the then "frontier" state of Paraná (especially the northern quarter of the state) between 1930 and 1970. During the decade between 1940 and 1950 the population of northern Paraná tripled from 340,000 to 1,029,000. Approximately 53 percent of the migrants to northern Paraná in the 1950s originated in São Paulo (Souza 1980:60). During the 1960s Paraná was the only area in Brazil to experience a net in-flow of migrants (Wood and Carvalho 1988:219). The third major movement has been from the South (especially Paraná) and Center-West to the Amazon and western Maranhão beginning in the 1970s. During the 1970s Paraná contributed fully 17.6 percent of the total number of rural migrants leaving their homes in Brazil. This dramatic turnaround, according to Wood and Carvalho (1988:219) "reflects the closing of the agricultural frontier in Paraná, and the expulsion of population due to changes in agricultural production, especially the expansion of labor-saving soybean production" (see chapter 6). Other regions of Brazil have also contributed migrants to the Amazon, especially the Center-West and Northeast, during the period between 1970 and 1990. Each of these broad movements has specific regional connections. For example, while migrants from Paraná have dominated the flow to Rondônia, the main flows of migrants to southern Pará have originated from the states of Maranhão and Goiás (chapter 7).

At the national level these major migratory movements have resulted in extraordinary rates of growth in the metropolitan populations.[3] Yet, the process of urbanization throughout Brazil has not been the simple result of direct rural-urban migration. Estimated net rural-urban migration rates grew from 35.6 percent during the 1960s to 42.2 percent during the 1970s (Wood and Carvalho 1988:217).[4] In terms of total migrations, however, these percentages become quite small. In 1970, for example, rural-urban movements represented only 5.6 percent of all movements between *municípios* (IBGE 1979a). Most of the metropolitan growth in Brazil has been due to stepwise urban-urban migration.[5]

Between 1970 and 1991 the population of the Brazilian North grew by 258 percent from 3.6 to 9.3 million. Interregional migration to the Amazon, the third main migratory wave in recent Brazilian history, diverges in two key respects from the national patterns. First, interregional rural-urban migration to Amazonia represented 11.9 percent of all migrations reported in the 1970 census, still a relatively small percentage but twice as high as the national average. Second, although much publicity has attended the agricultural colonization of the Amazon, interregional rural-rural migration to Amazonia, at 33.7 percent of total migrations by 1970, was considerably lower than in some other regions.[6] Urban-urban movements constituted the largest flow into the region, representing 41 percent of the total influx of migrants reported in 1970 and giving some demographic empirical support to Becker's (1990:46) assertion that the "Amazon was born urbanized."[7]

These official statistics, however, conceal a more complicated reality of urban growth in Amazonia that can only be discerned by examining local migration patterns *within* the Amazon. The intraregional movements are characterized by a high degree of circularity between rural and urban places, and multiple movements within and between urban centers—a spatial fluidity of the urban population.

PATTERNS OF MIGRATION TO THE AMAZON FRONTIER

Between 1960 and 1990 official estimates of the population of the Brazilian North region showed increases of more than 250 percent, from 2.5 million to 8.9 million (table 1.1). Since 1970 this population growth has occurred predominantly in urban areas of the region, except in Rondônia, which became predominantly urban only during the late 1980s. Our examination of the migration patterns of the urban residents surveyed in the two settlement corridors (Rolim de Moura–Alto Alegre in Rondônia and the Xingu corridor in Pará) reflects both the diverse directions of the interregional migratory flows to the Amazon and the long-term urban orientation of most migrants. In this section we examine both interregional (to the Amazon) and intraregional (within the Amazon) migration patterns of the more than 1,600 households surveyed.

Interregional Patterns

The contemporary human occupation of the Brazilian Amazon was never a smooth or even process. Rather it has been characterized by various ebbs and flows of migrants following different geographic trajectories. Rondônia's recent historic role as a labor safety valve for displaced rural workers from the South, Southeast and Center-West suggests that most of its current inhabitants have origins in those regions. Likewise, eastern Amazonia has long received streams of migrants from the Northeast and Center-West. These interregional patterns are illustrated in table 8.1.

Table 8.1 Region of last residence before arriving in Amazonia (percentage of households)

	RONDÔNIA			SOUTHERN PARÁ		
	RM	*SL*	*AA*	*XG*	*OU*	*TU*
South	44.3	37.0	38.6	2.7	0.6	9.9
Southeast	26.6	26.0	21.6	3.4	4.1	10.8
Center-West	19.0	24.1	26.2	53.8	44.4	39.5
Northeast	3.8	4.3	7.9	37.0	43.9	32.7
North (Amazon)	2.2	2.0	0	1.9	2.3	2.5
Foreign	2.9	1.4	4.5	0	0.6	0
No Information	1.2	5.3	1.1	1.2	4.1	4.6

Note:

RM = Rolim de Moura
SL = Santa Luzia
AA = Alto Alegre
XG = Xinguara
OU = Ourilândia
TU = Tucumã

Responses are based on the last state of residence before Rondônia and Pará, grouped by standard IBGE regions, of heads of household interviewed. Nonmigrants, natives of the states of Rondônia and Pará, are included in the figures of the North (Amazon) region. Columns may not sum to 100 percent due to rounding.

Consistent with other surveys showing a predominance of southern migrants to Rondônia, a weighted average of 67.1 percent of the residents surveyed in all three settlements comprising the Rolim de Moura–Alto Alegre settlement corridor reported their last residences in the South and Southeast, while only 3.75 percent of the total number of migrant households to the corridor had relocated directly from the Northeast. Natives of the Amazon region itself, however, accounted for the smallest proportion of the population—1.8 percent of the 713 residents surveyed in these three inland settlements, a sharp contrast to Rondônia's capital city, Porto Velho, on the Madeira River, where residents are largely natives of the North (chapter 5). Variations in these percentages between the settlements on each frontier generally were not statistically significant.

In the urban settlements surveyed in the Xingu settlement corridor in southern Pará, a different geographic pattern of origins emerges. Most of the population originated in the Center-West and Northeast, especially in Goiás and Maranhão. These states accounted for 90.8 percent of households in Xinguara, 88.3 percent in Ourilândia, and 72.2 percent in Tucumã. In contrast to Rondônia, in the Xingu corridor relatively small proportions of urban residents came from the South and Southeast. Because of its corporate institutional history, which encouraged an influx of southern farmers and bureaucrats (Butler 1985), Tucumã has the largest share of southerners in the corridor (about 10 percent) and another 11 percent originated in the Southeast. As in Rondônia, only a tiny fraction of the population of southern Pará is native to the Amazon.

Urban Origins of Interregional Migrants

Since virtually all the urban residents surveyed in both Rondônia and Pará were migrants from other regions of Brazil, it is instructive to examine the different types of migration contributing to frontier urbanization. Given that a growing proportion of total migrations has been between urban areas throughout Brazil, we can reasonably expect this urban-urban pattern to be dominant in the interregional migrations to the Amazon (IBGE 1977:211; Wood and Wilson 1984). Moreover, as predicted by Sawyer (1987), we expect the growth of urban areas to be generalized across the Amazon region and, therefore, we would not expect to find significant subregional variations in urban-urban migration rates due to the distinctive social and economic orientations of different subregional settlement areas.

The survey data from both Rondônia and Pará only partially support these expectations (table 8.2.). Urban-urban migration was the preponderant type of movement in both subregional areas. Weighted averages of 42.3 percent and 52.1 percent of all urban households surveyed in each state, respectively, reported urban origins outside the Amazon region. Only in Ourilândia do Norte was rural-urban migration slightly more important. However, the significant difference in these subregional rates can only be explained by the larger proportion of rural-rural and urban-rural migrants found in the Rondônia sample, further indicating the agrarian nature of this frontier area in contrast to that in Pará.

Beyond these general observations two interesting variations in migration patterns between the settlement corridors are apparent. First, in

Pará rural-urban movements were the second most frequent interregional migration type, accounting for a weighted average of 42.4 percent of households. In contrast, interregional rural-urban movements were comparatively less important in Rondônia, representing only 18.7 percent of urban household movements to the Rolim de Moura–Alto Alegre settlement corridor. In summary, in Pará 94.5 percent of all interregional household movements to the Amazon were directly to urban areas, in contrast to 61.0 percent in Rondônia.

Table 8.2 Interregional migration patterns (percentage of households by type of move to Amazonia)

	RONDÔNIA			SOUTHERN PARÁ		
	RM	*SL*	*AA*	*XG*	*OU*	*TU*
Urban-Urban	44.8	36.5	44.3	52.7	45.3	55.9
Urban-Rural	12.5	14.4	11.4	0.4	0	0.3
Rural-Rural	18.9	23.1	28.4	0.4	0	0.6
Rural-Urban	19.9	17.8	14.8	44.3	49.4	35.5
No information / does not apply	3.9	8.2	1.1	2.3	5.2	7.7

Note:

RM = Rolim de Moura
SL = Santa Luzia
AA = Alto Alegre
XG = Xinguara
OU = Ourilândia
TU = Tucumã

Second, whereas a substantial minority (34.2 percent) of current urban residents in Rondônia initially settled in rural locations of the Amazon (i.e., engaged in urban-rural and rural-rural interregional movements), such households were all but absent from the Pará sample.

The urban transformation of Amazonia displays divergent patterns. In Pará the urban transformation begins with the initial interregional migration of households to the Amazon. In Rondônia, to a significant degree, the urban transformation results from local intraregional migration within the Amazon. Indeed, the urbanization of Rondônia's migrant population appears to be a complex phenomenon, entailing multiple locational movements by different household members at various spatial scales, which invites the additional analysis we offer below.

Intraregional Migration Patterns Within Rondônia

Once in the Amazon, urban residents continue to move about in a variety of complex ways. Urban residency typically is unstable over time. Households often move between urban and rural places several times. The dynamics of resource extraction in various populist pockets of the frontier promote frequent household relocation as well. Moreover, significant intraregional differences in migration patterns, as between Rondônia and southern Pará, make sweeping generalizations problematic. While the surveys were able to capture the outlines of local migration patterns in Rondônia, in southern Pará the high frequency of household migration within Amazonia, especially among the gold-mining population of the Xingu corridor, defied systematic efforts at reliable classification.[8]

In the case of southern Pará, survey data collected about the last move prior to the current residence indicate a predominance of intraregional urban-urban moves, representing roughly 58 percent of the total number of last household changes of residence. Rural-urban migration was of secondary importance, comprising only about 27 percent of the total volume of previous movements. In southern Pará fewer urban settlers initially were able to obtain rural land, given the concentrated nature of land tenure in eastern Amazonia. It is reasonable to hypothesize that on the contested corporatist frontier, as in southern Pará, the main intraregional movements would be urban-urban. These findings from our southern Pará survey indicate that the urban transformation for most residents of the Xingu corridor occurred in the process of interregional migration to Amazonia. For a substantial proportion of Rondônia's respondents, on the other hand, the urban transformation occurred as a result of intraregional migration within Amazonia. Therefore, in this section we focus on Rondônia.

The importance of initial rural landings for current urban residents of Rondônia suggests that the recurring process of land consolidation and peasant expulsion is producing cities of landless peasants in the agrarian frontier, a prospect intimated by several notable scholars (Hebette and Acevedo 1979; Martine 1990; Wood and Carvalho 1988:231, to name a few). We acknowledge that this process (which we call involuntary urbanization) occurs throughout the Amazon, but it captures only one dimension of the urbanization experience. Indeed, more often than not, what superficially appears to be rural-urban migration within Rondônia on closer scrutiny is really one segment in a larger cycle of a circular urban-rural-urban migration pattern.

As predicted, when all intraregional movements among the sample of 713 household heads were summed and classified by type, the pervasive flow among Rondônia's urban populace was rural-urban, representing 42.1 percent of total intraregional household moves (table 8.3). The second most prevalent pattern in Rondônia was urban-urban movement, which occurred in 26.9 percent of the cases.

Table 8.3 Intraregional migration patterns, Rondônia (percentage of total household moves within Amazonia by type of move)

	ROLIM DE MOURA	SANTA LUZIA	ALTO ALEGRE	WEIGHTED AVERAGE
Urban-Urban[a]	25.6	29.0	29.0	26.9
Urban-Rural	14.3	14.5	13.3	14.3
Rural-Rural	18.4	15.9	17.8	17.7
Rural-Urban	43.0	41.7	38.5	42.1

[a]*Excludes crosstown moves within the same settlement.*

Three important nuances in the intraregional migration patterns in Rondônia require elaboration. First, while the analysis of the total number of household movements within Amazonia by type (excluding local crosstown moves) indicates a predominant rural-urban pattern as predicted (table 8.3), by examining the last two places of residence before the town of current residence we observe an important circular urban-rural-urban migration flow.[9] When only the last two intraregional movements are considered (among those moving two or more times), households were about evenly divided between rural and urban areas for the place of previous residence. A weighted average of 50.6 percent of all households indicated their last residence as a rural location, while 49.4 percent indicated an urban location (table 8.4). Of those indicating a rural place of last residence, however, only 27.3 percent indicated that their previous two residences were in rural areas; the balance, 23.3 percent, reported their second prior residence in an urban center. In other words, nearly half of all respondents indicating a rural place of last residence had previously resided in an urban area within the Amazon. Summing up, nearly three-quarters (72.7 percent) of all urban households in the Rolim de Moura—-Alto Alegre corridor had previously resided in an urban area, while only

27.3 percent were first-time urban residents. In summary, many rural-urban moves in Rondônia, over the longer term, are really one step in an urban-rural-urban migration pattern within the Amazon region. Most migrants interviewed have long been urban-oriented in their long-term local movements within the Amazon.

Table 8.4 Percentage of Rondônia households by type of previous two intra-regional moves

TYPE OF PREVIOUS MIGRATIONS	ROLIM DE MOURA	SANTA LUZIA	ALTO ALEGRE	WEIGHTED AVERAGE
1. Rural-Urban	48.9	55.0	48.8	50.6
1.1. Rural-Rural-Urban	25.9	30.3	27.4	27.3
1.2. Urban-Rural-Urban	23.0	24.7	21.4	23.3
2. Urban-Urban	51.1	45.0	51.2	49.4
Total Households	417	208	88	713

Note: Last two residences before town of curent residence. The above data thus exclude crosstown moves. Totals do not equal those indicated in table 8.3 because "total intraregional moves" slightly exceed "previous two intraregional moves."

The second nuance is that the dominant urban orientation of intraregional migration in Rondônia has led to the decentralization of the urban population. Whereas urban-urban migration is typically thought to occur in a stepwise progression of household movements to ever larger urban centers, our case studies suggest that the progression is reversed, as urban-urban migrants tend to move from larger to smaller urban places. In the Rondônia sample between 34 percent (Rolim de Moura) and 51 percent (Alto Alegre) of all current urban households had migrated directly from higher-order settlements. In contrast 15 percent (Rolim de Moura) to 6.9 percent (Santa Luzia) of households had followed the conventional pattern by moving to a larger urban place.

A third nuance concerns the geographic locality of the intraregional migration phenomenon. The majority (57.1 percent) of intraregional household moves among the respondents in Rondônia was within the *município* of current residence, i.e., typically within the same settlement (table 8.5). The proportion of such movements progressively decreases along the distance-age gradient of the Rolim de Moura——Alto Alegre corridor as the newer *município* are likely to capture a lower proportion of total intraregional moves. Not surprisingly, the vast majority (72.0 percent) of moves within a *município* are crosstown moves (within the urban area of current residence). We suggest that this local movement is due to the change in residence that most new migrants to an urban area undergo in the short-term as they move from a temporary landing residence (dwelling of a friend or relative or a hotel) to their own permanent dwelling.

Table 8.5 Intra*município* and crosstown household moves as a percentage of total intraregional moves, Rolim de Moura–Alto Alegre corridor

INTRAREGIONAL MOVE TYPE	ROLIM DE MOURA	SANTA LUZIA	ALTO ALEGRE	WEIGHTED AVERAGE
1. Intra*município* moves as a percentage of intraregional moves	65.8%	45.3%	39.5%	57.1%
2. Crosstown moves as a percentage of intra*município* moves	73.3%	72.9%	60.0%	72.0%
3. Total number of intraregional household moves	949	433	177	—
4. Average number of intraregional moves per household	2.27	2.08	2.01	2.20

While rural-urban migration accounts for the largest number of intraregional movements in our Rondônia sample, about half of these movements are but one segment in a larger, more complex pattern of urban-rural-urban migration. Ultimately, migration between urban centers is the dominant pattern of household mobility in both interregional and intraregional contexts. It would be inaccurate to conclude, therefore, that the lower echelon of urban settlements in Rondônia consists of "cities of peasants" or "centers of agglutination" of the rural poor. Although a significant peasant population is evident in these towns, frontier urbanization is largely the result of direct interregional urban-urban migration (in the case of both Pará and Rondônia) and intraregional urban-rural-urban migration (in the case of Rondônia).[10] Intraregional migration patterns are characterized by a preponderance of short-distance moves within a *município*, not long-distance moves across the region. Our survey data also suggest an inverted stepwise migration pattern in which migrants progressively move down the urban hierarchy, from larger to smaller frontier urban settlements in search of new opportunities, indicating a tendency toward population decentralization. The long-term stability of this decentralization pattern will depend upon the opportunity residency in smaller urban centers provides for socioeconomic advancement, an issue we address in the next section.

MIGRATION AND OCCUPATIONAL CHANGE

People typically move in response to changing occupational opportunities. Socioeconomic mobility is a critical dimension in our understanding of the patterns of internal migration and the process of frontier urbanization in the Amazon. Neoclassical dual sector migration models posit that migration is a response to expected income differentials between sending and receiving places. Migration, therefore, usually leads to upward social mobility (Lewis 1954; Todaro and Harris 1970; Todaro 1980). In contrast, critical political economists have argued that as capitalism advances in the periphery, peasants become alienated from the means of production, in this case land, and become progressively more dependent on urban wage and casual temporary employment (D'Incao 1975).

How has migration to and within the Amazon affected occupational change of migrants?[11] Interregional migration to the Amazon entailed at least one change in occupational category for three-quarters of the household heads surveyed in the Rolim de Moura corridor and for two-thirds of those in the Xingu corridor (table 8.6). For example, a residential move may indicate a shift in occupation from farming to gold mining or timber extraction. There were few significant differences in the number of occupational changes by workers residing in different settlements in the corridors. We conclude that migration to the Amazon generally is accompanied by high rates of occupational change among the migrant population.

Table 8.6 Number of changes in occupational class from last residence outside of Amazonia to current residence (percent of household heads)

| | RONDÔNIA | | | PARÁ | | |
Number of Changes	RM	SL	AA	XG	OU	TU
0	24.9	25.0	22.7	33.3	33.3	39.4
1	40.5	38.0	38.6	38.8	51.0	33.9
2	19.9	17.8	17.0	20.5	15.7	18.3
>2	11.7	11.5	19.2	6.6	0.0	8.3
No information	3.0	7.7	2.3	0.8	0.0	0.0

Note:

RM = Rolim de Moura

SL = Santa Luzia

AA = Alto Alegre

XG = Xinguara

OU = Ourilândia

TU = Tucumã

The central issues in this analysis are whether occupational changes associated with migration lead to a qualitative change in the socioeconomic position of migrants relative to capital and to corresponding increases in the generation of surplus household income. We compressed various types of socioeconomic relationships into six broad categories: autonomous worker, sharecropper/tenant farmer, temporary employee, permanent employee, unremunerated family labor, and proprietor.[12] The distribution of the labor force by socioeconomic position in July 1990 is represented in table 8.7.

Table 8.7 Socioeconomic position of the urban labor force (percentage of workers by socioeconomic position)[a]

	RONDÔNIA			PARÁ		
	RM	SL	AA	XG	OU	TU
Proprietor	7.6	12.3	15.8	10.6	11.5	10.6
Autonomous worker[b]	10.2	6.8	7.2	11.7	12.6	7.7
Permanent employee	32.3	26.8	19.4	20.5	18.8	26.5
Temporary employee	6.0	4.5	4.0	8.7	10.9	7.9
Tenant farmer/sharecropper	1.1	3.6	2.5	0.8	0.4	0.4
Unremunerated family worker	22.8	29.1	33.8	28.1	31.2	28.1
Other[c]	20.0	16.8	17.3	19.6	14.6	18.8
Total number of workers	1453	690	278	1336	506	840

Note:

RM = Rolim de Moura

SL = Santa Luzia

AA = Alto Alegre

XG = Xinguara

OU = Ourilândia

TU = Tucumã

[a] Socioeconomic position refers to primary economic activity. Between 6 percent and 10 percent of all workers in Rondônia also reported "secondary" jobs. The "labor force" is defined as all persons over ten years of age.

[b] Autonomous worker is an independent (i.e., nonsalaried) worker who is typically paid for services rendered; commonly thought of as a worker in the informal sector.

[c] Includes students older than age ten.

In the case of Rondônia, where the settlement frontier tends to be initially structured as a spatial continuum of hierarchically ordered central places, we would expect capitalist (wage) relations of production to be relatively more prevalent in defining the urban labor force of the dominant central place in the subregional settlement system (Rolim de Moura) and to become less prevalent in the smaller settlements located toward the fringe of the frontier. Indeed, the data confirm this expected pattern. Both permanent and temporary (i.e., formal wage) employment rates are highest in Rolim de Moura at 38.3 percent (combined) and become progressively lower farther out in the corridor. In Rondônia workers in the more developed core settlement, Rolim de Moura, are comparatively more homogeneous as wage workers than in the other settlements of the corridor. Conversely, unremunerated family labor becomes increasingly more important with distance from the core.

The opportunities to own means of production (e.g., land, businesses) are also greater in the periphery of Rondônia's settlement systems, and the percentage of the labor force classified as proprietors in Alto Alegre (15.8 percent) is twice that of Rolim de Moura (7.6 percent). While this tendency toward peripheral capitalization applied to both primary and secondary activities (all workers interviewed were queried about holding multiple jobs), it was particularly accentuated in the latter. While 25 percent of Rolim de Moura workers with two economic activities are proprietors in the secondary (part-time) activity, 56 percent of Alto Alegre's workers are so engaged. The frequency of ownership of a productive asset (e.g., urban shop or farm lot) that becomes a source of secondary income increases with distance from the core urban center. In essence this means that residents on the fringe are more heterogeneous in their relations to capital and pursue a greater diversity of economic activities to enhance survival.

Basically, as is predicted by recent adaptations of central place theory, the economy of core settlements is more formal, more structured, more intensively capitalized, while that of the periphery of the settlement subsystem is more familial, informal, and more extensively capitalized. In the populist frontier the urban transformation of the population gradually entails household decapitalization and the emergence of formal relations of production. The same, however, cannot be said for the contested corporatist frontier of southern Pará.

The Xingu corridor data turn the logic of progressive capitalist penetration found in Rondônia on its head. Wage relations of production are spatially inverted, such that Tucumã has the highest rate of formal employment in the corridor, perhaps an artifact of its corporatist origins. When permanent and temporary employment figures are combined, Xinguara (29.2 percent) has lower rates of wage labor than both Ourilândia do Norte (29.7 percent) and Tucumã (34.4 percent). Unremunerated family labor does not vary significantly among the settlements on the Xingu corridor. Within the Xingu corridor the opportunities to own the means of production do not vary much from the center to the periphery. Likewise, tenant farming and sharecropping are virtually absent among the urban labor force of both southern Pará and Rondônia.

The survey data on socioeconomic position in southern Pará fail to confirm assumptions of central place theory and reflect uneven patterns of regional development in a frontier affected by corporatist influences. Capitalism has expanded in an irregular, selective manner in eastern Amazonia. Market influences have advanced well beyond Xinguara into the resource-extraction areas of the Xingu region. Despite Xinguara's importance as a commercial center and its better developed physical infrastructure, the lower rates of wage labor probably reflect some degree of hollowing out of the frontier as postulated by our model of the corporatist frontier (chapter 4). Certainly the proportion of permanent wage laborers is much lower in Xinguara (20.5 percent) than in Rolim de Moura (32.3 percent), and the proportion of autonomous workers (a bellwether of informality) is higher in Xinguara (22.3 percent). Rates of proprietorship, on the other hand, are roughly equivalent in both subregional systems.

Paradoxically, the contested corporatist frontiers in southern Pará have not favored thorough proletarianization of the population. In fact, despite variations within each corridor, capitalist expansion has not succeeded in converting the bulk of the economically active population in either Rolim de Moura or Xinguara to steady wage labor. These empirical findings tend to support the recent critiques of contemporary studies of political economy (Cleary 1993) insofar as capitalism has not expanded evenly in Amazonia.

OCCUPATIONAL CHANGE AND
SOCIOECONOMIC MOBILITY

Migration and change of occupation typically entail some degree of change in socioeconomic position. We define socioeconomic mobility as a qualitative change in one's principal livelihood activity in relation to capital. This relationship to capital can be either direct, through the acquisition of means of production, or indirect, mediated through a mode of production (e.g., wage employment). We depart from rigidly dualistic categories of capital versus labor and accept the more relativist premise that all workers possess various degrees of both capital and labor. We propose a typology encompassing four different types of socioeconomic shift or directions of mobility. First, a worker's shift from any of the first four categories defined above (autonomous worker, employee, tenant farmer, and unremunerated worker) to proprietor (i.e., owner of some means of production) indicates a qualitative shift toward *capitalization*. Second, a shift in the opposite direction signals a move toward worker *decapitalization*. Third, the shift from unremunerated family worker, sharecropper/tenant farmer, and autonomous worker to employee (either temporary or permanent) suggests a move toward more formal wage relations, or *proletarianization*. Finally, the reverse shift indicates a move toward relatively more exploitative informal wage relations often associated with worker impoverishment, which we will call *pauperization*.[13]

Although three-quarters of the household heads surveyed in Rondônia indicated an occupational change associated with interregional migration to the Amazon (table 8.6), only slightly more than half indicated a shift in socioeconomic status. This pattern differs from that in southern Pará by degree, where 70 to 80 percent of household heads surveyed noted a qualitative change in socioeconomic status with migration to the region (table 8.8A). Thus, we can posit that interregional migration to the Amazon does generally result in a change of socioeconomic position, although significant regional variations exist.

Once in the Amazon, curiously, intraregional migration was not accompanied by correspondingly high rates of change in socioeconomic positions (table 8.8B). Nearly half of Rondônia's urban household heads reported no change in socioeconomic position between their first residence in Amazonia and their current residence. In Pará two-thirds of urban household heads reported no change in status between moves within the Amazon. Hence, among urban households in both the Pará and Rondônia study areas, interregional migration to the Amazon was associated with significant changes in socioeconomic position, while intraregional migration was not.

The directions of the observed changes in socioeconomic position are somewhat ambiguous. Three tendencies are nonetheless evident. First, a substantial proportion of those households undergoing a change of socioeconomic status in both study areas reported acquiring the means of production, that is, becoming proprietors since moving to the Amazon. This group, representing 11—16 percent of the Rondônia sample and 20—24 percent of the Pará sample, would be classified as becoming capitalized following their migration to the Amazon. Somewhat lower proportions (6—18 percent and 7—16 percent) reported a movement toward capitalization following relocation within the Amazon. Hence, we affirm that the rate of upward socioeconomic mobility stemming from interregional migration to the Amazon is higher than that associated with migration within the Amazon.

Table 8.8 Tendencies in the socioeconomic mobility of the urban workforce in Amazonia (percentage of household heads)

	RONDÔNIA			PARÁ		

A. *Changes in socioeonomic position from last residence outside of Amazonia to current residence inside of Amazonia*

	RM	SL	AA	XG	OU	TU
Capitalization	15.6	11.1	13.6	20.2	23.3	23.5
Decapitalization	7.9	2.9	8.0	4.6	8.1	3.7
Proletarianization	13.2	10.6	10.2	23.7	10.5	26.2
Pauperization	13.4	15.9	14.8	19.8	32.0	14.5
No change	44.4	48.6	48.9	29.0	21.5	27.5
No information	5.9	11.1	4.6	2.6	4.6	4.6

B. *Changes in socioeconomic position from first residence inside of Amazonia to current residence inside of Amazonia (where different)*

	RM	SL	AA	XG	OU	TU
Capitalization	6.2	14.4	18.2	13.6	15.7	7.3
Decapitalization	10.6	9.1	4.5	7.8	2.0	4.6
Proletarianization	14.1	11.1	11.4	9.3	0.0	6.4
Pauperization	8.6	5.8	5.7	7.0	9.8	6.4
No change	54.7	57.2	55.7	62.0	70.6	74.3
No information	5.8	2.4	4.5	0.4	2.0	0.9

Note:
RM = Rolim de Moura
SL = Santa Luzia
AA = Alto Alegre
XG = Xinguara
OU = Ourilândia
TU = Tucumã

Interestingly enough, in the sample in the Rolim de Moura—Alto Alegre corridor there is a positive distance gradient in capitalization rates arising from intraregional migration (but not from interregional migration), a pattern that is consistent with other settlement characteristics in Rondônia (chapter 6). The further away from the main central place in the urban system one moves, the higher the proportion of household heads moving into socioeconomic positions of proprietorship (see also table 8.7). Conversely, the likelihood of decapitalization is greater for those households choosing to locate in the more developed center (e.g., Rolim de Moura versus Alto Alegre). These findings appear at once contradictory: How is it possible for household capitalization to be more frequent in the more remote settlements where capitalist penetration is less pervasive and for household decapitalization to be more frequent in urban centers that are fully integrated into the larger capitalist economy?

The first tendency, that toward capitalization on the populist periphery, may be rather easily explained. Land (both urban and rural) at the fringe of the settlement corridor (e.g., Alto Alegre) in 1990 was still plentiful and comparatively cheap to acquire; indeed it may even be made available to claimants for free by government institutions. Capitalization, then, is achieved through the application of nonremunerated household labor to speculative and productive improvements to that land. The positive distance gradient characterizing agricultural employment in Rondônia's urban settlements, in which more remote smaller centers have higher rates of agricultural employment than dominant central places (chapter 6), may be taken to support this conclusion.

The second tendency, that toward decapitalization in the subregional core settlements of Rondônia, appears consistent with recent central place studies (see Meyer 1980). Petty merchants in towns with a larger market size (e.g., Rolim de Moura) face greater competition from more specialized vendors that arrive in the frontier in the later stages of its development. Unable to specialize (for lack of capital) and unwilling to relocate to the periphery where competition at lower levels of specialization is not so intense, these proprietors sell off their assets (decapitalize) and leave the frontier or join the urban labor force as proletarians.

The third tendency observed in the Rondônia data is that pauperization rates are higher following interregional migration than following household movements within the region, while proletarianization rates are not significantly different between the two migration categories. This tendency may be explained by the probability that interregional urban migrants relinquish formal sector jobs in moving to the Amazon and then for a time rely on irregular, informal work until they find new formal positions on the frontier.

Not surprisingly, the findings in southern Pará diverge significantly from those in Rondônia. First, as noted above, the rates of socioeconomic mobility arising from interregional migration to the Xingu corridor (71–79 percent) are significantly higher than in Rondônia (51–55 percent).[14] Changes in all categories of socioeconomic mobility, except decapitalization, were higher in the Xingu corridor as well. These data suggest an important difference between Pará and Rondônia. The interregional movement to the contested corporatist frontier is tumultuous in terms of occupation and entails greater opportunity for socioeconomic mobility among the urban population there. The interregional move to Rondônia, by contrast, has a relatively more stable effect on socioeconomic change, entailing greater continuity in class status.

More surprising socioeconomic changes occur in urban households in the Xingu corridor following intraregional movements. Three divergent tendencies appear. First, although the overall rate of capitalization is not significantly different between the Xingu and the Rolim de Moura—Alto Alegre corridors, it is spatially much bumpier within the former corridor. Ourilândia has a higher rate of capitalization following intraregional migration than either of the other two settlements in the Xingu corridor. Second, decapitalization rates in the Xingu corridor are lower than in Rondônia and do not display any spatial symmetry. Again, Ourilândia do Norte is a special case, having the lowest rate of decapitalization in the corridor. Finally, proletarianization rates are slightly lower in southern Pará than in Rondônia. Not a single household in Ourilândia do Norte obtained a wage-paying job following the last intraregional move, attesting to the high degree of occupational autonomy found in the gold fields that Ourilândia services.

Whereas opportunities for capitalization in Rondônia appear to be greater at the geographic fringes of the settlement system, proletarianization rates are rather uniform throughout the Rolim de Moura corridor for both interregional and intraregional movements entailing job changes (10–14 percent). Nevertheless, in Rondônia, a populist frontier, socioeconomic position shifts toward wage relations of production are generally of secondary importance. The plurality of migrants from other regions to the Rolim de Moura—Alto Alegre corridor fared well, becoming asset-acquiring proprietor-producers, at least for a time.

The same tendencies were not observed in southern Pará. In the contested corporatist frontier we would expect to find a significantly higher proportion of position changes toward proletarianization (formal employment). Indeed, about a quarter of all households shifted to proletarian positions in Xinguara and Tucumã. About 10 percent did so in Ourilândia do Norte, following migration to the Amazon, but a much smaller proportion of the labor force experienced such a shift following their internal movements within the Amazon. We find simultaneously high rates of socioeconomic shift toward both capitalization and pauperization in Ourilândia do Norte, suggesting a functional relationship of one shift to the other in the context of this goldmining community.

In summary, a comparison of the data sets from both study areas in table 8.8 highlights the difficulty in reaching unproblematic conclusions about the socioeconomic impacts of the process of regionwide urbanization in the Amazon. The Pará and Rondônia data suggest that these frontier urbanization fronts are differentially articulated to both larger national development processes and distinct local conditions. Nevertheless, three general conclusions should be offered. First, the data from all study sites suggest an increasing differentiation of owners and workers in Amazonia over time. While initial differences in capitalization and proletarianization rates following interregional migration are minimal, with a couple of local exceptions, the gap between these two socioeconomic groups uniformly widens following intraregional migration. Over time, socioeconomic class differentiation occurs among urban Amazonians generally.

Second, it might be argued that the shift toward wage employment and capitalization, representing the most significant combined shift, indicates a predominant tendency toward upward mobility among those experiencing a career change resulting from migration both to and within the Amazon.

Third, a closer examination of the anomalies (Santa Luzia and Ourilândia do Norte) accentuates the richness of diversity found in the urbanization experiences within these rainforest city systems. Santa Luzia had the lowest rate of decapitalization following interregional migration in the Rolim de Moura—Alto Alegre corridor, suggesting that conditions there differentially favored retention of means of production by migrants. We can only speculate about the possible causes of this anomaly but nonetheless point to it as evidence of local forces driving differential patterns of frontier urbanization.

Ourilândia do Norte is even more exceptional given the number of its anomalies. In contrast to Santa Luzia, with which it shares a cohort spatial position, the rate of decapitalization following interregional migration is the highest in the Xingu corridor. However, following intraregional migration, capitalization rates and pauperization rates become the highest, decapitalization the lowest, and proletarianization vanishes. The special case of Ourilândia do Norte aptly illustrates the disarticulated nature of settlement formation within the Xingu corridor.

INCOME GENERATION, SAVINGS, AND REMITTANCES

The high rates of occupational mobility associated with interregional migration to Amazonia and the generally downward shift in socioeconomic position over time for the majority of the urban population should be reflected in some fashion in household income generation and investment practices. How much income do urban households in Amazonia earn? How many borrow or save? How frequently do urban households remit or receive surplus income to or from distant (out-of-state) sources?

The average monthly incomes of the survey sample ranged from U.S.$248 in Xinguara, Pará, to U.S.$381 in Santa Luzia and are generally higher in Rondônia than in southern Pará (table 8.9). These transpose into average annual per capita household incomes of approximately U.S.$600 and U.S.$900, respectively, which are considerably below the Brazilian national average GNP per capita of U.S.$2,940 in 1991 (World Bank 1993:239).[15] Not surprisingly, about half of the respondents in both Rondônia and southern Pará claimed that they were unable to generate a surplus from their monthly income; all income was expended on subsistence. This, in part, explains the relatively low formal savings rates (14–26 percent of households) and the high rates of informal borrowing and indebtedness (30–59 percent of households), especially so in Pará.[16] Less than 20 percent of households surveyed in all Amazon sites claimed to have remitted any income to family or friends outside the state, and less than 10 percent acknowledged receiving financial aid from others outside the state.[17]

Several interesting features of the two survey areas warrant comparison. First, there is no linear progressive tendency in the distribution of household income by study site in either survey area as might be predicted in spatial economy models. In both areas, however, households in the middle node in the urban subsystem (i.e., Santa Luzia and Ourilândia do Norte) enjoyed significantly higher incomes than residents of the other towns. This convex spatial distribution of income cannot be attributed to any inherent quality of the urban subsystem as a whole. Rather, we suppose that this pattern is coincidental and reflects local idiosyncrasies of each middle settlement (probably the unusual political leadership of Santa Luzia and the greater reliance on the gold economy in Ourilândia).

Second, while households in the Rondônia sample indicated generally higher incomes than households in the southern Pará surveys, net (surplus) income generation was consistently lower in the former than in the latter. This apparent contradiction may be due to the somewhat higher cost of living in the more remote Rondônian frontier where consumers pay more for expensive interregional imports (chapter 9).

Table 8.9 Income generation, savings, and remittances (total monthly income and percentage of households)

	RONDÔNIA			PARÁ		
	RM	SL	AA	XG	OU	TU
Total household monthly income, June 1990 (U.S.$)	329	381	293	248	312	250
With no surplus income	57.4	50.5	51.1	43.5	41.6	42.0
With bank savings account	25.6	22.6	18.2	18.7	13.9	11.7
Borrowing informally	30.4	40.4	55.7	49.1	48.0	59.0
Remitting income out of state	17.2	18.7	5.7	16.2	14.5	19.8
Receiving income from out of state	9.1	5.8	9.1	6.0	2.9	3.4

Note:
RM = Rolim de Moura
SL = Santa Luzia
AA = Alto Alegre
XG = Xinguara
OU = Ourilândia do Oeste
TU = Tucumã

Third, the majority of households in both survey areas did not use formal savings instruments through commercial or state banks.[18] Despite the lower rate of surplus income generation in Rondônia, slightly higher proportions of households there used formal bank savings accounts. In both the Rondônia and the southern Pará survey areas the proportion of households with formal savings accounts progressively diminishes along the distance gradient of the respective corridors, reflecting the spatial distribution of bank branches in the frontier.

Fourth. informal credit transactions are a frequent feature of economic life throughout urban Latin America and no less so on the frontier. Rates of informal household borrowing were found to be higher in the southern Pará study area than in Rondônia, due, in part, to the greater degree of informality of the economy of the Xingu corridor as a whole.

In Rondônia, predictably, informal borrowing rates increase along the distance gradient of the settlement corridor in inverse relation to the spatial distribution of formal banking institutions (as reflected, in part, by lower formal savings rates).

Fifth, there is no significant difference between the two study areas in terms of income remittance rates. Alto Alegre has the lowest rate of households remitting income out of state, which is probably due to its lower than average household income. Overall rates of households receiving money from persons outside the state were lower than rates of remittances sent to such persons (except in Alto Alegre), suggesting that the frontier settlements are net exporters of disposable household income. The middle nodes in both settlement systems again distinguish themselves from the other settlements by having the lowest rates of receipt of external income support from outside the state.

Despite the various indicators of downward socioeconomic mobility, low income generation, financial debt, and economic hardship for older settlements in the Amazon's settlement system, we cannot easily embrace the conclusion that the majority of the region's urban residents perceive themselves to be worse off than they were before moving to the region or that most wish to return to where they had come from. Only in the Rondônia sample is there a clear and significant pattern consistent with the linear transition model regarding life satisfaction. As we would expect, Rolim de Moura, with higher rates of household decapitalization, lower rates of proprietorship, and a local economy in recession since the 1985 bust in the mahogany trade, was less popular among its residents than Santa Luzia and Alto Alegre were among their own. Indeed, whereas 22.5 percent of household respondents in Rolim de Moura indicated that the quality of their lives had worsened since arrival there, only 8.0 percent indicated the same in Alto Alegre (chapter 6). No such difference existed among the Pará respondents, more than one-fifth of whom uniformly asserted that their lives on the frontier were worse than before migrating to the Amazon (chapter 7).

Two conclusions follow from our analysis of income generation, savings, and remittances. First, urban residents of Amazonia generally are found within the lower-income stratum of the national population when comparing per capita household income to per capita GNP. While urban residents of Rondônia earn higher incomes than those of Pará, most do not accumulate any surplus income on a regular basis, but in Pará a slight majority does. Second, the patterns of household savings, informal borrowing, and income remitting and receiving are widely varied among the individual study sites. Generally, the data indicate a much higher reliance on informal financial relationships than on formal bank accounts. Moreover, with the singular exception of Alto Alegre, more households send money to friends and family outside the region than receive remittances from them, suggesting a possible net outflow of capital from the Amazon through informal channels. While progressive changes, such as those that might be associated with central place tendencies, are evident along the Rolim de Moura–Alto Alegre corridor (in terms of the distribution of bank savings accounts and informal borrowing rates), there is virtually no semblance of such ordering in the Xingu corridor. Local distortions in the spatial distribution of household financial practices are particularly evident in the middle node settlements of both systems. The variability observed in the distribution of these data cautions against sweeping generalizations and master principles of explanation that cut across the Amazon region as a whole.

CONCLUSIONS

A single portrait of the urban migrant to Amazonia cannot be accurately constructed. The interrelated processes of urbanization, migration, occupational mobility, socioeconomic change, and income generation in the Amazon are diverse and complex. Patterns, such as they are, vary widely between populist and corporatist frontiers. Even within these distinctive social spaces of Amazonia local diversity prevails.

Between 1960 and 1990 at least 6 million Brazilians moved to the Amazon according to official census data. The two most important streams of interregional migration have flowed into Rondônia and southern Pará, the areas of our comparative field research. We offer five general conclusions about these migrations. First, urban-urban migration has been the predominant type of movement to both Rondônia and Pará; 42.3 percent and 52.1 percent of our survey respondents, respectively, originated in urban areas outside the Amazon. In this sense the modern Amazon overall "is born urbanized" (Becker 1990) and the growth of urban areas is "generalized" across the region (Sawyer 1987). Contemporary images of rainforest cities as growing urban agglomerations of the dispossessed rural poor were not substantiated by our research findings, which instead point to more nuanced and complex patterns of frontier expansion and labor migration, such as the circular urban-rural-urban pattern noticed in Rondônia.

Second, once in the Amazon, urban workers continue (even accelerate) their household movements locally, shifting frequently between permanent urban dwellings, mining camps, and farmsteads, or between multiple urban dwellings within the same *município*. In southern Pará the
workforce is spatially so fluid in places like Ourilândia that the docu-
mentation of local migration patterns defies classification through standard household survey instruments. In contrast to many national studies
that depict the dominant migratory pattern as a progressive stepwise
migration from lower-order to higher-order cites, we observe the
reverse in the Rondônia analysis: an inverted stepwise migration pattern
down the urban hierarchy, from larger to smaller settlements, indicating
a tendency, perhaps only ephemeral, toward population decentralization.
This pattern, however, was not found in the southern Pará study, highlighting the heterogeneity of migration in the region and the perils of
reaching for generalities about this complex process.

A third conclusion concerns the relationships between migration,
occupational mobility, and change in socioeconomic position. We found
that interregional migration to both the Rondônia and the southern Pará
study areas entailed high rates of occupational change; between two-
thirds and three-quarters of all informants reported changing occupations. However, significant differences between the populist and contested corporatist frontiers emerged from the details of our survey data.
In Rondônia's populist frontier we found, as expected, that the labor
force is more homogeneous and formalized (in terms of wage employment) in the largest node in the urban system (Rolim de Moura) and
more heterogeneous and informal in the lower-order and more remote
settlements. Opportunities for petty ("primitive") capital accumulation
appear to be higher on the geographic fringes of the populist frontier,
while the chances of decapitalization are greater in the core centers.

By contrast, in the Xingu settlement corridor of southern Pará wage
relations of production were found to be spatially inverted; a greater proportion of the local workforce became engaged in formal wage employment following interregional relocation to the more remote settlement
(Tucumã) than to the core settlement (Xinguara) in the Xingu urban system. In other respects the distribution of the urban labor force by socioeconomic position was either convex or concave, in which the second
(middle) settlement (Ourilândia) displayed either higher or lower values
than did either of the endpoints (first and third settlements) of the Xingu
corridor segment of our study. An even gradient in the spatial distribution of socioeconomic characteristics was notably absent, illustrating the

differentially articulated nature of different nodes of the same settlement system in the contested corporatist frontier.

Fifth, we found diverse spatial patterns of income distribution and financial activity (borrowings, savings, and remittances). Oddly, in both the Rondônia and the southern Pará study areas, households in the middle node of the urban systems enjoyed significantly higher incomes than in the other settlements. We hypothesize that this apparent pattern of spatial convexity of income distribution is coincidental and reflects the distinctive local historical and economic characteristics of each middle settlement rather than a universal or generalizable tendency that might be applicable to the Amazon region as a whole. Formal (bank) savings rates are low in both study areas but are slightly lower in southern Pará, perhaps reflecting the parallel economy of the gold-mining fronts of the contested corporatist frontier. Similarly, informal (nonbank) borrowing rates were found to be higher among households of the Xingu corridor than among those of the Rolim de Moura–Alto Alegre corridor. In the latter, however, an increasing trend toward informal credit (mainly between shopkeeper and patron) was observed down the corridor; rates of informal borrowing increased with distance from the core. Finally, while there exists substantial research literature attesting to the economic importance of income remittances of migrants worldwide, our surveys found that less than 20 percent of households in all six Amazon study sites claimed to have remitted any income to family members or friends outside the state, and less than 10 percent acknowledged receiving financial aid from others outside the state. If a surplus extraction flows from periphery to center, then it is unlikely to flow through either informal remittances or formal bank savings, as we shall examine in chapter 9.

The wide array of spatial patterns (and idiosyncrasies therein) of labor migration, socioeconomic mobility, and income distribution reveal the Amazon to be a rich and colorful medley of social and institutional forces interacting simultaneously and sequentially in a mosaic of historically associated common spaces. Our observations of these characteristics drive us back to the central lesson of our study: We eschew claims that any single master principle of spatial structuration can adequately explain the systemic diversity and local anomalies of frontier urbanization observed within and between the sociospatial constructs we have denoted as corporatist and populist frontiers. Rather, a pluralism of concepts congregates, perhaps awkwardly at first, to simultaneously explain different aspects of these phenomena. Global, regional, and local influences interact, giving different places in the frontier different characteristics and defying universal statements. We suppose that critics may charge that we obscure the general patterns in the riot of particulars, that we fail to see the forest for the trees. However, if the variations we observe are understood as belonging within acceptable thresholds allowed by one universal theory or another, then would we not be compelled to question the universality of such a theory? Indeed, our concern is directed to the epistemic level, for we are not arguing the validity of one specific conceptual framework over another; rather, we call into question the entire class of theorizing that presupposes the macro-structuration of space and therefore of human social behavior around a single master principle.

Notes

1. For a sample of the demographic literature on Brazilian internal migration patterns, see: CEM (Centro de Estudos Migratórios 1986); Costa and Graham et al. (1971); IBGE (1979a); Merrick and Graham (1979); Souza (1980); and Wood and Carvalho (1988).

2. Afrânio Raul Garcia Jr.'s recent book *O Sul: Caminho do Roçado* (1990) provides an insightful account of the great migration from the Northeast to the Southeast, a well-researched analytical ethnography of peasant social reproduction strategies.

3. National urbanization rates grew from 44.7 percent in 1960 to 75.0 percent in 1991 (table 1.3).

4. Wood and Carvalho define the *net migration rate* (for 1970) as (net migrants 1960–1970/rural population 1960) x 100 (Wood and Carvalho 1988:217).

5. For example, in the state of Rio de Janeiro 75.7 percent of total migration (including migration to the state capital) was between urban areas (IBGE 1979a).

6. Interregional rural-rural migration to the Northeast, for example, represented 60.1% of total migrations and 55.5% in Paraná (IBGE 1977, vols. 1–5). The pace of interregional migration to Amazonia accelerated after 1970 because of the Transamazon colonization program and the opening of Rondônia.

7. In view of the rapid and dramatic transformation of the Amazon due to the influx of migrants, it is rather surprising that so few recent comprehensive analyses about the patterns of internal migration both to the Amazon region and within it have been published.

8. Testing hypotheses of intraregional movements proved to be unreliable in southern Pará due to the common frequency and irregularity of local migration patterns as illustrated in an ethnographic sketch of one particular individual. In early July of 1990 we met José de Conceição in Cuca, during a tour of the gold-mining areas south of Tucumã. José, then forty-four years old, was born near Codó, Maranhão. His current wife and two children lived in Tucuruí, where he planned to return soon, because Cuca did not prove to be as profitable as José had hoped. Already José could count some thirteen moves he had made in the eastern Amazon, not including previous movements in the state of Maranhão. José had lived in such frontier towns as Marabá, Itupiranga, Santarém, Tucuruí, Itaituba, Xinguara, and many other towns. Stories like those of José are commonplace in southern Pará and present significant difficulties for migration research.

9. In our Rondônia samples the average numbers of intraregional moves per household were: 2.27 (Rolim de Moura), 2.08 (Santa Luzia), and 2.01 (Alto Alegre). By examining the last two moves, therefore, we captured virtually 100 percent of all intraregional household migration within the Amazon.

10. Abers (1992) found a similar pattern of predominant downward urban-urban migration in the case of Boa Vista, Roraima, a gold and agricultural settlement frontier in the north-

ern Amazon, where 40 percent of the migrant population surveyed had previously resided in a large metropolitan center outside the region.

11. A typology of thirty occupational classes was developed from the survey data, in which over sixty different occupations were grouped. A change in occupation indicates only a change from one class of work to another, not necessarily a qualitative change in socioeconomic position relative to the ownership of capital or the means of production.

12. More specifically, these categories are autonomous worker (no wage relation, no capital ownership, e.g., rural day worker, peddler in the informal sector, etc.), sharecropper/tenant farmer (rural worker paid on the basis of production with access to land limited to household consumption), temporary employee, permanent employee (both defined by formal wage relations under legal employment contract), unremunerated family labor, and proprietor (an owner of the means of production, e.g., urban businessowner, farm owner-operator, etc.).

13. We recognize that these categories of change in socioeconomic position are idealized types and some of the terms used to describe specific shifts (e.g., proletarianization and pauperization) are loaded with specific ideological connotations beyond those intended in our usage. For instance, the shift out of formal wage relations of production to informal work (i.e., pauperization) does not always indicate impoverishment but may also reflect household life-cycle changes or other accumulation strategies that are consistent with a shift toward petty capitalization. Readers should regard this typology, therefore, as a rough heuristic of socioeconomic mobility.

14. Rates of socioeconomic mobility refer to the number of households in which the head of household experienced some change in socioeconomic position as defined above as a result of new activities undertaken following a migration event divided by the total number of households in the sample.

15. Survey respondents were asked to indicate their total household income for the preceding month (July 1990). We acknowledge two sources of potential bias from this approach. First, we found that the common respondent tendency to intentionally distort (both understate and overstate) income was more likely in interview situations where a male head of household was the sole respondent. More accurate estimates were reported, we believe, in the majority of cases where both female and male household heads (where applicable) participated as corespondents in the interviews. Second, to minimize effects of memory on internal validity, we limited our question to income earned from all sources in just the previous month (June 1990). We would expect household income to vary somewhat from month to month in relation to the agricultural calendar and between rainy and dry seasons. These are estimates of total household income, which typically includes the contributions of more than one income-earning worker, often working at more than one job. We recognize that GNP per capita is a poor indicator of personal or household income distribution in Brazil.

16. Several different mechanisms of informal credit are customary. We asked the question of whether or not anyone living in the household had an informal credit account in a local business. This, of course, merely indicates the number of households in the sample that periodically are indebted to local businesses, not the severity of indebtedness.

17. There is, of course, a bias in the question to answer in the negative, as respondents normally may not wish to acknowledge receipt of financial assistance, a bias that would understate the magnitude of informal interstate income transfers.

18. It is important to point out that just prior to the 1990 field surveys, then-President Fernando Collar de Melo implemented an eighteen-month freeze on savings account withdrawals as part of a national inflation management program. Reported savings rates may have been distorted by this policy. Reported rates may overstate the natural tendency of Amazonian urban residents to save by including households with sequestered accounts or understate that tendency by reflecting depositor distrust of the government.

9

Frontier Urbanization in the Global Periphery: Regional Disarticulation and Surplus Extraction

❦

The emergence of diverse urban systems in Brazilian Amazonia raises questions not only about the spatial organization of the frontier but also about its functional linkages to the global economy. In previous chapters the case studies of southern Pará and Rondônia have shown significant intraregional differences in the dynamics of urban system formation. Urbanization in Amazonia displays little of the symmetry presumed by central place theory and little of the hierarchy envisaged in world systems models (e.g., Armstrong and McGee 1985). We have argued that this variability in pattern can be traced in large part to diverse local histories, different institutional orientations of specific frontier zones, and heterogeneity in natural distributions of resource endowments. The multiple forces shaping these diverse urban formations suggest a pluralistic theory of frontier urbanization. This suggestion brings us into some contention with leading structuralist perspectives, especially with world systems theory, based on the notion of "surplus extraction" (Armstrong and McGee 1985; Bunker 1985).[1] This chapter takes up a central question of our study: what are the functions of rainforest cities in the global economy? Are they "theaters of accumulation" facilitating regional surplus extraction and global capital accumulation? Or are they spark plugs of autonomous economic development? We argue that frontier urbanization in Amazonia displays elements of both tendencies but cannot be fully apprehended by theories predicated on the master principles implicit in these questions. Our analysis starts with an examination of the surplus extraction argument in a world systems framework and draws upon both surveys of urban and rural households and key informants associated with various leading economic sectors of frontier settlements.

RONDÔNIA: DISARTICULATED URBANIZATION AND SURPLUS EXTRACTION IN AN AGRARIAN POPULIST FRONTIER

At least two conditions must be met to establish empirically the surplus extraction argument. First, a mechanism must exist that is capable of appropriating a producer surplus either as a function of some mode of production or at the level of exchange. This latter form of appropriation might arise from "unequal exchange" between interregional or international trading partners in which the value added to products traded is quantitatively different (Emmanuel 1972). For example, if the periphery sells a product with low value added to buy one with higher value added from the center, producer surplus is said to be appropriated by the latter.[2] Surplus extraction may also occur through oligopsonistic marketing practices by local merchant capital, taxes, profit repatriation (Armstrong and McGee 1985). Or, as in the Marxian analysis, it may arise from wage relations of production between various factions of capital and labor. Second, a system must exist that is capable of channeling this surplus, once appropriated, toward the highest profit-earning sector of the national economy. In both regulated and open market economies the banking system serves this purpose. "Financial capital performs a lubricating function in promoting the activities of commercial and production capital. It tends to enhance the power and importance of the largest cities and central subsystems within the urban hierarchy" (Armstrong and McGee 1985:58). Adopting a world systems framework is still more demanding and imposes a third condition: establishing the existence of a process by which local and national capital is transferred to the global sphere of circulation. Transnational corporations headquartered in world cities orchestrate this transfer through various processes, e.g., profit repatriation, external debt payments, and capital flight. As the dominant technological and financial control centers in the global economy, world cities also set global cultural standards of con-

sumption engendering materialist preferences that work to displace traditional (noncapitalist) production systems and values. In this section we
examine the empirical evidence for surplus extraction in Rondônia's
agrarian frontier by examining several key sectors in the local economy
in relation to these conditions.

The Grain Processing Sector: Food for Whom?

In his study of disarticulated accumulation under capitalism, Alain de
Janvry blends salient features of articulationist, capitalist penetration
and world systems perspectives to explain the agrarian crisis in Latin
America. In this hybridized formulation, dependent countries in the
transition to capitalism sacrifice food self-sufficiency to develop comparative advantages in export-oriented crop production that generates
foreign exchange (de Janvry 1981:152–175). Brazil's experience of
agricultural modernization provides a salient test of de Janvry's thesis.
The development of large-scale, mechanized soybean production for
export in the Brazilian South and Southeast (i.e., the center) during the
period from 1967 to 1976 led to the displacement of millions of rural
workers (see chapter 6). The production of food crops, notably black
beans, suffered domestic supply shortfalls, periodically compelling
Brazil to import staple foods. According to this thesis, by opening
Rondônia to settlement by displaced small-scale farmers the Brazilian
state could overcome the transitional crisis of capitalism associated with
agricultural modernization and solve two strategic problems. First, creating a populist frontier in Rondônia as a labor safety valve would obvi-
ate the social costs of urban alternatives for Brazil's displaced rural
poor. Second, by shifting domestic food crop production to the frontier,
more valuable land in the center could be released for conversion to
high-value export crop production.[3] Rural credits to small farmers on
the frontier, anticipated in the initial settlement plans for Rondônia,
would subsidize the higher costs of food crop production on the periphery, making it competitive with lower-cost food imports.[4] At the same
time, from the perspective of the state and national capital it would be
more efficient to produce low value-added food crops for national market consumption in Rondônia and higher value-added cash crops for
export in the South and Southeast. From the perspective of (household
welfare/survival maximizing) owners of small rural properties, it was

logical to trade their smallholdings in the South for considerably larger properties in the Amazon. Diverse rationalities, those of the state, the landless peasantry, and the family farming classes converged on the populist agrarian frontier solution.

However, the movement of a substantial segment of the agrarian (i.e., food producing) sector to the periphery imposed new marketing constraints on small producers arising from the sheer distance to national markets and their vulnerability to merchant capital and oligopsonistic marketing devices. Agricultural (cereal) processing companies *(cerealistas),* in conjunction with marketing middlemen, emerged in local market towns like Rolim de Moura as the principal agents of producer surplus extraction in the agrarian frontier.

In 1989 farmers in the Rolim de Moura *município* produced 43,141 metric tons of commercial food crops, mainly beans, rice, coffee, and corn (Rolim de Moura 1989). Most of this output was sold by farmers directly or through professional middlemen to twenty-seven different grain processing companies located in the town of Rolim de Moura or to the government's parastatal Company for Production Financing (CFP). These companies typically are private, legally incorporated but also frequently informal enterprises. Most are not subsidiaries of national food processing industries. The *cerealistas* minimally weigh and repackage, but often dry and dehusk commercial cereal crops (e.g., coffee, rice, beans), reselling this produce to truckers, wholesalers, and larger food processing companies under a wide range of contractual agreements.

Approximately 70 percent of Rolim de Moura's food output was destined for consumers in the South and Southeast, and 20 percent was shipped to supply Manaus and Acre in the Amazon, and 5 percent was distributed to the Northeast.[5] Only one company resold any locally produced food (5 percent of its total sales volume) to the local market in Rolim de Moura. None of Rolim de Moura's agricultural output channeled through these firms was destined for markets elsewhere in the state, despite its rapidly growing urban consumer population. The survey data, although taken from a small sample, tend to indicate a predominant national (i.e., southeastern) consumer market orientation of agricultural production on the frontier. An extraregional market orientation of frontier production is intuitive and does not necessarily indicate unequal exchange or surplus extraction benefiting global capital over local accumulation. But there is additional evidence to consider.

The Food Retail Sector

Under ideal conditions of autonomous regional development the agrarian frontier would become self-sufficient in food production, exporting only its surplus. However, the surplus extraction thesis situated in a world systems context argues that food production on the frontier supports the process of accumulation at the national and global levels, in part through an unequal exchange of low-value goods from the periphery for high-value goods from the center. We turn to the food retail sector to explore this proposition.

One of the few retail groups in Rolim de Moura showing considerable dynamism during the stagnant years following the mahogany boom was the national chain supermarket, growing from four to eleven establishments between 1984 and 1990 (table 9.1). Most young frontier towns depend upon a myriad of small-scale, part-time, family-owned and -operated corner grocers and dry-goods retailers *(comércio, mercenaria)*.[6] The number of such corner grocers, not surprisingly, declined from thirty-four to nineteen during this period and further shrank to twelve in 1992 (retail sector inventory, Rolim de Moura 1992).

Paradoxically, in both labor-based grocery and capital-based supermarket subsectors the majority of food products, including those derived from crops produced by Rolim de Moura's own farmers, were acquired from national suppliers outside of Rondônia (table 9.2).[7]

A few differences between large, corporate chain supermarkets and the family corner grocers are noteworthy. First, not surprisingly, the corner grocers, not being affiliated with large national supermarket chains, rely more heavily on local producers for stock. Thirty-one percent of all stock retailed by the corner grocers was directly supplied by farmers in the municipality, versus 15 percent for supermarkets. Local producers provided most of the fresh foods and basic staples (e.g., rice, manioc flour, and beans) that enter the local consumer market through corner grocers. However, corner grocers also depended on larger capitalist firms for most packaged goods. Nearly half of the corner grocers also said they patronized wholesalers located in Jí-Paraná, Rondônia's second largest city and main commercial center, representing 16.7 percent of all corner grocer supplies. Nearly half also depended on the seven local supermarkets for several merchandise items. Fifty-two percent of their stock is provided by national market suppliers (table 9.2B).

Second, in contrast to that of the corner grocers, 70 percent of Rolim de Moura's supermarket merchandise is provided by suppliers located in

Table 9.1 Principal businesses in Rolim de Moura (1984, 1990), Santa Luzia, and Alto Alegre (1990)

BUSINESSES	ROLIM DE MOURA	ROLIM DE MOURA	SANTA LUZIA	ALTO ALEGRE
	1984	1990	1990	1990
URBAN SERVICES	*145*	*158*		
Banks	7	5	X	
Mechanical repair[a]	23	40		
Hotels and restaurants	19	8	X	X
Bars and snack bars	38	59	X	X
Entertainment[b]	3	4		
Skilled services[c]	30	18	X	
Professional services[d]	11	12	X	
Other	14	12	X	
URBAN COMMERCIAL EST.	*180*	*222*		
Supermarkets	4	11	X	
Dry goods	34	19	X	X
Fresh fruit/vegetables	29	2	X	
Bakery	3	0		
Butcher	13	10	X	
Farm equipment	5	12	X	
Auto electronics	2	16		
Auto tires	4	10		
Clothing, boutiques	10	41	X	X
Pharmacies	17	22	X	
Gas stations	2	4		
Furniture	1	13	X	
Construction materials	7	7		
Auto parts	13	17	X	
Artesanal wares	2	0		
Other	34	38		
Total Businesses	325	380	—	—
Total Business Functions[e]	58	62	19	5

Note: X indicates presence of at least one establishment.

[a]Auto/mechanical includes businesses specializing in the repair of autos, bicycles, chainsaws, household appliances (radio/tv, etc.), and blacksmiths, locksmiths.

[b]Entertainment includes cinemas, discos, clubs.

[c]Skilled services include glaziery, drapery repair, barber, clothing repair, auto driving school, graphics, carpenter, photographic finishing, despachante, print shop.

[d]Professional services include veterinary clinic/pharmacy, public accountancy, dental clinic, funeral parlor, real estate office, laboratory, trading company, private offices.

[e]Total business functions refers to all different categories of business establishments and includes categories enumerated under "Other" and activities aggregated into larger categories. Hence, columns may not add up to totals.

Table 9.2 Origin of goods sold in Rolim de Moura by scale of supplier market

A. SUPERMARKETS

Product		Supplier market scale	
	LOCAL	STATE	NATIONAL
Coffee		xxxxx	x
Rice	xxx	x	xxx
Beans	xxx		xxx
Manioc flour	x	x	xxxx
Corn flour	x	x	xxxx
Popcorn		x	xxxx
Uncooked cornflakes		xxxxx	
Polenta			xxxxx
Potatoes			xxxxx
Soap		x	xxxxx
Sugar		xxx	xxx
Vegetable oil		xxxxx	
Powdered milk		xxxxx	
Macaroni		xxxxx	
Percent of stock	15.1	14.7	70.0

B. CORNER GROCERY STORES

Product		Supplier market scale	
	LOCAL	STATE	NATIONAL
Coffee	xx	xxxx	
Rice	xxxx		x
Beans	xxxxx		
Manioc flour	xxxxx		x
Corn flour	x	xx	xxx
Popcorn			
Uncooked cornflakes			
Polenta			
Potatoes			xxxxx
Soap		xx	xxxx
Sugar		xx	xxxx
Vegetable oil			xxxxx
Powdered milk			xxxxx
Macaroni			xxxxx
Percent of stock	31.0	16.7	52.3

Note:
x = 0–20 percent of product stock
xx = 21–40 percent
xxx = 41–60 percent
xxxx = 61–80 percent
xxxxx = 81–100 percent
Sample size: twelve corner grocers and six supermarkets

São Paulo and other national centers of food processing and whole-sale (table 9.2A). Even though an estimated 81.5 percent of farmers in Rolim de Moura produced rice (rural producer survey 1990), accounting for 31 percent of Rolim de Moura's total marketed grain output in 1989 (Rolim de Moura 1989), 40 percent of the rice retailed by the large supermarkets in 1989 came from the South and Southeast.[8] Similarly, while 59 percent of Rolim de Moura's farmers plant corn, the marketed output of which accounted for 35 percent of total local food grain production marketed, between 75 percent of corn flour and popcorn products and 100 percent of corn flakes and polenta originated from food processors in São Paulo, Paraná, and Rio Grande do Sul. Only one of the six super-markets surveyed carried locally produced manioc flour even though 55 percent of all local farmers produced and sold raw manioc root. In summary, most of the basic foodstuffs offered to consumers in agrarian frontier urban supermarkets are imported from the industrial core of the country.

Neither Santa Luzia nor Alto Alegre had any supermarkets in 1990. Three corner grocers in Santa Luzia were interviewed. Patterns of supply were different there than in the larger Rolim de Moura. Significantly, between 70 percent and 90 percent of merchandise sold in these shops was supplied by wholesalers in Jí-Paraná, Porto Velho, and Pimenta Bueno. Between 10 percent and 30 percent of stock was regularly acquired from the supermarkets in Rolim de Moura, or from wholesalers outside the state (Amazonas, Paraná, and Minas Gerais). Four shop-owners were interviewed in Alto Alegre, two of whom indicated that 100 percent of their stock originated from wholesalers in Jí-Paraná. In the third shop, 100 percent of the stock was bought from one supermarket in Rolim de Moura. Only the fourth shop had a diversified supplier network, with 30 percent of its stock acquired from one supermarket in Rolim de Moura and 70 percent from wholesalers in Jí-Paraná, Cacoal, and from outside the state. One shopkeeper in Alto Alegre remarked during our 1990 visit, "we are totally linked to Rolim de Moura. I buy in Rolim the goods that I sell here [in Alto Alegre], and my customers [all farmers] all sell their crops to *cerealistas* in Rolim." Unequal exchange?

The supermarkets of Rolim de Moura are clearly more effectively integrated into national food wholesale networks. Indeed, some supermarket chains (Pão de Açucar, SA) are not only vertically integrated into modern agro-industrial business but are also subsidiaries of Brazilian-based transnational corporations with diverse global holdings. In contrast, the corner grocers depend to a larger degree on family labor, local producers, and statewide wholesalers for the more limited range of products they offer. Food retailers in the outlying settlements depend on supermarkets in primary market towns like Rolim de Moura to supply much of their stock.

The local distribution of food products reflects regional patterns of disarticulated urbanization. Most of those retailers in the interior who rely upon state-based distributors cited the secondary city of Jí-Paraná, not the state's capital city, Porto Velho, as the wholesale center for the state, further highlighting the marginal economic importance of the state's metropolitan center in its the agrarian economy.

In summary, national supermarket chains have succeeded over time in penetrating the food retail sector of the populist agrarian frontier, displacing family-owned corner grocers (table 9.1). The supply linkages found within the food retail sector in the Rolim de Moura–Alto Alegre settlement corridor suggest that the provisioning of the frontier depends largely upon national chain suppliers and wholesalers in the South and Southeast. Supermarkets in the dominant center of this agrarian frontier (Rolim de Moura), in turn, supply a substantial proportion of the trade goods tendered by retail shops in the lower-order settlements of the corridor (Santa Luzia and Alto Alegre). The agricultural frontier presumably could be, but is not, self-sufficient in food production. These patterns would be consistent with the surplus extraction and unequal exchange arguments and de Janvry's concept of disarticulated development. Urban centers in the frontier are the arenas for this arbitrage.

Unequal Exchange in the Agricultural Sector

Two-thirds of the farmers in the Rolim de Moura–Alto Alegre corridor reported no positive net income from farming in 1989.[9] Three reasons help to explain the income squeeze on producers during the late 1980s. First, in 1989 interstate freight costs (from Rondônia to the Southeast) exceeded local producer prices, making commercialization of food crops from Rondônia uneconomical for farmers.[10] Second, while freight charges rose dramatically relative to producer prices, government agricultural supports (the crop minimum price purchasing program administered by Brazil's Production Financing Company) declined. In 1989, ostensibly under fiscal pressure from abroad, the CFP reduced its purchases of rice, beans, and corn in Rondônia to 18,000 metric tons, down from 107,000 metric tons the year before (World Bank 1991:54). In other words, the total amount of government grain purchases from the entire state of Rondônia in 1989 was less than half of the producer grain output from the Rolim de Moura *município* alone (43,000 tons). Third, faced with astronomical freight rate hikes and reduced government price supports, most farmers were forced to commercialize their output through private merchants. Reports of fraudulent and collusive marketing practices have been legion in the state for years (Browder 1986).

During the 1980s, colonists reported that private *cerealistas* and the government's quasi-public crop marketing agency (then denominated CIBRAZEM, the institutional predecessor to CFP) routinely refused to buy crops directly from farmers at harvest except at highly discounted prices, well below the minimum prices guaranteed by the government. Such refusals typically were based on the contrivance that the quality of the crops did not meet accepted standards (e.g, minimum moisture content, aroma, color, genetic purity, etc.). Farmers turned in frustration to petty merchants—anyone with a pickup truck able to haul crops to town—many of whom reportedly worked in collusion with the *cerealistas* and also paid producers less than the market price. In theory, government grain merchants and their middlemen accomplices could then split the difference between the actual price paid to producers and the minimum guaranteed price, but we found no irrefutable evidence of such collusion. By 1992 the producer price of coffee was so low that many farmers refused to harvest their crop, and others plowed under their coffee stands to plant pasture in their place (personal observations).

While producers endured severe constraints on income, *cerealistas* reported substantial profits of between 100 percent and 300 percent from rice, coffee, and beans in 1989 (1990 *cerealista* survey, Rolim de Moura). Was this global surplus extraction? Clearly structural factors affecting the frontier economy account for much of the income squeeze on producers during this period. World market fluctuations in commodity prices, inflation in the energy sector resulting in rate hikes in the trucking industry, austerity-induced reductions in the CFP's crop purchasing budget combined to increase pressure on the agrarian sector. However, it is questionable that these forces, each acting independently from the other, could be contrived as a coherent strategy of transnational capital to extract a producer surplus from the global periphery. Yet, the evidence of monopsonistic trading practices found even among CFP workers in collusion with petty merchants and grain processing firms suggests that some producer surplus was appropriated in the frontier at the level of exchange. Yet, such surplus appropriation by petty merchants in the periphery creates a bottleneck for capital accumulation in the world cities, running counter to the logic of national and global capitals, which presumably seek to maintain a steady flow of surplus out of the periphery. Moreover, surplus appropriation by petty factions of merchant capital in the frontier does little to legitimize the authority of the government at the national level since it siphons off surplus in the frontier that might otherwise be appropriated by politically more influential factions of capital at the national level. Producer surplus extraction by local petty merchants in the periphery itself would suggest local disarticulation from the grand strategy of global surplus extraction by transnational capital in alliance with the state, which is a conceptual cornerstone of world systems theory.

Financial Sector

The second condition for surplus extraction is the existence of a mechanism for transferring wealth, once extracted from the frontier, to other investment options elsewhere. The banking system, which we now examine, plays a potentially crucial role in channeling the regional surplus to the national and global levels of circulation, especially given the centralized structure of Brazil's banking system.

The 1980s brought economic havoc to Brazil in the form of recession, growing debt crisis, and a rapid succession of federal economic management plans. These were not exactly the most favorable conditions for expanding the financial sector in the Amazon. Yet the national banking system expanded dramatically in the Amazon region precisely during this period. According to the Brazilian Federation of Banks (Federação Brasileiro de Bancos), the number of private and state bank branches in the Legal Amazon doubled from roughly 200 in 1980 to 400 in 1990. But by the end of the decade the banking boom in the Amazon subsided. In a June 1990 interview, the regional representative of the Federation in Belém remarked that the big expansion of the 1980s was over, and banks in the Amazon were retrenching. "Private banks are taking money out of Amazonia and none of the branches will talk" (personal communication, June 1990). The force behind the expansion of the financial sector during the 1980s was more than an idle curiosity. The structure of the central banking system in Brazil serves to concentrate capital within the national financial system. The financial sector of Rolim de Moura illustrates how this process works locally.

In 1984, buffeted by the frenzy of the mahogany boom, Rolim de Moura received the services of eight different banks, all but two of which were branch offices of private national banks. In 1985, following the crash of the mahogany trade, three private banks closed their Rolim de Moura branches.[11] Incredibly, one of these banks recorded its second highest cash turnover rate among its network of thirty-two branch banks in the Amazon at its Rolim de Moura branch office.[12]

In a confidential interview, the manager of one of these private banks explained how the financial sector is structured to transfer wealth from the Amazon to Brazil's Central Bank. Brazil's private financial sector is closely linked to the national federal reserve bank (Banco Central do Brasil) through its compulsory savings policy. Deposits made in popular "overnight" accounts and other savings instruments are immediately transferred to the bank's state office in Porto Velho and from there to the bank's corporate headquarters in Curitiba, in the southern state of Paraná. In Curitiba, deposits from across the country are transferred to the Central Bank. At the time of our inquiry (1990) the Central Bank reportedly held 38 percent of these deposits in reserve in Brasília and returned 62 percent to the private banks for use in lending and investment. Only a fraction of the deposits made in the Rolim de Moura branch office returned as fundable capital to local districts. Private banks, however, captured a larger volume of federal reserve capital by participating in federal credit programs, a subject to which we now turn.

The most important profit opportunity for private banks in the 1980s was through federal credit programs for everything from annual crop production to export loans (see Browder [1988] for a partial review). The bank manager we interviewed estimated that 55 percent of his bank's current liabilities at the time were loans supported by federal credit programs. One illustrious example was the federal credit program established in the early 1980s to promote exports of industrialized products, a program that created a unique process of surplus extraction in the industrial wood sector of the frontier (see following section).

Approximately 80 percent of the unrestricted resources returned to our informant's bank in Rolim de Moura was lent to local urban businesses (stores, cerealistas, lumber companies), while 20 percent was lent to small farmers. All of these discretionary loans were short-term ones (i.e., twenty to sixty days). The bank manager commented that "not one house has ever been financed by [his] bank in Rolim de Moura."

The expansion of the financial sector into the frontier can be explained mainly by the surfeit of federal subsidies targeted to promote various activities in the Amazon. Private banks became important brokers for federal credit programs and clearly benefited from this role. The contraction of the financial sector witnessed in frontier towns like Rolim de Moura between 1984 and 1990 undoubtedly reflected the shrinking base of institutional rents provided to producers by the government in the form of subsidized production credits administered by the private banking sector.

It would be difficult to conclude that finance capital is successfully extracting the producer surplus from the Amazon's populist frontier. In our rural household surveys we found that a minority (about 27 percent) of farmers in Rondônia have formal savings accounts with local banks.[13] Only 23.5 percent of urban residents in the Rolim de Moura–Alto Alegre corridor hold formal bank accounts. The banking sector would not appear to be closely articulated to local processes of surplus production. Rather, federal credits and institutional rents, which represented more than half of the volume of business generated by banks in Rolim de Moura, provided the vital profit center supporting the expansion of the banking sector in Amazonia.

This evidence does not easily support a world systems explanation for the expansion of the financial sector. The highly centralized banking system is not well suited to support capital accumulation outside the national economy. Indeed, Brazil's national banking system explicitly functions to regulate the outflow of capital from the national economy. Although private banks still manage most of the financial resources in Brazil's economy, the Brazilian Central Bank is clearly the single most powerful institutional actor in determining the movements of Brazilian capital within the national economy.

Our argument thus far suggests that surplus extraction occurs only in contingent situations and, therefore, is not inherent or systemic to economic activity in the Amazon frontier generally. Moreover, our evidence does not lend much support to world systems perspectives on surplus extraction either. Surplus extraction, such as it exists, seems largely confined to diffuse factions of local capital. However, additional evidence from the industrial woods sector of the populist frontier suggests some caution in any hastily conceived revisionism of surplus extraction theory.

Unequal Exchange and Surplus Extraction in the Industrial Woods Sector

Given the importance of the Central Bank and its federal credit programs in directing capital throughout the Brazilian economy, it is instructive to examine how one federal credit program enabled a class of professional merchants to restructure relations of production so as to extract producer surplus in the Amazon's industrial wood sector during the early 1980s (Browder 1986, 1987, 1989a). This export promotion program, called CACEX Resolution 643 provided legally registered trading companies with comparatively advantageous loans.[14] Trading companies found in mahogany lumber (considered a "semi-industrialized" product) a small but lucrative export market with relatively low, short-term production costs. A short-term money market quickly emerged within the private banking sector to capture loan monies subsidized through Resolution 643 from exporters, and this practice constituted a novel form of social surplus extraction by financial capital.[15]

Three trading companies, each with financing through Resolution 643, moved into Rolim de Moura in the early 1980s to restructure local industrial wood production around a new export staple, mahogany. With this subsidized capital, averaging U.S.$1.1 million per firm, these exporters quickly scrambled to gain physical control of large tracts of open-access federal forestland containing mahogany timber. Whereas several independent lumber mills once contracted individually with loggers and colonists for timber, after 1980 the timberlands around Rolim de Moura became consolidated under the control of this oligopoly of federally subsidized trading companies. Small mills without global trading links that desired to participate in the lucrative mahogany export trade were typically compelled to engage in subcontractual relationships with one of the three trading companies dominating the mahogany timbershed. Three types of relationship emerged: advance purchase financing, production contracting ("jobbing"), and corporate acquisition or merger.

The most common relationship, the advance purchase agreement, was also the most predatory upon producer capital. In brief, traders advanced producers (lumber mills) an amount of cash (always drawn from Resolution 643 export loan monies) equivalent to the price of mixed-grade, export-quality mahogany lumber on the day the loan was executed. Financing terms were typically thirty to sixty days (well within the 180-day financing period for the Resolution 643 loan). Producers were usually obligated to purchase mahogany logs from trading company log suppliers. The export promotion programs created a strong incentive for loggers to "cream" the forest for mahogany (which has a typical distribution of one mature tree per hectare) as quickly as possible (see chapter 6). As mahogany became increasingly scarce and loggers cut deeper into the timbershed (up to 100 kilometers from Rolim de Moura's mills) to extract this export timber, log deliveries to contracted mills were increasingly delayed beyond the term of the advance purchase contract, and log prices skyrocketed. With mahogany roundwood prices increasing at an annual rate 2.3 times greater than the rate of interest on Brazilian treasury bonds (a surrogate indicator for inflation), loggers demanded price adjustments from producers, often in amounts exceeding the value of the original advance purchase contract for the finished lumber product. To avoid property condemnation that might arise from unfulfilled contracts, lumber producers were compelled to buy raw timber at prices that exceeded the value of the lumber they had yet to saw.

In 1980 none of the ten lumber mills operating in Rolim de Moura was engaged in any production contract for exporters. By 1984 the picture had dramatically changed, and 74 percent of the twenty-three mills operating in the town accepted advance purchase financing. During the five-year period from 1980 to 1984 during which Resolution 643 remained in effect, the average annual trading company profit rate from mahogany exports was 49.4 percent. The average profit rate of mahogany lumber producers was 3.4 percent, roughly equivalent to the estimated average rate of the producer's fixed capital depreciation in Rolim de Moura's lumber industry. From a cost-of-production perspective, the Brazilian mahogany trade was a losing proposition for the majority of Rolim de Moura's lumber mills, resulting in the appropriation of producer surplus and fixed capital investment by a Brazilian mercantile elite (Browder 1987).

Though our study suggests some evidence of surplus extraction, such processes do not occur in a contextual vacuum. Biophysical factors that explain the particular distribution of the *Swietenia* species combined with historically specific national policies (i.e., export promotion) promoted the rapid liquidation of this species. The growth in the number of local lumber mills, in part arising from the depletion of *Araucária* (Paraná pine) in the South, also constitutes an essential feature of any interpretation of surplus extraction during the mahogany boom on the frontier. Banks supplying trading companies with subsidized government financing and short-term, high-yield deposit products would not have located there either had it not been for the financial opportunities provided by the Central Bank's export promotion program. Political leaders in Rolim de Moura also played a role by providing free urban land in the town's so-called industrial sector to attract lumber mills as part of the local economic development strategy. In summary, there is no clear signal of a grand strategy of global capital to amass a surplus in the noisy riot of local, regional, national, social, economic, and biological forces at work on the frontier. Rolim de Moura provided an essential but episodic service in the multifaceted mahogany trade, serving as a convenient location for bankers, producers, and international merchants to interact with each other near this lucrative, hotly disputed, but short-lived resource front.

Spatial Interaction: Freight Flows and the Transport Sector

An important barometer of economic activity is the transport sector linking frontier towns to each other and to the industrial core of the country. In our spatial interaction analysis we sought to evaluate the changing economic orientation of the frontier on the basis of overland freight traffic. Indeed, direct observations of all vehicular traffic moving into and out of Rolim de Moura in 1984 and 1990 reflect the three important pattern shifts occurring in the town's economy before and after the mahogany boom (table 9.3).[16]

Table 9.3 Local direction of motor carrier freight movement in Rolim de Moura, 1984 and 1990 (number of motor carriers)

	1984			1990		
Freight Type	In	Out	Total	In	Out	Total
Timber Products	28	47	75 (76%)	6	10	16 (14%)
Agricult. Products	3	1	4 (4%)	30	8	38 (34%)
Dry Goods	4	2	6 (6%)	21	16	37 (33%)
Empties	8	6	14 (14%)	10	11	21 (19%)
Total	43	56	99	67	45	112

Note: The surveys were carried out by visual recognizance from posts on access routes on Linha 25 de Agosto and the Rua Norte-Sul, the only two routes directly linking Rolim de Moura to the BR-364 from 9:00 A.M. to 12:00 noon on August 16, 1984, and on July 24, 1990.

First, the economic contraction of Rolim de Moura during the period following the mahogany bust (1985) is reflected by the stagnant volume of freight. Between 1984 and 1990 the estimated overall truck traffic increased by only 13 percent or roughly 2 percent per year, which was significantly less than the population growth rates along the corridor of between 10 and 15 percent during this period. The dramatic loss in timber traffic, representing more than three-quarters of total movements in 1984 (dropping to 14 percent in 1990), was offset by increases in agricultural freight (from 4 to 34 percent) and dry goods (from 6 to 33 percent), reflecting the considerable expansion of the agrarian frontier along the Rolim de Moura–Alto Alegre corridor and the corresponding increase in crop production and consumer demand.

Second, disaggregating the crude truck flow data entering and leaving Rolim de Moura by direction, we found that frontier-bound flows moving southbound from Rolim de Moura toward Santa Luzia were distributed by type as follows: empty general freight backhauls and dry goods (53 percent of total southbound freight traffic observed); miscellaneous industrialized consumer goods and livestock (e.g., gasoline, tractor parts, supermarket products, bottled beer and soft drinks, and cattle—21 percent); and colonists in transit to the frontier aboard trucks (12 percent). In 1990, then, the still active frontier south of Rolim de Moura, generated a net demand not only for outbound agricultural freight carriers (reflected in empty trucks entering the corridor) but also for various inbound consumer goods as well.

The third pattern shift represented in table 9.3 is the reversal in the ratio of outbound to inbound traffic between 1984 and 1990, from 1.30 to 0.67. In 1984 inbound truck movements represented 43 percent of total movements. In 1990 inbound truck movements constituted 60 percent of the total. Given the very short observation period it is entirely possible that this pattern shift reflects nothing more than daily variations in traffic flows. But it is also likely that the directionality of traffic flow did fundamentally shift during this period. In 1984 Rolim de Moura was the center of a locally dynamic and rapidly expanding settlement frontier, as reflected by the observed net inflows. In 1990 Rolim de Moura and its immediate hinterland were contracting, as reflected by higher net outflows of produce originating from the new settlement frontiers well beyond its commercial market catchment area.

While there is some evidence of producer surplus extraction in the agrarian frontier, the specific forms such extraction assumes, involving local producers and petty merchants, professional merchant capital, and finance capital, are not altogether consistent with world systems prescriptions or any other single master theory. First, while the extension of the agrarian frontier created new opportunities for petty merchants to exploit farmers at the crop marketing stage, producer surplus extraction at local levels is not generally beneficial to the process of accumulation at either national or global levels. Second, professional merchant capital (i.e., trading companies) clearly appropriated some producer surplus through advance purchase financing during the mahogany boom. But this was almost incidental to the larger profit-seeking strategy of obtaining highly subsidized government export loans. Third, financial capital, in the form of the private banking sector, clearly saw Rolim de Moura as a quick profit center. However, given the low rates of local producer surplus savings, the opportunity for sustained capital accumulation by the banking sector depended mainly on the continuation of various federal credit programs and the possibility of capturing the institutional rents of the state.

We do not imply that rainforest cities operate in a global vacuum. Our retail sector survey, for example, indicates the important consumption function served by the frontier's urban markets. Rolim de Moura became a strategic location for the distribution and retailing of national consumer goods in the periphery. Dominated by national supermarket chains with direct linkages to agribusiness in the Southeast, Rolim de Moura's consumer market was flooded with national products, displacing even food items produced by the *municipio's* own farmers.

Rainforest cities on the populist frontier are not mere artifacts of the process of capital accumulation on a world scale. The hierarchical model of global capital accumulation postulated by Armstrong and McGee (1985), in which surplus capital flows progressively upward and ultimately to the level of world cities, is unrealistic and exceedingly reductionistic (fig. 4.1A). Rather, we suggest the process of capital circulation in the frontier is more accurately depicted as a series of circuits of financial flows between the government and diverse social classes (sociospatial centers of production), each circuit being locally contextualized by distinct historical, institutional, and cultural forces resulting, in the aggregate, in a regionally disarticulated pattern (fig. 4.1B). Different factions of capital (e.g., merchant, agro-industrial, and financial) clearly compete with each other for access to institutional rents from the government, the most important center of surplus available on the frontier. Producer surplus extraction on the frontier, such as it is, is a low-grade activity in the larger project of capital accumulation. There is little evidence that the processes of petty surplus extraction on the agrarian frontier lead to capital accumulation at the global level.

SOUTHERN PARÁ: UNEVEN URBAN DEVELOPMENT ON THE CONTESTED CORPORATIST PERIPHERY

We have seen in earlier chapters how the contested corporatist frontiers of southern Pará generally have led to more irregular spatial patterns than in the Rondônia case study. Compared to the relatively smooth gradient of central places in the Rolim de Moura corridor, the towns of the Xingu corridor exhibit more irregular demographic and socioeconomic patterns. Most important are the migratory pulls of the diverse corporate megaprojects of southern Pará, such as the Carajás mineral export complex, along with the small-scale *garimpeiro* gold-mining and timber-extraction operations. In addition, on a local level, the contemporary evolution of the towns in the Xingu corridor reflects the importance of road access for urban commercial expansion, land concentration in the rural sector, and changing patterns of resource extraction. All these forces operate in a regional context of competing social groups and institutions, whose battles over land and resources shaped the local social terrain. As rapid growth abated in the longer-settled areas of the Xingu corridor, new settlement and resource extraction generally moved to the areas to the west, near Ourilândia do Norte, Tucumã, and São Félix do Xingu. In a broader national context, processes of political-economic integration have affected both the immediate outcomes of local struggles and the long-term prospects for regional development. Now we turn to the sectoral interactions between the local, regional, and national levels of the Brazilian urban hierarchy exhibited by the towns in our southern Pará study area.

Internal Colonialism: Resource Extraction and the Industrial Woods Sector

As in Rondônia, valuable timber, particularly mahogany, provided the most lucrative form of resource extraction during the early phases of frontier settlement in the Xingu corridor. Lumber mills clustered around Xinguara from the town's earliest days, due to the availability of mahogany trees nearby. As local mahogany trees were cut during the 1980s, timber mills moved farther westward to new frontiers in the Xingu area. The ten sawmills operating in the urbanized area of Xinguara in 1978 produced an average of about 4,375 cubic meters of mahogany monthly and directly employed some 320 men in the milling operations, not including those extracting logs in the surrounding forests. By 1990 only three large wood processors, one lumber mill and two plywood-veneer plants, could be found in the town of Xinguara itself. Together these factories produced about 1,000 cubic meters of output monthly and employed a total of 400 workers, many of them now female, indicating the development of a feminized urban proletariat (Godfrey 1990).

Although the evolution from basic lumber milling to "beneficiated" wood processing reflected a maturation of the local economy, future prospects remained uncertain. Interviews with company personnel in 1990 indicated that the logs used in all three plants generally came from sites located 200 to 500 kilometers away. Although federal Brazilian forestry laws required reforestation projects for the granting of export licenses to firms, little if any tree planting has occurred near Xinguara. The local lumber mills tended to cut timber and plant pasture on large tracts of land in the westerly Xingu region, where small, showpiece demonstration projects were mounted and larger-scale afforestation promised for the future.

Another notable feature of the local wood products industry has been its control by national firms based in the Brazilian South. Although manual laborers are locally recruited in Xinguara, as in Rondônia, mill owners and plant managers generally come from the southern state of Paraná, where a lumber industry once thrived on the exploitation of Paraná pine early in the twentieth century. Virtually all lumber produced in the Xingu corridor is exported to the Brazilian South. For example, a 1994 survey of twenty-two industrial wood processors located in the three settlements of the Xingu corridor found that regions outside of the Amazon purchased 91 percent of the plywood and veneer output: the Southeast consumed 61 percent, the Center-West 16 percent, the South 11 percent, and the Northeast 3 percent (Mousasticoshvily 1994:28). These findings prompted forest economist Mousasticoshvily (1994:34) to conclude:

> Many of these industries, in spite of maintaining offices in the [North] region, are subsidiaries of financial groups in the south of the country. In this manner, the economic profits derived from this [industrial wood production] activity are channelized to other regions of the country, thus characterizing a process of "self-colonialism," where the region generating the wealth does not participate in the economic benefits flowing from the use of its natural resources.

These patterns of production, management, and trade create a dynamic of internal colonialism in the resource-extractive periphery of southern Pará.

The region's role as a peripheral resource-extractive frontier continued with the discovery of timber, gold, and other valuable commodities in the 1980s. As previously discussed, the history of Tucumã and Ourilândia do Norte puts into sharp relief the conflict between corporatist and populist forces in Amazonia. Despite the Andrade Gutierrez Corporation's plans for a directed program of small-farmer settlement at Tucumã, favored by fertile farmland and valuable mahogany, the influx of migrants moving into the Xingu region overwhelmed the corporatist plan in the 1980s. The discovery of gold in particular led to the emergence of expeditionary mining camp settlements and eventually to new towns at Ourilândia, Cuca, and elsewhere (see chapter 7). As the Xingu gold rush waned in the early 1980s, however, the region's continued peripheral status suggested the weakness of local central place functions independent of the resource-extractive functions.

The success of the local *garimpeiros* in overcoming the exclusionary designs of the powerful Andrade Gutierrez Corporation in the early 1980s recalls the earlier victories of the gold miners at Serra Pelada in confronting the state-supported mining concession, Companhia Vale do Rio Doce. Subsequently, the Kayapó Indians of the Xingu Valley successfully resisted incursions by goldminers, and in 1985 demarcated their 3.2-million-hectare reserve at Gorotire, south of São Felix do Xingu (Schmink and Wood 1992).[17] These cases point to the importance of often conflicting local forces in shaping national policy regarding access to land in Amazonia.

The Xingu corridor now resembles in many ways a hollow frontier rather than an example of sustained regional development. After an initial rush of in-migration and resource extraction, urban growth abates. In time, towns come to depend for their survival more on commercial success in serving interregional trade and less on resource extraction, agriculture, and ranching. Xinguara's strategic crossroads location favored its survival over other competing local towns in this regard. But the greatest dynamism in southern Pará has derived from the enormous capital investments in the corporatist enclaves that dot the region, the infrastructure they have left, and the linkages to the global economy that the corporatist frontier has forged. Commercial towns like Xinguara grow or contract in relation to the megaprojects described earlier in this work. Corporatist investments also have stimulated small-scale resource extraction, as illustrated by the *garimpeiro* gold-mining camps, and generally provoked a populist response contesting social control of the frontier, sometimes successfully.

Inequality follows in the wake of the historical cycles of resource extraction, both populist and corporatist, in Amazonia. As forests were cleared and cattle ranches were established, rural land in the municipality became concentrated into large landholdings reminiscent of traditional Amazon structures (Velho 1972): in 1985 registered latifundia and rural firms, which accounted for only 39 percent of the titled holdings, controlled 85 percent of the rural land in the municipality of Xinguara (Godfrey 1990). By the late 1980s the characteristic pattern of land occupancy in the vicinity of Xinguara was the large but increasingly degraded cattle ranch:

> Some of the holdings were so large that they stretched from the roadside clear to the horizon. Although only a few years old, many of the pastures . . . already showed signs of severe degradation. The small herds of zebu cattle . . . far exceeded the number of small farmers. . . . By then most of the *posseiros* who had pioneered the region years earlier had moved on to try their luck elsewhere. (Schmink and Wood 1992:191).

The dynamic settlement frontiers of southern Pará, which required abundant natural resources, steadily moved westward to the new areas in the Xingu corridor. The municipality of Ourilândia to the west, largely dependent on gold mining and large cattle ranches, now appears to be undergoing a transformation similar to that of Xinguara. The corporate orientation of Tucumã, present since the beginnings of the Andrade Gutierrez colonization project, serves to inhibit the rapid deforestation and land consolidation of the other two municipalities. Still, the general tendency on the Xingu corridor is for settlement to proceed toward a hollow frontier. In addition, as the Xingu region is selectively drawn into national commercial currents, the commercial viability of all of these towns remains in question.

Towns as Market Centers: Commercial Patterns in the Xingu Corridor

Xinguara, the highest-order commercial center in the Xingu corridor, grew as a regional market town oriented to the commercial traffic on the PA-150 and PA-279 highways. Due to its strategic location, Xinguara quickly prevailed in commercial competition with such nearby centers as Rio Maria, approximately 25 kilometers to the south on the PA-150, indicating the development of a regional urban hierarchy of places. In terms

of its economic evolution, Xinguara steadily lessened its initial dependence on natural resource processing and came to rely more on regional
business activity. Tucumã and Ourilândia, on the other hand, have
remained more dependent on extractive economies. Like in the towns
studied in Rondônia (table 9.1), the 1990 distribution of businesses in the
urban centers of the Xingu corridor confirms the most obvious tenet of
central place theory: the larger, higher-order settlement (Xinguara) in the
regional system supports a wider array of commercial functions and services than the smaller towns (Tucumã and Ourilândia) on the fringe of the
system. Yet the commercial viability of businesses in the Xingu corridor
remains highly dependent on outside forces. Many of the urban services
would disappear with the cessation of truck traffic to extractive sites in the
Xingu region or with the development of alternative transportation routes
(see "Transport Sector," below).

Despite evidence of emerging economic hardship (chapter 7),
Xinguara has witnessed a steady expansion of its commercial sector.
Indeed, as noted in chapter 7, local merchants constitute the social elite
of Xinguara. Besides their economic clout, this business elite enjoys the
prestige of national connections and often owns cattle ranches in the
region. In 1982 about 311 businesses could be found in Xinguara, according to Volbeda (1986), excluding lumber mills and other industries,
schools and government agencies, transport lines and telecommunications facilities.[18] Our own field surveys in 1990 indicated the presence of
392 businesses establishments along the main streets of the town. Most
numerous were the sixty-five bars and snack bars, which represented
16.6 percent of local businesses; the forty-seven clothing and shoe stores
(12.0 percent); the thirty-three fruit and vegetable, produce, bakery, and
butcher shops (8.4 percent); the thirty-two automobile and mechanical
repair shops (8.2 percent); the thirty-one miscellaneous businesses with
diverse consumer goods (7.9 percent); and the twenty-eight grocery and
specialized food stores (7.1 percent). Less numerous, but certainly
important for regional ties, were Xinguara's five commercial banks, nine
gold dealers, five agricultural supply and equipment shops, and various
professional and skilled services (table 9.4).

Distributions of businesses in the outlying settlements resembled that
of Xinguara in a general way, but some nuances are noteworthy. For
example, as a populist settlement originating from spontaneous in-
migration, Ourilândia has supported a disproportionately large number
of small-scale, informal businesses. Fifty-five bars and snack bars were
found in 1990, representing 21.7 percent of the local commercial estab

Table 9.4 Principal businesses in Xinguara (1982, 1990), Ourilândia do Norte, and Tucumã (1990)

BUSINESSES	XINGUARA		OURILÂNDIA	TUCUMÃ
	1982[a]	*1990*	*1990*	*1990*
URBAN SERVICES	134	180	120	77
Banks	1	5	1	2
Auto/mechanical repair[b]	32	32	9	6
Hotels and restaurants	23	22	22	13
Bars and snack bars	39	65	55	27
Entertainment[c]	2	8	6	7
Skilled services[d]	17	28	14	14
Professional Services[e]	14	14	11	6
Hospitals	6	6	2	2
URBAN COMMERCIAL EST.	177	212	133	126
Supermarkets/grocery stores	38	28	21	17
Fruit/vegetables, produce, bakeries, butchers	36	33	16	17
Misc. consumer goods[f]	17	31	21	26
Farm equipment, supplies	1	5	8	4
Auto parts, tires, electric	13	3	8	9
Clothing, shoes, boutiques	36	47	32	22
Pharmacies	18	18	12	10
Gas stations	2	3	2	1
Furniture, furnishings	5	13	—	3
Hardware and construction materials	7	8	6	12
Gold dealers, equipment	—	9	5	4
Beverage distributors	2	4	1	1
Cattle and grain dealers	2	10	1	—
TOTAL BUSINESSES	311	392	253	203

[a]*Based on data in Volbeda (1986).*

[b]*Auto/mechanical businesses specialize in the repair of automobiles, tires, bicycles, chainsaws, household appliances (radio/tv, etc.), and include blacksmiths and locksmiths.*

[c]*Entertainment includes cinemas, discos, clubs, Rotary Club, country clubs, and so forth.*

[d]*Skilled services include barbers and hair stylists, taxi dispatchers, carpentry, photographic finishing, despachantes, florists, and print shops.*

[e]*Professional services include veterinary clinics, public accountancies, dental clinics, funeral parlors, real estate offices, laboratories, trading companies, notaries, and office and clerical services.*

[f]*Misc. consumer goods include variety stores and shops specializing in tobacco, perfume, records, watches and jewelry, books and stationary, billiards, and magazines.*

lishments; in addition, twenty-two hotels and restaurants constituted 8.7 percent of the town's businesses. The relatively high number of hospitality services (e.g., hotels, bars, and snack bars) catered to transient goldminers, local travelers, and others. On the other hand, Ourilândia had fewer businesses requiring any significant degree of capitalization. Only one bank could be found in the town, the state-sponsored Banco do Brasil, which can be found in most municipal seats. Tucumã, on the other hand, had two banks. Clearly, Tucumã had taken the lead in the more capitalized firms: twelve hardware and construction stores (5.9 percent of local businesses), nine automobile parts stores (4.4 percent), and even four gold dealers (2 percent). In short, Tucumã had the more important central business district in an integrated market area with two twin cities.

To study the degree of external market penetration in the Xingu corridor towns, our 1990 survey teams examined the origins of common products sold in local supermarkets and corner grocery stores in the Xingu corridor, just as the Rondônia research teams had done. Although local farmers sold vegetables and fruits at informal outdoor markets, such produce fetched modest prices from consumers. The more lucrative commercial transactions occurred at established supermarkets and grocery stores, which sold packaged and nonperishable items. The Xingu corridor appeared less affected by national supermarket chains than the Rolim de Moura corridor in Rondônia: Xinguara had only two supermarkets, Tucumã had one, and Ourilândia had none. On the other hand, small corner grocery stores are more prevalent in southern Pará than in the Rondônia study sites.[19]

In terms of the origins of goods sold in the Xinguara supermarkets and small stores, we found that the Xingu corridor functions as a commercial outpost for national producers to an even greater degree than does the Rondônia study area. Nearly 85 percent of the goods in the Xinguara supermarkets were supplied by national producers; only 15.2 percent of the supermarket goods were produced locally or in the state of Pará (table 9.5A). In the Xinguara grocery stores, over 95 percent of the goods were produced nationally and less than 5 percent locally or elsewhere in the state (table 9.5B). In other words, although we expected the small stores to retain more ties to local producers, this did not prove to be the case in Xinguara. Indeed, the overwhelming national domination of local markets is the most important finding in southern Pará, as it is in Rondônia. The Xinguara supermarkets tended to sell a few more local and state products than the corner grocery stores, but in both cases, as in the Rondônia sites, national suppliers provided nearly all the available merchandise. Virtually all processed food, paper, clothes, construction materials, and manufactured goods were imported from the cities in

Table 9.5 Origins of goods sold in Xinguara by scale of supplier market

A. SUPERMARKETS

Product	Supplier market scale		
	LOCAL	STATE	NATIONAL
Coffee		x	xxxxx
Rice	xxxxx		
Beans	xxxxx		
Manioc flour	xxxxx		
Corn flour			xxxxx
Popcorn			xxxxx
Cornflakes			xxxxx
Polenta			xxxxx
Potatoes			
Soap		x	xxxxx
Sugar			xxxxx
Vegetable oil			xxxxx
Powdered milk			xxxxx
Macaroni			xxxxx
Percent of stock	9.1	6.1	84.8

B. CORNER GROCERY STORES

Product	Supplier market scale		
	LOCAL	STATE	NATIONAL
Coffee			xxxxx
Rice			xxxxx
Beans	xxxxx		
Manioc flour	xxxxx		
Corn flour			xxxxx
Popcorn			xxxxx
Cornflakes			xxxxx
Polenta			xxxxx
Potatoes			xxxxx
Soap			xxxxx
Sugar			xxxxx
Vegetable oil			xxxxx
Powdered milk			xxxxx
Macaroni			xxxxx
Percent of stock	4.5	0	95.5

Note:
x = 0–20 percent of product stock
xx = 21–40 percent
xxx = 41–60 percent
xxxx = 61–80 percent
xxxxx = 81–100 percent
Sample size: seven corner grocers and two supermarkets

Goiás, São Paulo, and Minas Gerais. Only basic foodstuffs, such as rice, beans, and manioc flour, were locally produced along with such perishables as fruit, meat, and vegetables. Suppliers based in Pará provided only coffee and soap. These results indicate nearly complete domination by national centers of production and weak representation of local and state producers in local consumption. On the contested corporatist frontier the national market has virtually taken over local markets, exchanging manufactured and processed goods from the center for natural resources extracted in the periphery that is southern Pará, a tendency that appears to be consistent with the surplus extraction thesis.

Interregional linkages

The weakness of Xinguara's ties to local central places and the extraction of resources and profits by agencies and enterprises in the Brazilian core regions suggest a form of urbanization increasingly dependent on national forces and largely disarticulated from autonomous regional market growth. Since the 1970s, when the Brazilian military launched a massive program of road construction in Amazonia, banking, telephone, radio, and television networks followed in rapid succession. The links between the Xingu corridor settlements and other regions of Brazil can be analyzed further in terms of the financial sector, transport, and communications ties. Overall, the region has seen increasing linkages to centers in southern Brazil and steadily weakening ties to the state capital region as southern Pará is drawn into the national city-system.

Financial Sector

Data on interregional financial flows in Amazonia are difficult to obtain, particularly in frontier settings. Interviews with officials in Belém and the Xingu corridor towns, however, indicate a period of growth in the financial sector in Pará during the 1980s similar to that in Rondônia. The federally chartered and supported Bank of Brazil, which operates with an explicit social mission of promoting local development, tends to be more decentralized in its operations than the private banks. The number of local branches of the Bank of Brazil in Pará increased from twenty-two in 1980 to forty-six in 1990.[20] Private-sector banks routinely transfer larger amounts of money to headquarter locations in southern Brazil. For example, Bradesco is a leading private bank with fifty-two branches in

Pará, including local branches in Xinguara and Tucumã in 1990. Although it was not possible to obtain precise figures on how much money Bradesco transferred from southern Pará to national centers, the expansion of the number of local branches suggests that the bank found the resource extraction associated with frontier expansion a profitable enterprise. The total number of local banking agencies in the state increased from about 120 in 1980 to 180 in 1990.[21]

Passenger Transport Sector

Bus service via the predominant Transbrasiliana Company, which links Amazonia to the rest of the country, indicates that predominant interregional patterns of passenger transportation mirror the flows of resources and capital from this regionally disarticulated frontier. Of the twenty-two bus lines passing through Xinguara daily in 1988, eight routes ran north to Marabá and Belém in the state of Pará; two led westward into the Xingu region, and twelve headed southeast to the states of Tocantins, Goiás, and on to the Brazilian South. The prevalent southerly direction of Xinguara's transportation connections reflects growing labor ties to the national urban system of Brazil (Godfrey 1990:115).

Freight Transport Sector

A careful monitoring of all vehicular freight movements on the various roads through Xinguara over a period of several days in July 1990 revealed significant patterns of regional interaction. During eight different three-hour observation sessions, researchers observed the direction of road transport entering and leaving Xinguara.[22] In terms of a breakdown of the trucks observed, those carrying logs, sawed timber, and mixed dry goods constituted most of the traffic. Agricultural products and cattle trucks were conspicuously infrequent. The results presented in table 9.6 divide these vehicular movements into two main directions.[23] Southeast of Xinguara on the paved PA-150, traffic flowed toward or from the national core, leading to Conceição do Araguaia and beyond to Goiás, Brasília, and Minas Gerais. On the other hand, traffic also flowed toward or from the Amazon periphery, either north on the PA-150 toward Marabá, or west on the unpaved PA-279 toward Tucumã and Ourilândia.

The flows documented in table 9.6 generally support the surplus extraction hypothesis. Trucks with timber logs, for example, usually

Table 9.6 Regional direction of motor carrier freight movement in Xinguara
(number of trucks)

Freight Type	Direction: National Core (southeast of Xinguara)		Direction: Amazon Periphery (north & west of Xinguara)		Total Trips	
	Into X	Out of X	Into X	Out of X	Into X	Out of X
Timber logs	3	22	10	0	13	22
Sawed lumber	1	7	0	6	1	13
Agricultural produce	0	0	3	5	3	5
Cattle	2	2	0	0	2	2
Dry goods	29	12	9	26	38	38
Empties	24	20	42	38	66	58
Total	59	63	64	75	123	138

Note: The surveys were carried out by visual recognizance from posts on access routes on the PA-150 and PA-279 highways around Xinguara during eight sessions in the morning, afternoon, and early evening, between July 14 and 17, 1990.

entered Xinguara from the periphery. A movement of sawed timber could be discerned leaving Xinguara, destined for both the national core and the regional periphery. Agricultural produce generally entered and left Xinguara in the direction of the Amazon periphery. In contrast, dry goods and mixed products generally entered Xinguara from the core and proceeded on to the Amazon periphery. In short, Xinguara serves as a way station in the national long-distance commerce, linking sources of raw materials in Amazonia with manufacturing centers in the Brazilian core regions.

Communications Sector

The diffusion of modern communications in southern Pará also reveals telling patterns of regional interaction. In the late 1980s Xinguara and Tucumã obtained "TelePará" telephone posts, linked directly by satellite to the other centers of Brazil. Our 1990 survey indicated that only 6.7 percent of the overall population of the three towns had telephones at home: 9.1 percent of the households in Xinguara, 7.4 percent in Tucumã, and 0 percent in Ourilândia. Therefore, the overwhelming majority of the population used the local telephone posts in Xinguara and Tucumã. A regional breakdown of telephone calls between Xinguara's TelePará post and outside areas on a single day—Monday, August 1, 1988—revealed prevalent patterns of regional interaction. The state of Goiás (which then included Tocantins) received 43 percent of the calls, 25 percent went to the rest of Pará, 8 percent connected to the state of São Paulo, 6 percent rang up Brasília, and the rest were directed to other Brazilian states. These communications connections reflect the continuation of traditional migrant ties to Goiás and other sending regions, relatively weak ties to centers in Pará state, and significant linkages to São Paulo, Brasília, and other metropolitan centers in the Brazilian South (Godfrey 1990:115).

In addition to interregional transportation tying the Xingu corridor to other regions of Brazil, television is another important means of national integration. Indeed, the popular film *Bye-Bye Brazil* cites the appearance of television antennas in remote towns (the dreaded "fish bones") to explain the declining importance of regional Amazon culture. As the masses were captivated by nationally syndicated television programs, interest in local entertainment pastimes, such as the traveling circus depicted in the film, disappeared. Television also serves as a mechanism for the diffusion of consumer trends from the metropolitan centers of Rio de Janeiro and São Paulo and propagates images of modernity to the backward interior through national broadcasts (Godfrey 1993). By the late 1980s the Xingu corridor towns were all retransmitting television broadcasts, captured via satellite, to local markets. Our 1990 survey indicated that a weighted mean of 45.0 percent of the households in the three towns had television at home: 53.3 percent in Xinguara, 42.3 percent in Tucumã, and 30.6 percent in Ourilândia. Even among those without television in the home, the ubiquitous influence of TV could be experienced in local bars, restaurants, and stores.

In southern Pará the national Globo network dominates the airwaves and includes little local content for television consumption. The national focus fostered by television, along with other patterns of regional communications, puts the prospects for the maintenance of a distinct regional culture in Amazonia in serious doubt. Although local programming exists in some broadcast markets, national programs, originating from São Paulo and Rio de Janeiro, tend to enjoy prime-time slots regionwide (Roberts 1995).

REGIONAL CONTRASTS: AUTONOMOUS REGIONAL DEVELOPMENT OR GLOBAL SURPLUS EXTRACTION?

The case studies of Rondônia and Pará indicate the complexities and local variations in patterns of interregional linkage. Both regions are linked to national capital in different ways. For example, the data on timber, commerce, communications, transport, and banking indicate increasing interactions of the towns in the Xingu corridor with the Brazilian national cultural and economic center. Although those towns retain links to the state capital, increasingly their connections are to Marabá in southern Pará, to centers in Tocantins and Goiás, and to Brasília, São Paulo, and other parts of the Brazilian southeastern core region. Virtually all merchandise sold in the Xingu towns is imported from outside the region, and the most valuable resources tend to be sent to the southern centers. Profits from cattle, lumber, and other operations presumably are remitted to company headquarters in the Brazilian South. In the contested corporatist frontier of southern Pará, commercial linkages reflect a pattern of unequal exchange between the Amazon periphery and the national core. Though the regional export of natural resources has led to local growth in the Xingu corridor, the predominant extractive patterns reflect centralized control of key profit-generating sectors. Arguably, the urbanization of southern Pará is largely driven by forces of surplus extraction from the peripheral centers of southern Pará and the diffusion of market exchanges and investment decisions from the higher-order centers of the Brazilian core region (Godfrey 1990).

The populist agrarian frontier offers some interesting contrasts to the corporatist frontier in its interregional linkages. Although agricultural producers in Rondônia are chronically cheated out of their profits by local merchants and their oligopsonistic marketing practices, there is little evidence that this form of petty extraction in the frontier has served global accumulation processes. Arguably, the appropriation of producer surplus by petty merchant capital in the frontier runs counter to the logic of capital accumulation at the national and global levels. Periodically, national capital interests directly participate in local processes of surplus extraction. For example, during the mahogany boom, trading companies based in São Paulo, themselves subsidiaries of major banks, shopping mall chains, and industrial transnational corporations, embellished their profits by exploiting local industrial wood producers through advance purchase financing arrangements. But this low-grade form of surplus extraction must be seen in the larger political economic context of Brazil's export promotion program, which provided enormous institutional rents to these diverse factions of national capital. Once the export credit program was terminated, the national corporations withdrew from Rondônia's mahogany trade, and local lumber producers resumed business as usual.

Although the Brazilian banking system is structured to concentrate capital in the national core, there is little evidence that banks in either corporatist or populist areas are mobilizing significant amounts of local producer capital or enhancing global accumulation. While banks proliferated in relation to national credit programs, recently the Amazon's financial sector has contracted and stabilized. Low rates of formal savings by Rondônia's urban and rural producers further suggest that banks are not siphoning off the frontier's producer surplus. Rather, the financial sector depends upon the continuation of federal credit programs and the opportunities they provide to capture institutional rents of the state in the form of subsidized capital, which banks can easily transfer to subsidiary corporate activities around the world.

Surplus extraction in the populist frontier occurs in fits and starts, not as a homogeneous characteristic of capitalist expansion on the frontier. Indeed, it is axiomatic (at least among Marxists) that the absence of surplus extractive relations of production signals the incomplete penetration of capitalism into the frontier, a central point we made in chapter 2. In one sector of the frontier economy, however, national capital almost completely dominates: the formal retail sector. The Rolim de Moura corridor in Rondônia offers a better example than does the Xingu corridor of self-sufficiency in food production, but even here, where agriculture dominates the local economy, national supermarket chains play the dominant role in the frontier food supply.

The regional differences in how the frontier functions in the national economic context suggest that no single theory of regional development adequately explains the variations. True, the development of the frontier is constrained by global and national forces, but local factors also explain a great deal of the variation in regional patterns. In this chapter's exploration of the surplus extraction thesis, central to the world systems position, no unambiguous and uniform pattern of surplus extraction supporting the accumulation of capital at the global level is evident. At least in Rondônia's populist frontier rainforest cities are not unambiguously "theaters of accumulation." Southern Pará, dominated by corporate megaprojects like Carajás and Tucuruí, generally shows more coherent support for the extractive hypothesis, but even here local groups serve as vital intermediaries in processes of capital accumulation. For example, small-scale *garimpeiros* at Ourilândia and other populist groups elsewhere in the Xingu corridor successfully confronted the ambitious corporate colonization program of the Andrade Gutierrez Corporation, just as occurred at the Serra Pelada. Ultimately, neither populist nor corporatist frontiers are experiencing a process of autonomous sociospatial development as envisaged in many spatial economic formulations (especially central place theory). The fractal social landscape of the Amazon is volatile, changing more rapidly in some areas than others in response to the particular mix of social, ethnic, and economic class factions that, for historical and institutional reasons, congregate in a given area.

Notes

1. Armstrong and McGee (1985:58) define surplus extraction as "value extracted by capitalists (private and state) from wage labour, rent, interest and unequal exchange." While acknowledging that the process of surplus extraction "is subject to historical cycles of expansion and contraction and is exceptionally uneven in its class and spatial impacts" (p. 58), Armstrong and McGee argue that the process invariably leads to "the upward flow of surplus through the urban hierarchy and rural areas" (p. 59) to the national metropolis and from there to global cities (fig. 2.4).

2. In Emmanuel's initial formulation (1972), an unequal exchange arises from trade between countries in which the wage paid to labor is quantitatively different, such that countries with low wage rates transfer more labor value congealed in the products sold to countries with high wage rates than they receive in the form of products purchased from those countries.

3. Interregional rural land price differentials favored the structural shift in production leading to the settlement of Rondônia. From 1966 to 1987 the average real price of crop land in Mato Grosso do Sul (on the western fringe of Brazil's agricultural heartland) rose by a factor of 14.2, while that of Rondônia grew by a factor of only 6.9 (World Bank 1991:39).

4. During the period of most intensive smallholder settlement in Amazonia (1973 to 1983), approximately U.S.$147 billion were distributed to agricultural producers throughout Brazil from the National Rural Credit System. Between 1977 and 1983 producers in the North Region received only 2.4 percent of national rural credits (Browder 1988). Beef, the most important Brazilian staple, is a special case in point. Credits to promote beef production on corporate ranches were particularly encouraged by SUDAM in Operation Amazonia and the First Amazon Development Plan (Browder 1988a; Hecht 1985). Smaller producers also received significant subsidies through regular credit lines that were used to develop cattle operations in the Amazon. The economic costs of subsidized rural credits to promote Amazonian cattle production, however, vastly exceeded the opportunity costs associated with foreign beef imports. Between 1971 and 1982 one metric ton of Amazon beef (before processing) embodied about U.S.$4,000 in social costs. In stark contrast, Brazil paid on average U.S.$1,086 per metric ton for imported foreign-produced and processed beef during the same period (Browder 1988b).

5. Following our July 1990 census of all grain processors in Rolim de Moura, resulting in the identification of a population of twenty-seven, a random sample of six was selected for interview. These six *cerealistas* together had moved 15.6 percent of the total volume of food crops produced in the municipality of Rolim de Moura in 1989. Time constraints and high respondent refusal rates prevented us from interviewing the entire population. A trained and experienced Brazilian postgraduate research associate administered a standard questionnaire to the owner or plant manager of each firm selected. Given that respondent refusal was higher among larger firms, our sample, although randomized, is likely to be biased by over-representing smaller, locally based merchant capital. This bias, we speculate, would probably

understate the relative proportions of Rolim de Moura's agricultural output destined for national markets.

6. There is no completely comprehensive analytical definition of corner grocers, but commonly these commercial establishments share the following characteristics: They are typically family-owned and -operated, often using nonremunerated household labor. They have low capital entry requirements and are often physically set up in or adjoining the family's dwelling unit. They maintain a minimal stock of a small range of low-priced, high-turnover goods, many perishable and locally produced. In towns without supermarkets, they are seemingly ubiquitous.

7. In Rolim de Moura six supermarkets and twelve corner grocers were surveyed in 1990. The place of origin (manufacture or production) of fourteen basic food products was ascertained by examining product packaging labels or by interviewing the store manager/owner.

8. Personal communication, manager, Supermercado GB, July 29, 1990.

9. In our 1990 rural producer survey of 115 systematically selected farmers along the Rolim de Moura–Alto Alegre corridor 69 percent overall indicated no positive net income from agriculture during the preceding year (unpublished survey). Our rural producer survey was undertaken only in Rondônia in tandem with the urban household surveys in June and August 1990 by a separate team of Brazilian researchers. A sample of systematically selected rural households were visited and male and female household heads were interviewed using a standardized questionnaire. These surveys were designed, in part, to ascertain land use, crop output, and household income (see technical appendix).

10. As a percentage of wholesale prices transport costs rose 44 percent for corn transport and 35 percent for upland rice. In the case of corn, the amount of the freight rate increase alone exceeded its local producer price, making the commercialization of corn from Rondônia uneconomical for farmers (World Bank 1991:54).

11. In 1984 six private national banks had branch offices in Rolim de Moura (Bamerindus, Itau, Bradesco, Banco Mercantile do São Paulo, Finanza, Comind). In addition, the federal Banco do Brasil and the official State Bank of Rondônia (Beron) had offices operating in the city. In 1985, after the collapse of the mahogany trade, only Bamerindus, Itau, and Bradesco and the two government banks remained.

12. Personal communication, Celso de Freitas Barbosa, manager, Rolim de Moura branch bank, Bamerindus, July 20, 1990.

13. In the 1990 rural household survey conducted in the Rondônia study sites we found the following rates of formal savings among small producers: Rolim de Moura, 27.7 percent; Santa Luzia, 30.3 percent; and Alto Alegre, 23.5 percent.

14. These export credits were relatively long-term 180-day loans in amounts equivalent to 50 percent of the applicant's "liquid assets" (defined as value of its exports in the preceding year plus current cash on hand). The nominal interest rate of 40 percent charged to such companies conferred a substantial subsidy of approximately 75 percent of the true cost of capital during this period. Qualified producers were also eligible for export financing but under terms from the Central Bank that were less favorable than those accorded to the professional export merchants (Browder 1986, 1987).

15. Bamerindus of Belém, for example, offered depositors a special six-month "supplemental financing plan," in which the bank retained 60 percent of the amount of the loan from the Central Bank in a money market account bearing an annual interest rate of 178 percent, while releasing only 40 percent of the loan amount to the trading company for purposes of financing exports. Moreover, most trading companies could easily schedule the disbursement of their loan monies to the final thirty days of their loan periods, allowing these resources to draw interest in overnight accounts for the initial 150 days. It is estimated that 73 percent of

all Brazilian mahogany exporters participated in the central bank's financing program (Browder 1987).

16. On August 16, 1984, a research team directed by John Browder surveyed all vehicular traffic entering the city of Rolim de Moura along Avenida 25 de Agosto and Rua Norte-Sul from the north and west (the only points of access to the BR-364 interstate highway at the time). On July 24, 1990, the survey was repeated except this time traffic enumerators were positioned at all four gateways into the city in order to capture through traffic leaving Rolim de Moura to the east and south en route to more peripheral destinations in the frontier settlement corridors.

17. The Kayapó Indians of Gorotire provide one of Amazonia's best documented cases of indigenous resistance to land incursions. Indeed, the Kayapó have had spectacular success using traditional dress and rituals in the modern media to obtain strategic advantages in negotiations with Brazilian authorities (Schmink and Wood 1992:253–275).

18. These data on the evolution of businesses in Xinguara are based on Volbeda (1986) and subsequent field surveys by our teams in 1990. The 1982 figures cited here differ slightly from those in Volbeda because of the different categories used to classify businesses.

19. Our research teams visited Xinguara's two supermarkets, and Tucumã's sole supermarket along with several small grocery stores in both of these two settlements during June and July of 1990. Standard questionnaires on product origins were administered in both southern Pará and Rondônia.

20. Personal communication, Júlio César dos Mendes Lopes, Assessor Técnico, Bank of Brasil, Belém, Pará, June 15, 1990.

21. Personal communication, Dr. Aleijandrinho Moreira, director of the Federação Brasileira de Bancos, Belém, Pará, June 22, 1990.

22. The surveys were carried out by visual recognizance from posts on access routes on the PA-150 and PA-279 highways around Xinguara during eight sessions in the morning, afternoon, and early evening, July 14–17, 1990.

23. Given the absence of competing central places near Xinguara due to the hollow character of the urban system of the Xingu corridor, local vehicular movement generally indicates regional direction.

10

Frontier Urbanization and Environmental Change in Amazonia

❧❦

Amazonia's most dramatic contemporary environmental spectacle, tropical deforestation, has been steadfastly portrayed by researchers and the popular press as the result of rural class struggle (e.g., cattle ranchers versus rubber tappers), misguided rural development policies, or the displacement of rural peoples by the penetration of global capital. The urban dimension of environmental change in Amazonia has passed virtually unnoticed in a literature now spanning hundreds of popular and academic titles in several languages. We submit that the popular focus on rural-based agents of deforestation in the Amazon misses the important roles that diverse urban interests play in processes of environmental change. Moreover, while deforestation is clearly an important dimension of the regional environmental change in Amazonia, it is by no means the only important one.

How has the urban transformation of the Brazilian Amazon affected the "natural" environment? In this chapter we explore four different aspects of the urbanization-environment relationship in Amazonia: first, the deforestation and land use implications of the transference of rural farmlands to absentee urban residents; second, the social and environmental impacts of large-scale hydroelectric developments that largely serve metropolitan populations; third, the social and ecological implications of the mining frontier, which has engendered particular urban forms in the rainforest; and finally, the deterioration of the urban living environment in Amazonia over the last twenty years. Each of these environmental issues is linked to the larger processes of regional urbanization occurring in Amazonia. To fully explain the processes of environmental change underway in Amazonia requires an understanding of each of these urban-environmental linkages.

MODERNIZATION, FRONTIER URBANIZATION, AND TROPICAL DEFORESTATION

Modernization theory predicts that as developing societies urbanize, agricultural production becomes more intensified; traditional low-input, consumption-oriented cultivation gives way to higher input cash cropping (Hirschman 1958; Lewis 1954, 1955; Rondinelli 1986; Todaro 1969). This predicted tendency arguably materialized in the case of the United States during the twentieth century. While the modernization hypothesis refers to fundamental societal transformations, not necessarily to changes on the regional or household level, these broader mutations cannot be without local correlates. For instance, in a frontier setting undergoing rapid urbanization it would be logical to speculate that urban-resident farmholders (i.e., farmers who reside in urban areas) would have greater access to information, technology, and financial resources that would make them more likely to engage in intensive commercial cropping strategies than rural-resident farmholders who visit urban centers infrequently. Modernization theory, then, postulates significant differences between urban- and rural-resident farmholders in their land use and relation to capital. From this postulate follows this basic argument: As regional populations congregate in urban areas, and as rural populations decline, rural land use intensifies, and pressures on remaining natural resources, such as tropical forests, are reduced.

Based on rural and urban household surveys undertaken in the Rondônia study sites in 1990, we explore three questions in this section. First, do urban- and rural-resident farmholders pursue distinct land-use strategies on their rural holdings? Second, has the urbanization of the Amazonian frontier induced greater intensification of agricultural production consistent with modernization theory? Third, does the recent tendency toward urbanization of rural property ownership lead to lower levels of deforestation? To address these questions we consider the different land-use characteristics of rural-resident and urban-resident farmholders.

In 1990 fifty-six farms in the Rolim de Moura colonization area were visited and their residents interviewed using a standardized questionnaire. The vast majority (86 percent) of these residents owned legal title to the farm properties on which they resided. In addition, of the 418 randomly selected urban households interviewed in the town of Rolim de Moura, 98 (23 percent) reported owning rural lots in the surrounding colonization area. Two interesting observations emerged from the survey. First, 28 percent of the rural households surveyed indicated they owned urban properties, while 23 percent of urban households surveyed reported owning rural properties. We surmise, then, that approximately one-quarter of all residents in Rondônia own both rural and urban properties. Our analysis compares the land-use differences observed in 1990 between the 23 percent of urban-resident farmholders and the 72 percent of rural-resident farmholders that do not also own urban properties. Second, while the percentage of the urban population holding farms dropped between 1984 and 1990 from 44 percent to 23 percent, the general increase in the urban population of Rolim de Moura from approximately 15,000 to 30,000 during this period meant that the number of urban-resident farms slightly increased from 1,320 to 1,380. As the populist frontier has become increasingly urbanized, more rural properties have come under urban-resident ownership.

Several land-use differences between these two groups should be highlighted (table 10.1A). First, urban-resident farms tend to be smaller than rural-resident farms. While the mean value of rural property size for the urban-resident group (136 hectares) is 44 percent larger than that of rural residents (94 hectares), the distribution of urban-resident farms is skewed by three outliers each exceeding 500 hectares (table 10.1B). Indeed, 61 percent of urban-resident farmholdings were 50 hectares or smaller while only 14 percent of rural-resident holdings were found in this size range. Rural-resident farms appear normally distributed around the mean of 94 hectares, their original size.[1] The varied size distribution of urban-resident farms suggests some greater degree of heterogeneity among urban-resident farmholders than among their rural counterparts.

Second, contrary to the trend predicted by modernization theory, the urbanization of rural property in Amazonia appears to be associated with higher, not lower, levels of deforestation. The average natural forest area cleared on the sample of urban-resident farms was 41 hectares, 24 percent higher than on the rural-resident farms, where 33 hectares of natural vegetation had been cleared by 1990. While it is uncertain from the

ondary growth or fallows, which typically generate no income. The survey data reveal that rural-resident farms are more intensively utilized than urban-resident farms. The former have more than twice the area in the most (labor and capital) intensive uses, coffee and annual crops, and just over half of the area in the lower intensity landscapes of pasture and secondary growth, than do the latter. In sum, 35 percent of the area on rural-resident farms is in higher intensity uses, compared to only 12 percent of the area on urban-resident farms.[4]

Three other observations about land use warrant mention. First, a higher percentage of urban-resident farmholders (15.5 percent) exclusively practice pasture-based (i.e., cattle) production on their farms than did rural-resident farmholders (9 percent). Second, a slightly lower percentage of urban-resident farmholders extracted timber from their farm lots, possibly indicating a lack of interest in managing farm properties for income.[5] Third—and this is perhaps the most striking difference between these two groups—while not a single rural-resident farm lies completely idle, fully one-fifth of all urban-resident farms in the sample were not producing anything whatsoever. For a significant proportion of urban-resident farmholders, rural property ownership is a speculative, rather than a productive investment, a form of passive savings and a weekend retreat.

Returning to the three suppositions of this inquiry: First, do urban and rural-resident farmholders pursue distinct land-use strategies on their rural properties? The findings of the survey indicate potentially significant differences in the rural landscapes of urban and rural residents of the Amazon. With regard to the second, modernization hypothesis: Has the urbanization process in the Amazon induced greater intensification of agricultural production? Our survey data suggest just the opposite. The urbanization of rural property ownership has led to less intensive, less productive uses of rural land. Finally, has the progressive urban transformation of the Amazon frontier led to lower levels of natural forest conversion? On this point the survey data are equivocal. On the one hand, urban-resident farmholders have cleared more forest, on average, than rural-resident farmholders. On the other hand, urban-resident farms tend to be smaller.

All of this begs a few questions about the so-called urban transition in Third World agriculture and the incorporation of the rural sector into the global economy. Are urban and rural farmholders fundamentally different social groups? Or are they different factions of the same social group at different stages in the structural transition to advanced capitalism? The answer to these questions may depend on how urban residents come to acquire rural properties. Clearly there are key differences between these groups. It is likely that urban-resident farmholders, by virtue of their location in urban areas, pursue multiple income opportunities that do not so easily offer themselves to rural households, whose survival depends predominantly upon agricultural production.[6] For urban residents, farmholding may be just one of many elements in the household's accumulation strategy. For rural residents, the ability to survive off the land is more important than off-farm opportunities (Browder 1995).

Yet, our survey data suggest some danger in dichotomizing rural and urban groups. A substantial proportion of the urban population in Rondônia owns rural land.[7] This urban population, however, is difficult to classify regarding its relation to capital. Most are likely drawn from the petty urban bourgeoisie, which is not consistently associated to any significant extent with either merchant or agro-industrial capital, the principal agents of modernization and structural transformation. Yet, powerful urban groups are also present in the rural frontier. To characterize the process of the urbanization of rural property ownership as sequential or transitional, in which powerful urban social groups appropriate farmland from impoverished farmers only after their labor power has been expended in clearing the land, would be only partially accurate. Urban interests are also very much at the frontline of frontier expansion from the start, not at its closing, as the case study below indicates.

A Day-Trip to the "End of the Line"

"On August 2, 1990, as our urban survey crew was finishing up its work in Alto Alegre, I took the Toyota and our rural survey team on a day-trip to the end of linha 176, 15 kilometers south of Alto Alegre. The road abruptly ended against a wall of tropical forest where the bulldozer turned back. Even at this, the most remote terminus of the frontier in Rondônia, the whining noise of a poorly oiled chainsaw, followed intermittently by the howl of falling timber, could be heard in the forest background. Where the road ended, the trail began and as the Toyota could go no further, we proceeded on foot toward the ruckus. In a short time we came upon a makeshift lean-to constructed of rough wood beams, palm fronds, lumber and metal scraps, and cardboard tied together by pieces of rope, wire, and liana vines. Here, we found Maria Moreira da Costa, who told a lurid story of bondage, disease, and despair at the end of the line (plates 10.1 and 10.2).

Plate 10.1. The Moreira da Costa homestead, Alto Alegre, Rondônia, 1992, near the ``End of the Line'' *Photo: John O. Browder*

"It was a familiar frontier story. Maria and her husband, Miguel, had been recruited by an urban businessman, 'Dr. Astir,' to clear an area of forest on 365 hectares of land he had 'acquired' from a *grileiro* (professional land speculator) in Alta Floresta *município*. Miguel was to receive a salary for his services and the right to plant beans, maize, and rice partly for his own use, after the costs of all consumer goods supplied to his family by Dr. Astir had been deducted. As we watched and heard Miguel clearing forest over the ridge, Maria described their situation. The food crops Miguel and Maria complained that 'Until now we have not received any payment, because the *padrão* furnishes all of our food supplies and at the end of each month we keep owing him more money than we have earned. After clearing these 7 alqueires (about 17 hectares) of forest for the *padrão* we will still owe him 12,000 Cruzeiros.'

"The economic impoverishment resulting from their de facto indentured servitude, combined with their isolation and lack of regular transport, had clearly taken its toll on these migrants to Amazonia. Maria and all four of her children, ages 2 to 12, were suffering from malaria, malnutrition and various skin ailments and intestinal disorders. Frequently, their food stores ran empty and the family was forced to rely solely on wild palm heart. They were sick, remained in debt, and continued to clear tropical forest. These are the people pushing the edges of the frontier forward. Interestingly, the key figure in the background of this story, Dr. Astir, reportedly owned 43 urban properties in Alto Alegre, Alta Floresta, and Rolim de Moura including a lumber mill, several residences, a 'party house' (*casa de festa*), and several vacant urban lots.

"The pathetic story of the Moreira da Costa family at the end of the line captures the essence of the urban center-frontier periphery relationship. The frontier is often pushed forward not by small farmers acting alone, but under the sponsorship of urban-based interests in the region and beyond. The causes and dynamics of tropical deforestation cannot be fully grasped by examining only the accessible victims of development, the landless farmers nibbling away at the forest margins. We need to con-

Plate 10.2. The "End of the Line" (southern terminus of the Linha 36 in the Rolim de Moura–Alto Alegre corridor) at the forest edge, Rondônia, 1992
Photo: John O. Browder

sider the remote fringes of the frontier as extensions of the urban social and economic space and the urban-orientation of the forces that propel frontier expansion" (John O. Browder, adapted from logbook, 1990).

POWERING UP THE AMAZON: HYDROELECTRIC DEVELOPMENT AND ENVIRONMENTAL CHANGE

To meet Amazonia's growing urban and corporate demand for energy, Brazil's federal government has turned to the immense hydroelectric potential of the Amazon River Basin. Draining an area of approximately 350 million square kilometers, the Amazon Basin's hydroelectric potential is a staggering 100,000 megawatts, equivalent to 100 nuclear power stations (Bunyard 1987). Developing this potential through the construction of a network of large dams, however, entails significant social and environmental costs. In this section, we review the record of urban and corporatist-oriented hydroelectric projects in Brazilian Amazonia.

The environmental and social impacts of dam construction in the tropics are hotly debated issues. Some 63 percent of the Third World's hydroelectric potential is found in thirty countries containing all of the world's tropical rainforest (Mougeot 1990:94). In the South American tropics only Brazil and Paraguay are net energy importers. Domestic production provides for 73 percent of Brazilian national consumption; in Paraguay, the corresponding figure is only 12 percent (WRI 1987:300). To reduce dependency on foreign sources, these countries have aggressively promoted hydroelectric development. The Itaipú dam, a joint Brazilian-Paraguayan venture, is one of the world's largest hydroelectric complexes. In Brazil, where one-quarter of the foreign debt is attributed to energy imports, such hydroelectric projects have played a central role in national energy policy.[8]

The most significant environmental impacts of hydroelectric development in Brazil are directly linked to the rapid urbanization of Amazonia. In the early 1980s Brazil generated only 65 megawatts of electric power from the tributaries of the Amazon River, but current plans for forty-seven large and medium-sized hydroelectric power plants are expected to boost the region's contribution to 22,000 megawatts, providing for 40 percent of the country's projected demand by the year 2000 (Caufield 1983:60).[9] Brazil's federal power utility company for the Amazon, Centrais Elétricas do Norte do Brasil (ELETRONORTE), has identified eighty potential dam sites in the Amazon Basin, the impoundments of which would flood an estimated 100,000 square kilometers of forestland (Fearnside 1989:403; Mougeot 1990:95). Reservoirs at just three active dams (Tucuruí, Balbina, and Samuel) have flooded 6,430 square kilometers, accounting for the destruction of over 10 percent of the total area of Brazilian Amazonia deforested by 1990.[10]

But deforestation is only one environmental impact of hydroelectric development to serve the Amazon's growing urban population. Numerous social and public health problems are also linked to hydroelectric development and, as such, to the urban transformation of Amazonia. Moreover, hydroelectric projects are often the first stage in more complex land distribution and development schemes that influence the region's spatial organization and settlement structure by attracting unemployed workers, displacing resident populations, and flooding highways and existing towns and villages. A brief review of three hydroelectric projects illustrates the range and complexity of issues associated with powering up urban Amazonia.

Tucuruí, Pará

The Tucuruí dam, planned by ELETRONORTE in the 1970s, was constructed on the Tocantins River between 1980 and 1984 to supply electrical power to the Greater Carajás development pole in southern Pará, to metropolitan Belém, and to the national power grid. Brazil's huge Camargo Correa construction firm was the principal contractor for the project. Both the Tucuruí and Carajás projects, driven by links to national and global capital, were instrumental in establishing a corporatist social formation in eastern Amazonia. Expected to generate between 3,960 and 7,300 megawatts, the Tucuruí project is the world's largest hydroelectric dam in a tropical forest environment, initially flooding an area of 2,430 square kilometers (Adam 1988; Hall 1989). The dam flooded about 100 kilometers of the original route of the Transamazon Highway between the towns of Marabá and Altamira—a road planned and constructed only a few years before—attesting to the conflicting social goals of successive development programs in Amazonia. Superseding the previous populist emphasis on small-farmer settlements of the 1970s, the huge Tucuruí hydroelectric complex tangibly symbolizes the corporate megaprojects of the 1980s (plate 10.3).

When construction activity began at the hydroelectric site in the late 1970s—pictured, by the way, in the popular film *Emerald Forest*—the small village of Tucuruí, situated below the dam, suddenly exploded into one of southern Pará's most visible boomtowns. Attracted by the lure of construction work at the dam, or by the associated service-sector activity nearby, migrants poured into the town from surrounding areas. During visits in 1978 and 1988, a decade apart, we observed the transformation of a ramshackle river town, similar to Itupiranga below, into a modern dual city with an improved infrastructure and notable extremes of wealth and poverty. In the late 1970s the town consisted of unpaved streets teeming with vehicles and a rapidly expanding commercial sector of small shops, restaurants, hotels, and lodging houses to serve the burgeoning population of construction personnel. A decade later, in the late 1980s, Tucuruí was much larger and boasted carefully maintained squares and paved streets in the central areas. The informal sector was still quite apparent in the lively street fairs and ambulatory vendors, however, as well as in the peripheral shantytowns.

For project professionals and their families, ELETRONORTE planned and built an enclosed company town, designed in a curvilinear suburban fashion and located on a ridge overlooking the teeming informal settlement along the river below (plate 7.3). The hybrid morphology of the planned and unplanned towns of Tucuruí reflects the differentiated nature of the corporatist frontier of southern Pará, as at Carajás (chapter 4). Overall,

Plate 10.3. View of the massive hydroelectric works at Tucuruí, Pará, 1989
Photo: Brian J. Godfrey

the urban population of the municipal seat of Tucuruí grew from 5,673 in 1970 to 16,687 in 1980 and to 46,011 in 1991 (IBGE various years).

Besides the explosive urbanization of the town of Tucuruí (plate 7.3), the hydroelectric project has had two major social and environmental impacts on the regional population distribution. First, the dam's impoundment area inundated seventeen towns and villages, displacing between 25,000 and 35,000 people, many of them indigenous, only a fraction of whom received any compensation from ELETRONORTE (Bunyard 1987; Hall 1989). Many small towns along the Tocantins River between Marabá and Tucuruí were flooded by the dam's artificial lake. For example, Itupiranga had emerged as a small fishing village and collecting point for forest products about 40 kilometers north of Marabá during the early twentieth century. Reminiscent of Wagley's "Amazon Town" in its cohesive *caboclo* community, Itupiranga was forced to relocate about 15 kilometers to the west in the early 1980s. Although the municipal seat of Itupiranga roughly doubled in population between 1980 and 1991, reaching a population of about 9,225, the new planned town bore little resemblance to the former sleepy river village (IBGE 1991). Such traditional riverine populations, concentrated in channel settlements, are the most vulnerable to dislocation by hydroelectric development projects. Given the economic logic of their settlement location, based on the extraction of river resources (e.g., fish, turtles, manatees) and forest products (e.g., rubber latex and Brazil nuts) and the necessity of river transport, these "traditional" peoples are among the first to be dislocated in the process of river impoundment. While many of those who actually received dislocation relief moved to other rural areas, most of the dislocated undoubtedly ended up in Pará's burgeoning towns and cities, involuntarily contributing to the region's urbanization.

A second environmental consequence of Tucuruí arose from ELETRONORTE's belated decision in 1979 to selectively remove the existing vegetation from the planned reservoir area prior to impounding the Tocantins river.[11] Hoping to cut costs, the Ministry of Agriculture announced in 1978 an open competition among logging companies to salvage an estimated 13.5 million cubic meters of exportable hardwood timber from the impoundment area over a three-year period (*Gazeta Mercantil* 1978). A private company, Agropecuaria CAPEMI (Caixa de Pecúlio dos Militares), was awarded the contract. However, by the end of 1982, after a U.S.$100 million loan from Banco Lazard Frères to finance the salvage operation was canceled, Agropecuaria CAPEMI went bankrupt, owing more than U.S.$4 million to its 3,000 workers (Goodland 1985; Bunyard 1987). Agropecuaria CAPEMI had cleared less than one-tenth of the area it was contracted to clear, and it had cut fewer than 100,000 cubic meters of timber, leaving journalists to speculate publicly about the volume of wood that must have been sold clandestinely (Pinto 1984).[12]

In March and April of 1984 Belém's two major newspapers, *O Liberal* and *A Província do Pará* published front-page stories revealing that CAPEMI had indiscriminately dumped large quantities of the defoliant dioxin (Agent Orange) throughout the Tucuruí area since 1980, exposing thousands of rural inhabitants to this and other lethal toxins.[13] Both ELETRONORTE and Dow Chemical Corporation, the manufacturer of the herbicides, acknowledged the use of these plant killers in 1982. An international controversy ensued. *Newsweek* called it a potential "environmental disaster for generations to come" (Smolowe and Margolis 1984). *O Liberal* referred to the Tucuruí area as an "ecological cemetery" (*O Liberal* 1984).[14] João Batista Bastos, secretary of agriculture of Pará, blasted the federal government, calling the ecological disaster "a direct consequence of the absence of a rational policy for the occupation of Amazonia and of the hydroelectric construction project . . . that is totally oriented to serve large projects in the area . . . , exclusively benefiting foreign groups" (Bastos 1984).[15] This local rhetoric accentuated the rift between Amazon state politicians and the federal bureaucracy that first developed in the period of the New Republic under President Getúlio Vargas. In this era global capital shifted its alliance from regional to national elites, a spatial realignment of power beginning with World War II but largely resulting from twenty years of military dictatorship during which the urban transformation of Amazonia occurred.

The environmental modifications induced by the Tucuruí project have led to other public health concerns, especially the expected rise in malaria cases in the immediate area where already one in five people have been found to be infected.[16] Schistosomiasis, a serious problem in Brazil affecting 14 million people, is eventually expected to become endemic to the reservoir (Bunyard 1987). Onchocerciasis (river blindness) is carried by blackflies, which are locally abundant in Tucuruí, and leishmaniasis, a fly-borne disease, could also become a serious problem in Tucuruí lake (Lemos de Sá 1992).

A final concern is declining fisheries. Many important fish species found in the rivers prior to impoundment are disappearing due to degrading water quality both upstream and downstream from the hydroelectric plants and due to the physical impediments to fish migration presented by dams (Lemos de Sá 1992; Smith 1985). It is presently unclear what measurable threat hydroelectric development of the Amazon Basin might pose to the sustainable potential of the basin's total fishery, which is estimated to be on the order of 320,000 to 350,000 tons per year (Junk 1984).

Balbina, Amazonas

The Balbina dam project makes Tucuruí seem like a success story in wise environmental management. Planned in 1975–76, Balbina floods approximately the same amount of rainforest (2,360 square kilometers) to generate 2 percent of Tucuruí's output (an average of only 112.2 megawatts) (Fearnside 1989). The relatively flat terrain of the impoundment area, the shallow depth of the reservoir, and the long average residence time of water in the lake as well as the availability of alternative energy options for Manaus made the Balbina option monumentally ill-advised from the start on technical, financial, and social grounds. For example, Fearnside notes that Balbina sacrifices thirty-one times more forest per megawatt of generating capacity installed than does Tucuruí (Fearnside 1989:407). Only 50 square kilometers, or 2 percent of the submergence area, was cleared of vegetation with little done to salvage valuable timber and fuelwood. The total mass of potential fuelwood alone is estimated to be 63 million metric tons for the entire 2,360 square kilometer area, equivalent to 39.4 gigawatt-hours, 161,000 barrels of crude oil, or U.S.$3.2 million (Fearnside 1989). Moreover, the construction cost of the dam, which doubled from initial estimates of U.S.$383 million, amounted to about U.S.$3,000 per kilowatt, 4.4 times higher than that of Tucuruí, total construction costs of which reached U.S.$675 per kilowatt (Fearnside 1989).

The Balbina reservoir is a labyrinth of canals among an estimated 1,500 islands. Consequently, the extremely long water residence time of 11.7 months (in contrast to 1.8 months at Tucuruí), combined with slowly decomposing residual vegetation in the submergence area is expected to lead to higher than normal maintenance costs. In the first six years, cumulative maintenance totaled U.S.$2 million or U.S.$16,600 per installed megawatt per year—seventy times the cost per megawatt of maintaining a comparable dam in the semi-arid northeastern part of Brazil (Fearnside 1989). Although the social impacts, in terms of people dislocated, of the Balbina project were thought to be considerably less than those at Tucuruí,[17] the economic waste has been appalling, prompting Hugh W. Foster, U.S. alternate executive director of the World Bank

Board of Executive Directors to describe Balbina as an example of "totally unacceptable investments" (quoted in Fearnside 1989:415). Most disturbing of all were the environmental losses. Paulo Nogueira Neto, former president of the Special Secretariat of the Environment (SEMA) described Balbina as "the greatest ecological disaster ever provoked by a reservoir" (quoted in Fearnside 1989:417). This inefficient and poorly planned energy project will "never supply 50 percent of the Manaus demand" (Fearnside 1989:413).

Fearnside (1989) speculates about the various explanations for the Brazilian government's counterintuitive decision to implement the Balbina plan. Two conclusions are clear. First, corporatist interests associated with ELETRONORTE, including its own immediate constituency of professionals (*barrageiros*) have a particular vested class interest in building dams regardless of their economic, social, and environmental trade-offs. In other words, ELETRONORTE's constituency helps to shape the corporatist frontier and landscape (chapter 4). Second, the Balbina dam project, indeed the urban economic expansion of Manaus itself, illustrates subsidized growth for the benefit of transnational capital:

> The power from Balbina will largely benefit the international companies that have established factories in the Manaus Free Trade Superintendency Zone (SUFRAMA). That power will be subsidized for these firms at the expense of residential consumers throughout the country is an irritant to many Brazilians. (Fearnside 1989:418)

The Manaus Free Trade Zone and SUFRAMA were political concessions made by the military authoritarian regime in 1967 to compensate western Amazonia (and ensure its continued political support of the regime) for the concentration of fiscal incentives in the eastern Amazon through SUDAM in Belém (see chapter 3). The contemporary growth of Manaus and the environmental trade-offs it has entailed can only be attributed to the political decisions of an authoritarian regime seeking to legitimize its power by subsidizing national and global capital investment in economically questionable development projects.

Samuel, Rondônia

As mentioned earlier, ELETRONORTE's penetration into Rondônia's populist frontier had the predictable effect of enclosing a large area of tropical forest and creating a controlled bedroom community (Vila

Eletronorte) designed on strict social class divisions on the fringe of Porto Velho in stark contrast to the populist peri-urban squatter settlements now enveloping the state capital (chapter 6).[18] More importantly, perhaps, the Samuel dam project illustrates the flexibility of the government-corporatist relationship. Closely linked to global capital, ELETRONORTE has maintained its own institutional viability by serving the energy needs of both corporatist and populist frontiers.

The inadequate supply of electrical energy has long been considered a severe constraint on the development of Rondônia (World Bank 1981). Two energy production companies are responsible for generating electricity in the state. The Centrais Elétrica de Rondônia, SA (CERON), a largely state-owned utility, produced 172.5 megawatts in 1986, 95 percent of which was generated from burning diesel fuel, and distributed them to thirty-one localities in the interior of the state (FIERO ca. 1993). In 1981 ELETRONORTE began generating electricity in Rondônia from diesel oil. By 1986 it was producing more than half of Rondônia's total energy output (192.5 megawatts). Prior to the closing of the Jamarí River at the Samuel dam in 1989, Rondônia had been a deficit energy consumer producing only 45 percent of its energy consumption. The balance was acquired through the Petrobrás network from outside of Rondônia with considerable subsidization (FIERO ca. 1993).[19] The Samuel Dam is devoted to supplying the state capitals (Porto Velho and Rio Branco) and the larger urban centers along the BR-364 corridor (from Ariquemes to Jí-Paraná).

The social benefits of the Samuel dam project have been arguably immediate and impressive in scale. In 1990, one year after beginning operation, Samuel provided 36 percent of Rondônia's total power supply and as much as 58 percent in 1996. The number of households in Rondônia provided with electrical service grew from 40,419 in 1983 to 132,157 by 1990, predominantly (42 percent) in Porto Velho (FIERO ca. 1993). Virtually the entire urban area of Porto Velho, except for several shantytowns on the metropolitan fringe (chapter 5), is on the municipal power grid (Prefeitura Municipal de Porto Velho ca. 1994).

The Samuel dam reflects ELETRONORTE's increased attention to some of the environmental concerns expressed during its earlier dam-building projects at Tucuruí and Balbina. Although the 217-megawatt, five-turbine plant inundated a relatively small area (560 square kilometers) by Balbina and Tucuruí standards, considerable efforts were expended to rescue wildlife and protect adjoining forests at the Samuel Biological Reserve.

For nearly five months following the closing of Samuel's flood gates, ELETRONORTE sponsored an animal rescue operation, evacuating some 16,000 animals (Lemos de Sá 1992).[20]

Unlike the Tucuruí and Balbina experiences in the central Amazon, two additional points merit mention in the case of Samuel in the southern Amazon. First, because of the distinct seasonality of rainfall in the southern Amazon, the influence of the dam and its reservoir on local ecosystems is not a continuous or one-time phenomenon. During the dry season (May to October), it is estimated that the reservoir contracts to 40 percent of its capacity. Approximately 300 square kilometers of nonsedimentary (e.g, *non-várzea*) soils will be exposed each year during the dry season cycle (Lemos de Sá 1992), potentially altering adjacent biological communities and fostering growth of mosquito populations and with them increasing the prospects of malaria in a region already known as the world capital of malaria (Sawyer and Sawyer 1987). The Samuel dam, while providing tangible benefits to Porto Velho's urban population, has entailed serious environmental and public health trade-offs, especially for local inhabitants in the vicinity of the dam.

The second environmental effect of the Samuel dam includes the larger social and historical context: In the Jamarí River Basin deforestation rates due to rapid colonization by farmers have risen sharply. According to one study, the forest area cleared in the basin from 1978 to 1983 rose from 5 to 1,654 square kilometers (Graham 1986). The estimated change in sediment load in the river due to the soil erosion in the Jamarí watershed ranges from 25,417 tons (delivered sediment) to 54,643 tons (Graham 1986). The impacts of continued growth in sediment load due to upland deforestation on the economic longevity of the dam, estimated to be one hundred years (ELETRONORTE 1979), remain disputed and unclear.

The economic costs of powering up the Amazon's rainforest cities are another question. In 1975 a nationwide power facility financing system, the Global Guarantee Reserve Fund (Fundo de Reserva Global de Garantia), was established drawing upon the profits of financially successful utilities throughout Brazil. A common tariff was established for all utility consumers nationwide regardless of production cost differentials.[21] Consequently, hydroelectric development in the Amazon, where costs are higher than in the more densely populated Southeast, has been heavily subsidized. However, the system is financially unstable due to the reduced number of profitable utilities throughout Brazil and their inability to support the fund with sufficient resources to offset the effects of hyperinflation during the 1980s.[22] To help offset the shortfall in capital needed to complete and maintain the hydroelectric system of the Amazon, ELETROBRÁS, the national parent company of ELETRONORTE, secured a U.S. $241 million loan from the World Bank in 1983. From this loan CERON obtained a contract for U.S. $29 million, the equivalent of only 18 percent of its 1984–1988 development budget (FIERO ca. 1993). Other credits have been provided by European manufacturers of machine parts used in dam construction (Fearnside 1989).

The development of Brazil's hydroelectric system in the Amazon is thus highly dependent on both national and global capital. Besides relying heavily on global capital for financing, hydroelectric development in Brazil also depends upon imported technology and foreign-made generators and machine parts. In both Balbina and Samuel, the high costs of replacing and maintaining foreign-manufactured equipment have added significantly to the total costs of these projects, resulting in chronic service interruptions to consumers and industries alike. A logical tension has emerged in the debate over the Amazon's hydroelectric future between local industrial elites (e.g., FIERO) and the transnational financial sponsors of Brazil's electrification program (e.g., World Bank). For example, FIERO's own constituency is local industrial capital oriented to the domestic market. Echoing the concerns of local politicians (e.g., Bastos) and researchers (e.g., Fearnside), FIERO has publicly criticized the whole financing system for "subsidizing projects oriented to export, some making highly intensive use of electrical energy" (FIERO ca. 1993:113).

Urbanization, Global Capital, and Amazonian Hydroelectric Development

After reviewing the three case studies above, it is useful to consider the social contexts of hydroelectric development in the Amazon. Hydroelectric development serves Amazonia's growing urban centers and influences the distribution of the region's human population and associated settlement patterns in two principal ways. First, during the construction phases these megaprojects attract unemployed workers from around the country creating rings of new makeshift shantytowns around the project sites, as described for Tucuruí (see also chapter 4). Second, the water impoundments created by some of the dams have inundated areas already occupied; some existing settlements were washed away, resulting in population displacement. Mougeot (1990:97) estimates that if currently contemplated dams in the Araguaia-Tocantins Basin and major southern tributaries of the Amazon were constructed, the total area of tropical forestlands inundated would be about 81,850 square kilometers and would displace between 85,000 and 156,000 residents. Clearly these involuntary population relocations have immediate local effects on urban population growth in the Amazon.

How is hydroelectric development related to the various corporatist and populist frontiers of Amazonia? As indicated above, ELETRONORTE, a quasi-corporatist entity, has appeared in both social spaces of the frontier to offer its dam-building services with seeming equanimity. The impacts of hydroelectric development are mediated by the social context of the specific frontier spaces in which ELETRONORTE operates. Even though the dam-building enterprise is a corporatist initiative, the hydroelectric development of the Amazon assumes the social character of the specific frontier spaces in which it occurs. Moreover, the environmental and social impacts of hydroelectric development differ depending on preexisting microregional and biophysical characteristics; they are differentially articulated to local historical and environmental features.

Global capital, while supporting a general plan for Amazon urban-oriented hydroelectrification, clearly favors those specific energy projects that directly support export-oriented development projects (e.g., Carajás). Overall, global capital only selectively penetrates the Amazon, reinforcing the corporatist character of specific frontier zones. A general process of globalization is simply not evident in the energy sector. Global capital no longer requires regional metropolitan centers as administrative "landings" from which to direct the restructuring and flow of economic activity in the frontier. Working through national economic and political elites in São Paulo and Brasília, transnational capital physically bases itself directly on project sites in the Amazon's rainforests, engendering a corporatist character in those areas of the frontier it selectively invests in.

URBANIZATION AND THE GOLD-MINING FRONTIER

We have contended that the social orientation of Brazilian frontier expansion may be conceptualized as a continuum, ranging from corporatist to populist ideal types. Distinctive patterns of urbanization derive from the particular mix of socioeconomic forces driving the expansion at any given place. The Amazon, as we have repeatedly indicated, cannot be neatly divided into mutually exclusive social spaces; such false dichotomies simply do not exist. Nowhere in the Amazon are the boundaries between corporatist and populist forces more ambiguous than in the region's boisterous mining frontiers.

The Amazon's small-scale mining sector (*garimpo* in Portuguese) became increasingly important in the 1980s. The extraction of gold in particular provoked a frenzied in-migration reminiscent of the great gold rushes of North America. Amazonia's contemporary gold rush already ranks in importance as equal with the great previous regional extractive boom, the rubber trade. Indeed, in one of the few contemporary studies on the subject, Cleary (1990) argues that the contemporary gold rush in Amazonia is one of the largest in modern history: though the informal and often clandestine nature of gold mining makes it a particularly elusive subject of study, probably over 100 metric tons of gold, worth some billion dollars, now are produced annually in the region. It is often assumed that these gold-mining activities are ephemeral and leave no significant imprint on the regional settlement structure other than environmental problems such as mercury contamination. Yet gold mining has become one of Amazonia's principal economic activities, at times surpassing even the combined ranching-agricultural sector in importance. Probably at least half a million goldminers now operate in Amazonia, not including the merchants, restauranteurs, prostitutes, and others linked to mining. In areas of southern Pará, Roraima, parts of Rondônia, and elsewhere in the Amazon, gold mining has become the leading economic sector (Godfrey 1992).

The Amazon Basin boasts extensive, varied, and still largely unexplored mineral riches. In geological terms, the *várzea* lowlands along the main Amazon watercourse exhibit a wide plain of comparatively recent origin, comprised largely of unconsolidated gravels, sands, clays, and silt. These Amazon lowlands lie between two immense highland formations: to the north the Guiana Shield and to the south the Brazilian Shield. These highlands, set on Precambrian basement complexes of ancient crystalline rocks, contain enormous concentrations of valuable minerals, such as gold, cassiterite, bauxite, copper, zinc, iron ore, and others. Many of the largest and most valuable discoveries so far have taken place in relatively accessible parts of eastern Amazonia, especially in the south of Pará between the Araguaia and Xingu rivers. For example, the Carajás mountains contain the world's largest known deposits of high-grade iron ore, the third largest reserve of bauxite, and large quantities of manganese, nickel, copper, gold, and other minerals (Becker 1990:64–65; Roberts 1991).

Mining activities are not new to the Amazon. Historically, gold and precious stones have been extracted since colonial times in a series of periodic booms, especially in western Maranhão and other parts of eastern Amazonia. Mining was a way of life practiced by runaway slaves and others escaping government regulation, and this clandestine tradition continues to the present (Cleary 1990:27). After the rubber boom ended by World War I, the regional population clustered along the major watercourses of the region engaged in the extraction of other resources, such as Brazil nuts, animal skins, and minerals; diamond mining became a particular focus of activity on the Tocantins and other rivers after the late 1930s (Velho 1972:69–72).

The scale and capitalization of Amazonian mining activities increased dramatically after World War II, leading at times to bitter social conflicts among the diverse public and private mining interests. Large-scale mechanized mining, authorized by the Brazilian government in a concession to foreign capital, began with the production of manganese in Amapá in the 1940s. The mining of cassiterite by small producers became widespread during the 1950s in Rondônia and on the Tapajós River in Pará. Since the late 1970s the massive iron-ore project at the Serra dos Carajás, in southern Pará, owned and operated by the state-owned Vale do Rio Doce (CVRD) Company, most vividly dramatizes the clash between corporatist and populist sectors. As chronicled in chapter 7, a bipolar urban system emerged, indicative of the corporatist frontier, as the spontaneous settlement of Parauapebas grew to provide subcontracted services and informal activities for the nearby company town at Carajás. Parauapebas, located just outside the entrance to the Carajás iron-ore project, has grown at annual rates of nearly 20 percent in recent years and doubled its population in just four years (Roberts 1991, 1992).

Corporatist-driven urbanization at Carajás has remained largely disarticulated from the populist forces of expansion surrounding this industrial mining site, as in the infamous case of Serra Pelada. In 1979 a fantastic gold strike at the Serra Pelada near Carajás ignited a huge infusion of independent miners to the region. The settlement at Serra Pelada reached a peak population of over 100,000 at the height of the gold extraction in 1983 (plate 10.4). The miners resisted government efforts to mechanize production and regulate the gold mines at Serra Pelada, as it had done at the tin mines of Rondônia more than a decade earlier. The miners came to form a potent political force in the region, but with the exhaustion of accessible gold, mined manually and separated by indiscriminate applications of mercury, production dropped, and this mining town shrank to a population of about 30,000 by 1988. By this time other mines in southern Pará attracted tens of thousands of miners (Godfrey 1990, 1992; Schmink and Wood 1992).[23]

Gold mining also emerged in other regions of the Amazon during the 1980s (see Feijao and Pinto 1990 for a comprehensive listing). River dredging operations multiplied on the Madeira River of Rondônia. After the discovery of gold in northerly Roraima in 1987, thousands of miners encroached on the traditional native lands of the Yanomami Indians; at least 10–15 percent of the approximately 10,000 Yanomamis subsequently died of disease, hunger, and

Plate 10.4. Miners emerging after a day working in the pits at Serra Pelada, Pará, 1989 *Photo: Brian J. Godfrey*

violence. In early 1990 the new president, Fernando Collor de Mello, heeded international protests and ordered the closure of the *garimpeiro*'s landing strips in the area, but thousands of miners still remained there. This "forgotten frontier" of northern Brazil (Rivière 1972), long dominated by cattle ranching, has been overwhelmed by this recent mining boom in Roraima, and this remote state's population is now about two-thirds urbanized. The state capital, Boa Vista, has grown from a population of 43,000 in 1980 to an estimated 165,000 in 1989 (Brooke 1990a, b; Sawyer et al. 1990). In this section, we focus on the expansion of the mining frontier along the Xingu corridor.

Populist Mining Activities in the Xingu Region, Southern Pará

As recounted in chapter 7, gold was discovered in the São Félix do Xingu region in the early 1980s. A gold-mining camp at Vila do Cuca sprang up about 25 kilometers directly south of Tucumã, inside the area claimed by the agricultural colonization project of the Andrade Gutierrez Company. The town of Ourilândia (literally, "Goldland") formed nearby, composed of miners excluded from the lands of the private colonization project. Tensions between the private company and the hostile local population of miners ultimately led to the demise of the colonization project (Butler 1985; Schmink and Wood 1992).

Unlike the highly capitalized corporatist forms of mining, such as those used at Carajás, mining along the Xingu corridor has been predominantly populist in nature, challenging the corporatist control of this space. During the 1980s gold mining developed as an important economic sector with vital local linkages in the Xingu region. In our 1990 survey of households in southern Pará, we found that significant percentages of the labor force in urban residence actually worked in nearby gold fields: 10 percent of the economically active population in Xinguara was engaged in gold mining, 11 percent in Tucumã, and 24 percent in Ourilândia do Norte (Godfrey 1992). The value of local mineral production is difficult to estimate, since most gold is flown out as contraband from local landing strips. In a personal interview in 1990 the municipal prefect of Tucumã claimed that gold mining had surpassed the declining timber industry and stagnant agriculture as the municipality's leading sector. But he argued that gold mining had brought many social and environmental problems and few long-lasting benefits to the area (Silva 1990, personal communication).

By 1990 the town of Cuca itself had a population of about 2,500 residents, who worked in scores of businesses and enjoyed telephone service, television, and other modern conveniences. The town had become a service center for other *garimpos* in the region, places like Vila Bananal, Serrinha, and Rio Branco in the municipality of Tucumã. The emergence of the nearby Rio Branco gold mines in early 1990 provided an opportunity to witness firsthand the formation of a new *garimpo* in southern Pará (Godfrey 1992).

The settlement at Rio Branco grew quickly as word spread among miners dissatisfied with other mines in early 1990. By late July the population of Rio Branco was estimated at 850; along the main road, leading into the mining pits about one hundred businesses (half of them bars) served the local miners (plate 10.5). In the adjacent alluvial gold mine, ore was pumped out of the river bed and separated by mercury in boxes. The owner of a claim generally took 70 percent of the profits, but he was also responsible for providing food for the group of five to seven miners and the price of the hydraulic machines and other machines to separate the ore. Workers divided up the remaining 30 percent of the profits from

Plate 10.5. A mining camp at Rio Branco, Pará, 1990. The hoses criss-crossing the foreground were used in the hydraulic mining operations. *Photo: Brian J. Godfrey*

all gold found. This gold provided an unsteady source of income, but at least it did not require that the miners go into debt to pay for food, housing, and so forth.

In July of 1990 our team of eight researchers administered seventy questionnaires on migration and socioeconomic status to the miners working at Rio Branco. Over three-quarters of the miners had been raised in the Brazilian Northeast, a traditional migratory path into southern Pará; in particular, they hailed from the neighboring states of Maranhão (55.7 percent) or Piauí (17.1 percent). This predominance of northeastern states indicates a continuation of a long-term migratory trend. Asked about their father's state of birth, over 82 percent responded with a northeastern state, particularly Maranhão (35.7 percent) or Piauí (28.5 percent).

The vast majority (78.5 percent) had their first mining experiences in Pará. The most important initial sites for the local gold miners were Serra Pelada (16 percent), Cuca (11 percent), Rio Branco itself (10 percent), and Cumarú (7 percent). While most (52.8 percent) reported previous rural residences, significantly almost half (45.7 percent) had previously lived primarily in cities, challenging the prevailing view that these migrants are frustrated peasant farmers. Social mobility was common, if not universal, among the miners. Fully half of the men had bought some important capital goods or land from profits earned in gold mining. In contrast to the populist agrarian frontier, a thriving remittance economy was indicated by nearly a third (30 percent) of the miners sending money to someone in another state; however, hardly any miners (5.7 percent) reported receiving money. Overall, nearly a quarter of the miners (24.2 percent) described themselves as richer since arriving at the mine, slightly less (21.4 percent) felt poorer, and nearly half (44.2 percent) considered themselves in the same status as before.

The most alarming environmental health problems at Rio Branco and other *garimpos* stem from the widespread mercury contamination. The liquid mercury is applied to the recycled water, pumped from large pools, to separate gold particles from river sediments. Mercury bonds with the gold in sluice boxes before the amalgam is heated in pans; the mercury eventually burns off and leaves pure gold. Probably about half of the mercury escapes as vapor, inhaled directly by the miners, and the residues are dumped back into the rivers or into the stagnant pools of water at the mines. Miners use as much as two pounds of toxic mercury to produce one pound of gold. Estimates of the amount of mercury dumped in the Amazon Basin range from 100 to 235 tons annually (Brooke 1990c; Hoffman 1994). Resultant problems are reported to include mercury pollution of the water sources for communities downstream, the extinction of aquatic fauna, and health effects on local populations.

Serious injuries to the health of miners and local people already have been documented.[24] Tests of hair, blood, and urine in several communities in this region found high levels of mercury contamination; about a third of the miners had mercury levels over the tolerable limits set by the World Health Organization. A study on mercury contamination at four sites in the Brazilian Amazon found that

> evidence of excessive exposure to mercury vapour . . . could give rise to adverse health effects in a proportion of the workers . . . sampled. . . . In addition there is strong evidence of environmental contamination with mercury giving rise to high blood levels in a proportion of residents and fishermen . . . with high weekly intake of fish, presumably with high methyl mercury levels acquired from contaminated waterways. (Thornton et al. 1990:)

Particularly disturbing has been the contamination of nearby native communities not themselves directly involved in mining. For example, about a quarter of the Kayapó Indians of nearby Gorotire have been exposed to excessive levels of mercury (Couto et al. 1988). In Roraima's Yanomami indigenous area illegal gold mining was reported to be responsible for the deaths of an estimated 15 percent of the Yanomami population between 1988 and 1990 (McMillan 1993). Mercury poisoning has numerous short-term health effects, but even more serious are the long-term prospects of birth defects in affected populations.[25] Still, miners themselves overwhelmingly regard malaria, borne by anopheles mosquitoes in the stagnant local waters, as a more serious and immediate problem for their livelihood. One study in 1988 found that 87 percent of the miners regarded malaria as their primary health problem, as opposed to only 6.2 percent who thought mercury contamination and the general absence of sanitary facilities were as important (Couto et al. 1988).

Populist gold-mining activities raise vexing problems of public policy. Despite serious environmental and health problems, *garimpo* mining employs hundreds of thousands of people and therefore cannot realistically be shut down by authorities in Amazonia. Indeed, given the importance of gold mining by private prospectors in contemporary Amazonia, the independent mineral sector is unlikely to disappear even in the face of official constraints or political repression. The diversion of labor into the mining sector may even prevent the extensive deforestation that would occur with alternative populist activities, such as farming and cattle ranching. Rather than trying to eliminate spontaneous migration to the gold-mining areas, as federal authorities attempted in the case of Serra Pelada, public policy would be better directed at ameliorating acute environmental problems in the mines, at controlling the incursions of miners into native lands, and generally at providing better social services in the frontier areas. Recent experience indicates that the resilient sector of independent miners cannot be eliminated, even by force, as it provides a necessary avenue of social mobility for hundreds of thousands of migrants lacking other opportunities elsewhere in Brazil. Still, the benefits of populist *garimpo* mining are debatable, as Schmink and Wood (1992:245) note in their analysis of southern Pará:

> The *garimpo* was a mixed blessing for the southern Pará region. The gold economy contributed to local development, but it also undermined other activities. If mining provided men a much-needed source of income to help finance their family's economic future, their absence left women and children to fend for themselves. The lure of the gold fields often undermined the already precarious stability of family relations and stimulated prostitution. Similarly, mining provided an avenue for upward mobility and served as a refuge for the landless and unemployed, but it did little to resolve the underlying problem of unequal access to land.

In terms of frontier urbanization, the serendipity of independent *garimpeiro* mineral strikes adds significant but volatile elements to the urban transformation of Amazonia. Irregular local patterns and strong external linkages characterize the mining settlements of disparate areas in southern Pará, Roraima, and Rondônia. Though some mining camps, like Cuca, eventually evolve into local service centers, generally the boom-and-bust cycles of growth and decline hinder the creation of a stable, sustainable urban system. Consequently, mining-driven settlement generally contributes to the regionally disarticulated nature of frontier urbanization in Amazonia.

URBAN ENVIRONMENTAL DEGRADATION

The rapid growth of the urban population of Amazonia and the horizon-tal expansion of urban areas have raced far ahead of the capacity of most municipalities to supply these new urban areas with adequate physical infrastructure (e.g., housing, streets, water, sewer, energy, schools, and clinics). As Sawyer notes:

> A great part of the urban population on the frontier lives in the most precarious conditions. [Typically] their housing is self-built using easily available materials: rough wood, sawmill refuse, mud, thatch (*palha*). Floors are customarily compacted soil. Black plastic sheets ("zinc of the poor") serve as roofs. Sometimes, for water and sewage, wells and latrines are dug. Other times, streams are used. . . . Dirty and stagnant water accumulates inside the urban area, pol-luting the water courses that are used for drinking, bathing, and washing clothes. (Sawyer 1993:163)

We estimate that between 60 and 80 percent of the growth in Amazonian urban centers has occurred in self-built shantytowns located mainly on the periphery of existing towns and cities, well beyond the reach of established utility and public transport lines (see fig. 5.1). This rapid growth has increased traffic congestion, and air quality has percep-tibly deteriorated in all metropolitan centers. As accessible and suitable vacant lands fill in, ecologically unstable but unoccupied areas are even-tually taken over by squatters, especially areas prone to flooding (plates 10.6 and 10.7). In Porto Velho large tracts of open land well within the urban area stand vacant, owned or claimed by local politicians, the mili-tary, and ELETRONORTE, while the metropolitan fringe expands at break-neck speed outward from the city limit (Pinheiro 1994). In 1990 only one-third of Porto Velho's urban population was connected to the city's water supply system, and 1.9 percent of its 67,266 dwelling units were connected to the municipal sewage disposal and processing network (Barreiros da Silva 1993; SEPLAN 1991). Compacted soil streets, intran-sitable much of the year, discourage regular services of refuse collection, potable bottled water distribution, and emergency evacuation as well as police and fire protection services. The municipal planning department of Porto Velho estimates that more than 35 percent of the garbage produced each month in the metropolitan area (some 1000 metric tons) is never collected (Prefeitura Municipal de Porto Velho 1990). As one community leader in the spontaneous peri-urban neighborhood Ulisses Guimarães lamented:

> The situation here is critical, principally in health, and everybody complains a lot about this and the problem of trash, the intransitable streets, streets on which even bicycles cannot pass. People throw garbage in the middle of the street, the garbage truck comes by once per week to collect it, and the rest of the week the people just dump their junk in the street or any place. ("Francisca" 1994)

The situation in Manaus is somewhat better. Here, according to one study, 21 percent of households are hooked up to the city's sewage disposal system (Lacerda de Melo and de Moura 1990). But much of the expansion in informal housing has occurred along streams (*igarapés*) and seasonally inundated drainage ditches, which are easily contaminated by human waste, trash, and industrial toxins (Fearnside 1989). Most of the expansion of Belém has occurred in the lowlands (*baixadas*) south of the city. Informal shanties, built precariously over heavily polluted streams

Plate 10.6. View of peri-urban slum, Porto Velho, Rondônia, 1991. Self-built housing sits on drainage ditch with pirated electrical power lines.
Photo: John O. Browder

and adjacent to the city's overflowing municipal landfill, house growing numbers of migrants from traditional extractive areas in the Bragantina and the Lower Tocantins regions (Mitschein et al. 1989).

Remarkably little research has been conducted on the public health impacts of the degradation of the Amazon's urban environment. Official census indicators measuring the incidence of various illnesses are woefully inadequate in representing trends in health indicators in most states of Amazonia. It is widely acknowledged by researchers that the worst impacts of environmental degradation disproportionately affect the urban poor who cluster in the slums on the metropolitan fringes (Mitschein et al. 1989). However, urban environmental deterioration is by no means limited to the Amazon's large metropolitan centers. In chapters 6 and 7 we outlined some of the public health perils of living in the frontier towns of Rondônia and Pará.

Squalid living conditions promote infectious diseases and chronic long-term ailments. Malaria, amoebic dysentery, and a bewildering host of gastrointestinal maladies are pandemic in urban Amazonia. The typically unsanitary environmental conditions created by rapid urban expansion combined with the low incomes, poor diet, and large families of

Plate 10.7. View of peri-urban slum, Porto Velho, Rondônia, 1991. Intransitable mud streets. *Photo: John O. Browder*

those who have come to live in the region's towns and cities have given the urban transformation of Amazonia a despairing side; unimaginably, it is an existence preferable to that many migrants left behind in their former homelands in rural Brazil.

CONCLUSIONS

Despite the general focus on tropical deforestation in rural Amazonia, several processes of environmental change are directly or indirectly related to regional urbanization. First, while a diminishing proportion of the urban population in Amazonia owns rural real estate, an increasing proportion of tropical forest real estate is coming into urban-resident ownership. These urban-oriented landowners are typically involved in numerous urban economic activities, unlike their rural-resident counterparts who typically invest more heavily in farming. Not surprisingly, the land-use patterns of urban-resident landowners are different from those of the rural-resident farmers.

How much tropical deforestation can be attributed to urban-based interests in the Amazon? The urbanization of the Amazon's rainforest has ambiguous environmental impacts. Based on our comparative surveys in Rondônia, we found that while smaller forest areas are cleared by urban-based farm owners, the more casual and unproductive forms of land use they pursue may entail opportunity costs that lead to unnecessary forest clearing elsewhere. In general, the urbanization of the region has not been accompanied by increases in agricultural intensification and productivity. Stated differently, the urbanization of the Amazon's tropical forestlands is disarticulated from the process of agricultural modernization as observed, say, in the history of the United States.[26]

Urban-based interests are also indirect agents of deforestation. In contrast to popular portrayals of the rainforest as besieged by small farmers and shifting cultivators, individuals who are clearly urban-oriented in their economic activities but with mixed relations to capital are also directly poised on the cutting edge of the frontier. Through various social relations of production, small farmers are often only the pawns in the deforestation spectacle dominated by urban groups.

A second connection between urban and corporatist interests and the environment is the regional energy policy of the Brazilian government emphasizing the development of the basin's considerable hydroelectric potential. Dam-building directly results in extensive forest conversion but also entails various social, public health, and economic impacts. Dam-building is both a direct response to and an influence upon urbanization. As Amazonian urban populations grow, so too does the necessity for energy. Yet, dam-building displaces large populations, contributing to involuntary urbanization, just as it draws population to dam-building sites to find short-term construction jobs, sometimes creating enclave boom-towns with separate corporate and popular (and largely informal) sectors, as at Tucuruí and Samuel.

Dam-building is certainly a corporatist undertaking, and ELETRONORTE has contributed to the corporatist segmentation of frontier space. However, despite the enormous amount of capital, both domestic and multinational, involved in the hydroelectrification of Amazonia, these projects eventually assume the distinctive social character of the frontier in which they are situated. Local political figures and development financiers capitalize on the grand public investments to promote populist (as at Samuel) or corporatist (as at Tucuruí) interests. Capital is thus flexible in adjusting to the diverse social orientations of Amazonia, accommodating both corporatist and populist interests in the frontier. The political-economic control of both ELETRONORTE and decision making over dam construction remain divisive social issues. Hydroelectric conflicts highlight the tension between national elites in the southeastern core and the more diffuse local elites spread across the Amazonian periphery. The situational logic and resulting incongruities in regional pattern inherent in the Amazon's hydroelectric program underscore the differentially articulated nature of Amazonian development, leading, in the aggregate, to a fundamentally disarticulated composite.

A third dimension of the urbanization-environment nexus is the expansion of populist and corporatist mining frontiers. The informal, often illegal, and largely uncontrollable phenomenon of popular gold and tin mining across the basin stands as a vivid contrast to the highly organized, controlled corporatist mining landscape at Carajás. The contrasting social organizations of mining activities engender strikingly different and often conflictive urban forms. Spontaneous mining towns emerge at crossroads locations near gold fields, serving both to market the mineral discoveries and to supply local miners with victuals and materials. The corporatist urban center, designed to accommodate professionals and permanent workers, is closed to the seeming melee of the populist mines, though the adjacent popular settlement and its burgeoning informal sector make for a dual city, as at Carajás-Parauapebas. Overall, the mining sector gives urbanization in the Amazon region a dynamic but essentially irregular and regionally disarticulated aspect. The dramatic contrasts and vicissitudes of Amazonia's mining frontiers further argue against the validity of master theories of autonomous regional development or dependent development in a world system.

Finally, most of the expansion of the urban-built environment in Amazonia has occurred on the periphery of existing towns and cities. Unregulated, self-built shantytowns wedged on the precipices of streams and drainage ditches and other unserviceable spaces have rapidly appeared in the urban landscape. Without power, clean water, or adequate sanitation, the urban environment of Amazonia is under siege, exposing a growing proportion of the region's public to various health risks. The sprawling metropolitan shantytowns outside Belém, Manaus, and Porto Velho are the poorer in-laws of the great slums of Lima, Peru, the satellite settlements ringing Brasília, and the Zona Norte of Rio de Janeiro. While lacking many of the metropolitan amenities of the older cities, the Amazon's regional centers lack none of their problems. Though still largely unrecognized by academics and policymakers, the problems of the Amazon's burgeoning urban centers are well known to local planners and merit higher priority in regional development policy. As Amazonia rapidly urbanizes, the region's many environmental and health problems increasingly demand effective public policy responses.

Notes

1. When the Rolim de Moura settlement area was laid out in the late 1970s, INCRA fixed farmlot sizes at 100 hectares, except for lots located on the main road link (Linha 25), which were typically set at 250 hectares.

2. In the first of three surveys (1985) of seventy rural households in Rolim de Moura's colonization area (Browder 1995) the average period of tenure on the farm was 5.2 years. Thirty (45.5 percent) colonists inherited some forest clearing on their farm lots from previous occupants, averaging 8.6 hectares. The 1990 survey included thirty-five farms (or 50 percent) from the 1985 survey. Of this group of thirty-five farms surveyed five years apart, twenty-seven (77 percent) were still owned by the same rural resident, two (6 percent) had new owners, and six (17 percent) had been abandoned (i.e., were vacant) in 1990. Given the low turnover rate (23 percent) found in the Rolim de Moura settlement area since its original colonization around 1979, it is unlikely that the difference in forest clearings between urban and rural-resident farmholders can be totally attributed to previous land occupants. Based on the findings of chapter 8, we might reasonably speculate that many urban-resident farmholders were previously rural residents in an earlier stage of circular migration between urban centers.

3. Obviously, this is a rather crude and unreliable indicator of capital intensity, but it provides a reasonable scale for the ranking of labor intensity, which is indisputably the more important factor in colonist agricultural production.

4. We should point out that this ranking scheme simplified a much more complex reality. For instance, coffee and annual crops are often interplanted on the same area for several years, pastures may or may not be stocked with cattle, and crop fallows and numerous other secondary forest tree species are often used for household consumption.

5. Some lots may have been previously logged out of commercial timber species before coming into urban-resident ownership.

6. In our 1990 rural household survey only 18.5 percent of the respondents in the Rolim de Moura colonization area had engaged in off-farm work during the twelve months preceding the survey.

7. It is interesting to note that 30 percent of the respondents to a 1984 urban household survey indicated that their migration decision to Rondônia was primarily motivated by the desire to secure farmland on the frontier. This finding suggests that a significant proportion of the urban population is mobile, either using the town as a temporary urban "landing" and engaging in whatever economic opportunities present themselves until a rural plot can be secured or combining rural and urban investments in a single household strategy.

8. Brazil is already a major producer and consumer of hydroelectric energy. In terms of consumption, hydroelectric energy represents 33.6 percent of total primary energy consumed in Brazil, in contrast to the United States (4.6 percent), Japan (6.0 percent), Canada (28.2 percent), and Argentina (10.5 percent) (FIERO ca. 1993:113).

9. Mougeot (1990) estimates Brazil's total hydroelectric potential at 213,000 megawatts, only 14 percent of which has been tapped. Brazilian Amazonia is believed to hold 97,800 megawatts of potential power.

10. Estimates of Amazon deforestation rates have varied widely. We estimate that by 1990 as much as 12 percent of the original tropical vegetation of the Legal Amazon had been significantly modified by human actions since 1960 (see also Mahar 1989).

11. Bunyard (1987) reports that ELETRONORTE originally had no plans to clear the forest from the impoundment area of the Tucuruí dam even though past experience in Surinam and at the 20-megawatt Curúa-Una dam near Santarém indicated that the decomposition of intact vegetation increased water acidity, causing the premature corrosion of metal casings on the dam's turbines while emitting noxious sulfurous gases. Other problems frequently associated with hydroelectric projects include sediment accumulation due to soil erosion in the dam's hydrologic catchment area and the proliferation of turbine-choking water vegetation (e.g., floating fern, water hyacinth) that reduce the effective economic life of dams and dramatically increase their maintenance costs.

12. Journalist Lúcio Flávio Pinto (then reporting for *O Liberal*) noted that the original estimate of marketable roundwood volume from the Tucuruí impoundment area was officially reduced several times. While CAPEMI's own forest inventory suggested 6 million cubic meters, the Banco Lazard Frères estimated only 3 million cubic meters of any "real commercial interest." Indicating a density of about 14.3 cubic meters per hectare, this latter estimate is not unreasonable given the results of forest inventories in the region undertaken by EMBRAPA/CPATU in Belém.

13. Various toxic chemicals were alleged to have been spread by ELETRONORTE in the Tucuruí area in unconfirmed quantities. Lúcio Flávio Pinto reported 262 oil drums (some 40 tons) of pentachlorophenol ("China dust"), although CAPEMI had registered only 500 kilograms of this herbicide, plus unconfirmed quantities of other fungicides used to preserve cut timber (Pinto 1984). Some commentators were not so convinced that CAPEMI's dioxin dumping was *indiscriminate*. Thousands of Brazil-nut trees were reportedly targeted for poisoning in forests managed by traditional populations to drive them out of the impoundment area, pre-

sumably so they would be unable to demand compensation from ELETRONORTE for displacement costs (*A Província do Pará*, March 24, 1984; *O Liberal*, April 16, 1984). Other roughneck
tactics were also reportedly used by ELETRONORTE. For instance, Alano Penna, Bishop of
Marabá, in an 1979 interview with the journal *Estado do Pará* testified that "the management
of ELETRONORTE is using soldiers from the Military Police as vigilantes to practice great violence against the homesteaders who have occupied for several years now the land the company wants vacant for the Tucuruí dam" (Scherl and Netto 1979).

14. While only five confirmed human deaths were directly linked to dioxin exposure by
March 24, 1984, the 12 million milligrams of sodium pentachlorophenol also reportedly
dumped by CAPEMI were believed sufficient to kill 20,000 adult persons (Pinheiro 1984;
Província do Pará 1984). By April 1994 more than one hundred residents of the Amazon had
filed a class action suit against ELETRONORTE demanding reparations for loss of cattle, crops,
and human life (Smolowe and Margolis 1984).

15. Altogether, ELETRONORTE constructed three main transmission lines from Tucuruí: The
largest line (at 500 kilovolts) runs to the national grid 385 kilometers east of Imperatriz. A
second 500-kilovolt line provides power to the alumina smelter under construction at Vila do
Condone, 280 kilometers north of the dam. A smaller, 230-kilovolt line runs from there to
Belém (Goodland 1985). While it might appear from these data that the distribution infrastructure established would give residential users at least equal access to power generated at
Tucuruí, the rate charged to Alunorte/Albrás (a consortium of Japanese firms and Brazil's
Companhia Vale do Rio Doce), at U.S. $0.01 per kilowatt, is roughly one-third the rate paid
by residential consumers throughout the country and one-sixth the cost of generating the
power at Tucuruí, a subsidy to corporations paid by the Brazilian populace through their taxes,
home power bills, and public debt payments (Fearnside 1989).

16. Bunyard (1987:66) also notes that the reservoir's estimated 12 meter drawdown will
reduce the surface area of the lake from 2,160 to 1,260 square kilometers, thus exposing a
900-kilometer area of ideal mosquito breeding habitat.

17. Fearnside cites a report of a survey by three organizations opposed to the dam, which
concluded that 217 families (about 1,000 people) would be directly affected by the Balbina
dam project. However, two of the ten remaining villages of the Waimiri-Atroari tribe would
be flooded, dislocating 29 percent of the tribe's population, now totaling 374 individuals
(Fearnside 1989:410).

18. Vila Eletronorte, located on the perimeter of Porto Velho on the BR-364 toward Rio
Branco, is an internally stratified dormitory village with security gates and a social club for
the company's employees working at the Samuel project.

19. The FIERO source indicates that 55 percent of Rondônia's total supply was acquired at
well below cost from outside sources. The subsidy rates ranged from 68 percent for diesel fuel
to 86 percent for services, suggesting that the operating costs of energy provision to
Rondônia, based on imported diesel, were heavily subsidized by the general Brazilian public
during the pre-hydroelectric years.

20. According to ELETRONORTE, these animals included 6,590 arthropods, 3,729 mam-
mals, 3,504 reptiles, 2,099 amphibians, and 78 birds. Of the 16,000 animals rescued, 11,417
were sent to research institutions, 1,729 were sacrificed for research or for museum speci-
mens (including those that died during the rescue operation), and only 2,854 were released
inside the Samuel Biological Reserve (Lemos de Sá 1992:3). It is widely conjectured that rescue operations that result in the reintroduction of displaced animals into existing (and biologically occupied) forests place increased population stress on those ecosystems, probably
resulting in the loss of those animals anyway (cf. Fearnside 1989; Lemos de Sá 1992).

21. Local utilities, such as that in Porto Velho, are able to assess a flat 10-percent tariff on private consumers for public street illumination, even for consumers who do not receive street light service.

22. The proportion of Brazilian utilities contributing profits to the fund dropped from 12.4 percent in 1975 to 6 percent in 1986 (FIERO ca. 1993).

23. For a detailed history and comparison of the Serra Pelada and other major contemporary gold-mining strikes in southern Pará, including Cumarú and Cuca, see Schmink and Wood (1992:219–249).

24. The toxicity of mercury depends on the particular chemical form it takes. According to Cornell University chemist Ronald Hoffman, the worst offenders are organomercurials—compounds such as salts of methylmercury, CH^3HG_+. "These are stored in fat tissue, are easily transported across membranes, and cause nerve damage. Ionic mercury, Hg_+ (in the form of $Hg^2{}_{2+}$) is not very toxic, for it forms the stable Hg^2Cl^2 (calomel) with stomach chloride. Mercuric ion, Hg_{2+} is dangerous (it can be methylated, forming CH^3Hg_+, by bacteria) but not easily transported across biological membranes. Elemental mercury is very toxic when inhaled. Pascal, Faraday, and especially the great German inorganic chemist, Alfred Stock, suffered a painful irritation of the nose lining and bladder, as well as memory loss and impaired mental processes, as a result of inhaling it" (Hoffman 1994).

25. In a study of 311 hair samples from *ribeirinha* residents on the Alto Madeira of Rondônia, 53 percent of the female population of reproductive age had been exposed to environmental concentrations of mercury higher than 10 parts per million. Mercury concentrations ranged from 7.0 to 303.1 parts per million in this survey (Boischio and Barbosa 1993).

26. One policy implication of the urbanization of rainforest ownership suggests that conventional rural extension approaches be partially redirected to urban centers.

11

Patterns of Development and Urbanization on the Global Periphery

❦

Our quest for patterns of development on the global periphery too often is framed by deterministic theories of social and economic change that pivot on the unproven validity of master principles. Regarding the case of one such model, John Friedmann (1986:30) writes that "despite the seeming inevitability of global economic forces, 'fate is not traced before-hand.'" In this spirit we call for a pluralistic conceptual framework to interpret patterns of regional development and urbanization in the world's last great rainforest frontier, the Brazilian Amazon. Such a frame-work begins with one fundamental premise: the possibility that a priori multiple explanations of a phenomenon coexist simultaneously.

Looking beyond the seeming bedlam of frontier expansion in specific places to the larger global dimension of the urban transformation of Amazonia, we question whether these rainforest cities have served any useful functions in the global accumulation of capital. We doubt, there-fore, that the world cities hypothesis provides an appropriate initial struc-turation principle for interpreting regional change in the Amazon. In this study we acknowledge that forces emanating from global centers of investment and exchange indirectly influence the spatial organization of human activity on the frontier. Global forces, however, constitute only one of the threads weaving together a larger tapestry, which also includes the actions of local peoples, their organizations, and their politics and the fixed locations of valued natural resources. Global political-economic currents and national policies set the broad frameworks within which selected Amazon settlement frontiers expand and contract, while other (arguably most) frontier spaces go untouched by global forces; nevertheless, even in these affected zones local levels of society ultimately mediate the processes of regional change. Local peoples of Amazonia alternately adopt, adapt to, and resist external pressures. Our case studies of Rondônia and southern Pará show how the variable agency of local peoples, along with the diverse distributions of natural resources, moderates the regional impacts of national and transnational forces. The history of these complex, multilevel interactions becomes geographically etched into the evolving cultural landscapes and spatial structures of Amazonia.

During the colonial period the spatial organization of the Amazon's settlement system was largely limited to the rivers, the main transportation thoroughfares, although some overland trade routes were also traversed. Small riverbank settlements, like Charles Wagley's "Itá" (Gurupá), together formed a dendritic trade network typically converging upon the major trading centers on the Atlantic littoral or the Amazon gateway city of Belém. During the rubber boom years (1850–1920) this extractive mercantile settlement system was greatly extended upriver to include vast and remote reaches of the Amazon Basin. Manaus, Porto Velho, and Belém grew in commercial importance and reigned over a myriad of smaller upstream settlements, trading posts, and labor control points in the extractive rubber economy.

However, the Amazon's contemporary urban system did not directly evolve from the preexisting spatial network of the rubber economy. Rather, the shift from river transport to upland roads beginning in the late 1950s abruptly thrust the modern network of human settlements onto the relict rubber network. New modes of regional production supplanted the older ones. New forms of spatial organization of human activity superimposed themselves over the older ones.

Urban systems are physical manifestations of the prevailing mode of economic production. Hence our study concerns more than just rainforest cities; it speaks to the broader issue of regional development and global integration. What is the prevailing mode of economic production in Brazilian Amazonia today? Our study suggests there is not just one but rather numerous economic systems operating simultaneously and shaping spatial organization in Amazonia. The economic space of the Amazon is fractured by numerous divergent modes of production, some directly linked, for a time, to the "technopoles" and world cities of the "Global North," while others remain isolated pockets of traditional noncapitalist society. The predominant tendency structuring the Amazon's economic space has been the selective integration of the region into the national economic space of the Brazilian southeastern metropolitan core.

Yet, no uniform process of capitalist expansion is occurring in Amazonia. Rather, different social groups that come to dominate local culture and politics in the Amazon today are linked to capital in different ways. We have found it useful, as an analytic tool, to conceptualize the frontier region as a highly dynamic sociospatial continuum in which different social groups, defined by their respective orientations to capital and labor, interact to give different spatial points on this continuum distinctive socioeconomic characters. We broadly classify two important sociospatial ideal types as "corporatist" and "populist." These are not treated as deterministic models of spatial organization, but rather as probabilistic spheres. Once a space in the Amazon has been socially established as a corporatist space, for example, the chances are high that the prevailing economic structure, the spatial configuration of human activity, and the function of the space in the global economy will follow from the logic of corporatist forms of capitalist production. Since corporate interest in the Amazon is almost entirely resource-extractive, corporatist settlement systems tend to be short-term, locally closed, and externally controlled. The populist frontier, although also largely extractivist in economic orientation, by contrast, tends to be more socially open and its settlement system more diffuse; its general pattern of spatial organization resembles a loosely packed system of central places. Each social space organizes economic opportunity differently, connects differently to national and global forces of change, and produces distinctive patterns of spatial organization.

Diversity, not uniformity, prevails in contemporary Amazonia. Found in the region's populist frontiers are pioneer settlements not unlike those that dotted the prairies and badlands of the North American West during the nineteenth century. Here, on the populist agrarian frontier, farmers clearing tropical forest, produce for a time the conditions required for autonomous regional development. The marketing principle of spatial organization takes hold, and an incipient system of central places emerges, as in the Rolim de Moura-Alto Alegre case study. This incipient process of spatial structuration of the populist agrarian frontier was reinforced by its leading institutional orientation; INCRA and its social equity theme and national security ideology dominated political life during the formative years of Rondônia's modern settlement. But even here the pace and direction of local development can suddenly shift with the discovery of valuable natural resources, such as mahogany timber or gold, and the episodic burst of new extractive activity—such as occurred in Rondônia's 1980–85 mahogany boom or in southern Pará's gold-mining boom in the late 1980s. Local development can also shift suddenly with the sudden introduction of new economic groups, such as Brazilian trading companies or multinational corporations. Our evidence suggests that in Rondônia the essentially populist quality of the frontier, with its archetypal yeoman pioneer farmer, is a transitional phrase toward an ever more differentiated rural population with emergent tendencies resembling corporatist social interests. The physical manifestation of the clashing logics of frontier expansion is a settlement system that is "differentially articulated" to different modes of production, to different principles of spatial organization, and hence "disarticulated" from any single master principle of national development.

Elsewhere in the Amazon, corporate "stealth cities" are constructed seemingly overnight to house in relative comfort the professional employees of state enterprises and their transnational contractors who are engaged in the megadevelopment projects for which the Brazilian Amazon is famous. Provisioned through their own closed economy, and ensconced behind guarded security gates, these spaces in the Amazon have an unmistakable corporatist character. The logic of spatial segmentation in the corporatist frontier, based on centralized planning, capital-intensive resource exploitation, and social control, engenders a decidedly different form of urban development than that found in populist frontier areas. In contrast to Rondônia's populist agrarian frontier, southern Pará displays the fragmented dynamics of a corporatist frontier, institutionally mediated by SUDAM, challenged by populist impulses and driven by strong extractive impulses. With SUDAM's developmentalist ideology, a framework for spatial organization vastly different from that of INCRA in Rondônia was established in southern Pará. Yet, time fades the rigid boundaries that separate the tidy corporatist bungalow cities from the shanties that inevitably envelop them. The social character of this corporatist frontier space tends to shift toward a more populist orientation, as described in the Tucumã-Ourilândia case study.

The process of sociospatial change in the Amazon's rainforest frontier works in both directions simultaneously, toward corporatist and populist poles, and in sequence, as one develops into the other over time. This dialectical quality of frontier change, the importance of emergent properties, novel generative structures and local contextuality, and the ability of different social spaces within the frontier to transform into each other, bear witness to the necessity for an interpretative framework based upon conceptual pluralism.

The development of the Amazon has not been a smooth and progressive process but has moved unevenly, in fits and starts. An unlikely array of production modalities alternately coexist, conflict, and coevolve, fragmenting and melding space. Far from being a homogeneous social space with predictable hierarchical settlement patterns, the human organization of the Amazon is riddled with fissures; its volatile development history of boom-bust cycles has engendered a region with polymorphic and disarticulated configurations of rainforest cities.

WHITHER THEORY?

While eluding any single hegemonic theory, our efforts to situate the phenomenon of frontier urbanization in relation to different conceptual frameworks is not without reward. We now briefly consider the comparative validity of the five major frameworks we reviewed in chapter 2 (see summary in table 2.1).[1] Then we close by revisiting the basic principles of disarticulated urbanization outlined in the introduction and chapter 4.

Central Place Theory

Originally developed to explain the spatial distribution of market towns emerging in medieval Germany (and subsequently applied to the United States Midwest and other world regions), central place theory has enjoyed a long-standing eminence in the modern fields of spatial economics and location theory. The master principle driving central place theory is the "market principle." In the abstract, the human landscape is conceived as a homogeneous surface over which goods and workers flow without impediments. Market development is endogenic and sequential in relation to the settlement of the plane. Over time as population increases, so too do agricultural land and commodity prices, changes that induce innovation, urban growth, and productive and commercial specialization. As urban settlements grow, they specialize in the provision of different services to the surrounding agricultural hinterland. The relative location of settlements, their respective sizes, and the range of functions they provide all depend on the size and density of their market catchment area. Eventually a symmetrical, hierarchical system of settlements emerges.

In our study of the Rolim de Moura–Alto Alegre settlement corridor in Rondônia we found that each of the three settlements displayed different key characteristics associated with age, location, and economic functions that reflect incipient central place tendencies. Rolim de Moura, the oldest, largest, and most central of the corridor's settlements, displayed more advanced physical infrastructure, higher nonagricultural employment, and higher material development than the peripheral settlements in its urban system. Moreover, we found that the distribution of some employment characteristics (e.g., agricultural employment) loosely followed a positive distance gradient from the center, as might be predicted by central place theory. But this incipient pattern is not without certain idiosyncrasies. Several bellwether indicators (e.g., incidence of human diseases, employment in extractive and industrial sectors) did not vary significantly with distance from the core, urban size, or degree of physical development among the three settlements. Finally, we noted that in general terms, over time, each settlement's respective market envelope shifted as the settlement frontier expanded, underscoring the saliency of the market dimension in the formation of urban settlement systems.

Despite the general relevance of the market principle to the evolution of Rondônia's populist agrarian frontier, the recent contraction of the economy in Rolim de Moura calls into question the long-term, progressive growth tendencies implicitly postulated by central place theory. The area's chronology recalls some elements of Amazonia's familiar boom-and-bust cycles. Rolim de Moura as a pioneer service center dominated the corridor's emergent urban system in the provision of all goods and services during the initial years of settlement (1977–1984). As the corridor extended and the market area expanded, other towns (Santa Luzia and, later, Alto Alegre) emerged to provide more efficiently a variety of basic services to the agrarian population that hitherto had been provided by Rolim de Moura. By 1990 Rolim de Moura felt the woes of market area contraction.

The spatial organization of Rondônia's populist frontier has been structured largely by shifting market forces, which have engendered some degree of regularity in its urban system morphology. The same, however, cannot be said of the contested corporatist frontier of southern Pará. The Xingu corridor in our study provides a chronological parallel to the Rolim de Moura-Alto Alegre corridor. A growing number of landless migrant farmers drove initial expansion of southern Pará's agrarian frontier, as in the case of Rondônia. Settlement of Xinguara began with a populist announcement of rural colonization programs by the state of Pará and then by the national agency INCRA. The subsequent concentration of

mercantile theory increased system integration induces competition and specialization among stockholders in more central settlements, forcing further outward expansion of the system. Any industrialization that develops is concentrated in the "gateway city" at the central terminus of the mercantile system.

Our study illustrates that the mercantile model of regional development more closely corresponds to the form of spatial organization characterizing the rubber boom epoch than to the contemporary period in Amazonia. Currently, the location of settlements has not been limited to natural transportation corridors, most notably the rivers, and resource extraction is only one of several important economic activities; others include agriculture and services. The spatial configuration of the Amazon's regional settlement systems only vaguely resembles a dendritic pattern, since numerous local twists and turns do not conform to the master principle of trade efficiency maximization.

While mercantilism is clearly not generalized as the dominant mode of production in the Amazon today, mercantile activity does occur as natural resources are systematically extracted from both corporatist and populist frontier spaces and transported to the national economic core. We find that mercantile activity tends to be limited to specific commercial commodities, such as gold, mahogany, bauxite, diamonds, and fish. The exploitation of these natural resources often influences the spatial organization of Amazonia's settlement systems, as in both the gold mining of southern Pará and the mahogany cutting of Rondônia. Other resource flows, such as those of capital and labor, are only partially oriented toward the Brazilian southeast and as such are not as fully explained by mercantile theory as by structuralist theories.

Intersectoral Articulation Theory

Drawing upon both Marxian modes of production and spatial dualism models, the master principle of intersectoral articulation theory is "functional dualism." According to this principle, capitalist growth is class specific. The specific material expansion of the capitalist classes depends upon modes of production that tend to impoverish the working class. To continue to grow, however, the working class, both rural and urban, must be preserved at the minimum level of material consumption. Intersectoral articulation theory postulates that the frontier is, above all, a labor safety valve. The progressive occupation of the frontier enhances the capital's class control of labor, frees up prime agricultural farmland for consolidation under profit-maximizing corporate agribusiness, and exploits landless rural workers to subsidize the urban industrialization of luxury goods through protective tariffs. Explicit in articulationist formulations is the alliance between capital and the state, which expands to serve the interests of the former. This process promotes the accumulation of capital in the national center. Market expansion is uneven and disarticulated from national development. Although the implications for urban system morphology are unclear, arguably articulationists would envisage an irregular system of cities which, among other things, serve as unemployed labor reserves. Not surprisingly, in this formulation the urban space is dichotomized into formal (capitalist-dominated) and informal (populist-dominated) sectors.

Our comparative case studies suggest that the Amazon, writ large, cannot be summarily reduced to a labor safety valve, although clearly large pockets of the region have served this purpose for a time (e.g., Transamazon, Rondônia).[2] National and some transnational capitals are clearly dynamic elements helping to shape spatial organization of the Amazon. Nor is it obvious that capital's class control of labor is enhanced by the latter's diffusion throughout the frontier. The frontier is rife with rural labor union movements.[3] With the restoration of civilian rule beginning in 1985 and the devolution of political power to newly formed Amazon states, capital has ostensibly given over political power to local populist classes in sizable areas of the Amazon (e.g., Rondônia). Others have pointed to the limitations of the articulationist's dichotomization of the urban social space and to the rigid functional dualism characterizing the relationship between so-called formal and informal sectors. Finally, while unemployed labor does periodically "pile up" in Amazon cities, such cities do not serve primarily as reservoirs for exploitable labor as articulationists suppose. Rather, lower-order settlements in the frontier often serve as temporary landings for settlers seeking land.

Our analysis of migration suggests that the rural/urban divide is a false dichotomy in Amazonia. A substantial proportion of the urban population periodically resides in nonurban areas and engages in extraction and agriculture, as well as in urban-based activities. Our survey of employment patterns in urban households indicates that many households have members simultaneously employed in both formal and informal sectors. This may seem to be perfectly consistent with the functional dualism principle of intersectoral articulation theory except for two significant breaches: First, the state's role in orchestrating frontier expansion has not been homogeneous, but ambiguous, favoring different social groups at different times, further disarticulating regional development from any unified vision of national development. Second, the expansion of the urban network does not appear to be contingent upon the preservation of an urban or rural underclass in the frontier. Urban dwellers, while generally among Brazil's poor, are socioeconomically mobile and differentiated into multiple class factions. Yet, the articulationist's emphasis on class polarization can hardly be ignored in our analysis of frontier urbanization. A powerful capitalist elite class, in alliance with the state, drives the expansion of corporatist frontiers throughout the Amazon, while its role in structuring social life on the populist frontiers is more ambiguous and nuanced than is typically allowed by intersectoral articulation models.

Capitalist Penetration Theories

Among the more popular structuralist perspectives, capitalist penetration theories tend to view the frontier as a transitional zone of contestation between capital and labor. The master principle is the inexorable expansion of capitalism, seen as progressing through a series of predictable linear stages. Peasants open the frontier with their labor. Capitalist and speculative expansion follows, eventually consolidating the frontier and transforming the peasantry into a semi-proletarianized class of itinerant, landless day workers who congregate in frontier towns and cities, creating a "false urbanization." Frontier urbanization is explained, then, by changes in the agrarian sector—especially the expansion of export-oriented cash cropping and modern agricultural technologies—induced by the penetration of capitalism. Frontier towns simultaneously become centers of agglutination for the dispossessed peasantry and service centers for the mobilization of agro-industrial capital.

Our analysis indicates a more complex reality. We find that the Amazon's urban centers are not predominantly "peasant cities," where landless rural workers accumulate. Rather, with some significant variations between corporatist and populist frontiers, the preponderance of the urban population has a long history of urban living elsewhere in Brazil. The patterns of frequent local migration, especially in southern Pará's gold-mining areas, indicate an active labor force on the frontier for whom urban centers facilitate various processes of occupational mobility. Indeed, our study suggests that rigidly inelastic categories of capital versus labor are anachronistic; all workers possess varying degrees of both capital and labor. Finally, we find little evidence of modernization in Amazonian agriculture and hence little evidence that capitalism has penetrated the frontier as predicted by modernizationists or capital penetrationists.

Patterns of socioeconomic mobility vary significantly between corporatist and populist frontiers. In our surveys, interregional migration to the former entailed changes in socioeconomic status for more than 70 percent of the urban population; rates of capitalization and wage employment (proletarianization) were roughly comparable. In the populist frontier, slightly more than half of the urban population reported a change in socioeconomic position associated with the interregional migration to the Amazon. This change was about evenly divided between capitalization, proletarianization, and dropping out of the formal economy (pauperization). Indeed, for the majority of those households reporting a change in socioeconomic position following interregional migration, that change reflected some degree of upward mobility, defined as the socioeconomic shift from informal to wage employment and from other forms of employment to ownership of the means of production (capitalization). Yet, most of the household movements recorded in our surveys occurred within the Amazon region. For the majority of households these intraregional movements entailed no change in socioeconomic position. Notwithstanding the significant variations between corporatist and populist frontiers, observed patterns of frontier migration and socioeconomic mobility are more nuanced and less uniform than those envisaged by most capitalist penetration theories.

Capitalist penetration theories cannot be rejected outright. They do provide a basis for classifying the frontier into distinct social spaces, and they accurately predict one important trend, the tendency toward socioeconomic differentiation among the working class. Whereas traditional formulations of this hypothesis focus on the differentiation within the rural population, we observe differentiating processes occurring among the urban population between urban business owners and workers over time.

World Systems Theories

World systems (i.e., world cities, dependency, and globalization) theories tend to adopt a spatial framework, and most formulations postulate a dualistic world order. Typical polarized categories include: center/satellite; core/periphery; global North/global South. Evolving from both neo-Marxist and Latin American structuralism, many writers in the traditional world systems genre base their work on the master principle of surplus extraction, differing in the relative emphasis given to its particular form (e.g., through unequal terms of trade versus wage relations of production). Third World production is seen as being driven by transnational corporations, which operate globally by restructuring local modes of production. Both the center and the periphery maximize the extraction of labor's productive surplus, concentrating it in world cities. World cities are "basing points" in the global project of capital accumulation. Most world cities (e.g., New York, Tokyo, London) are located in the advanced, postindustrial countries of the Global North. Yet, by drawing upon capitalist penetration ideas, world system advocates suggest that transnational capital extends itself as a multitiered urban system throughout the periphery right down to the local market town (Armstrong and McGee 1985:56–58). Each town and city in the global system serves to "bulk" the surplus value extracted from workers in its hinterland, from whence it inevitably and irreversibly flows to its ultimate destination in the transnational capitalist sector. The urban systems supporting this world accumulation of capital are hierarchical in vertical organization and distinctly linear in spatial distribution (Friedmann 1986:72).

A meticulous examination of the economic transactions between key sectors in the local economy of the agrarian frontier found scattered evidence of surplus extraction at both levels of production and exchange. Agricultural producers are routinely subjected to unfair marketing practices that enable local merchant intermediaries to extract a petty surplus from producers. There is little evidence, however, that these practices directly or indirectly support some global project of capital accumulation in world cities. In the case of Rondônia's mahogany boom, we found that national capital (in the form of Brazilian trading companies), in alliance with the national state (through its national export promotion program), effectively appropriated local capital (in the form of small, often family-owned lumber mills). Yet, in only two of the eight cases surveyed were the corporate parents of those trading companies engaged in any contractual arrangement with a non-Brazilian transnational corporation.[4] The financial sector, the principal institutional mechanism through which local capital is concentrated at the national level, underwent unprecedented expansion in the Amazon during the 1970s and 1980s. One regional hub of heightened financial activity was the unlikely *município* market town of Rolim de Moura. Here one major Brazilian commercial bank reported its second highest Amazon earnings in 1982. Yet, most of the *município*'s farmers failed to earn a financial surplus. So who was supporting the expansion? We conclude that only the federal subsidies to merchant capitalists (i.e., mahogany exporters), channeled through the private banking sector, could have fueled this frenzied growth in financial activity. After the mahogany boom, Rolim de Moura's banking sector contracted; the possibility of windfall profits through the recirculation of the national surplus ended. The institutional mechanism of capital accumulation at the global level is only vaguely present in this, the clearest case of an alliance between national capital elites and the national state.

We argue that convergent processes are not generalized, but selective, and that, on the whole, heterogeneity prevails. One example of convergence, however, is found in the food retail sector in the Amazon, which is closely integrated into national production and supply networks. National supermarket chains have succeeded in penetrating the food retail sector in both the populist and corporatist frontiers, displacing family-owned corner grocers and increasing the dependency of Amazon consumers on interregional food imports. The extension of commercial capital through the food retail sector suggests the importance of the frontier as an extension of the national consumer market and a form of disarticulated development as postulated by de Janvry. Selective sectors of the frontier are integrated into the national economic complex, while other sectors remain indigenous or localized in their orientation.

Our analysis does not dispute the widespread observations that transnational capital often determines the terms of trade around the world and selectively effectuates structural changes in local modes of production. Rather, we question the implications that transnational capital provides the "structuration" of the frontier and that global integration is spreading uniformly throughout the Amazon and with predictable effects everywhere. Admittedly, we find numerous cases of surplus extraction, some crude (e.g., oligopolistic crop marketing) and others more subtle (e.g., siphoning off local savings through the formal financial sector). Yet such local extraction often works against, not in favor of, capital accumulation at the global level.

So, whither theory? Contemporary thinking about urbanization in the forty-odd years following Wagley's classic *Amazon Town* has moved through a series of master principles and conceptual frameworks, from central places to class conflict to world systems. The hegemonic debate among the various paradigm proponents seems irreconcilable. Is a final theory of urbanization in sight?

In his provocative book, *Dreams of a Final Theory*, Nobel-Prize winning physicist Steven Weinberg writes: "Our present theories are of only limited validity, still tentative and incomplete. But behind them now and then we catch glimpses of a final theory, one that would be of unlimited validity and entirely satisfying in its completeness and consistency (Weinberg 1992:6)."

Weinberg's dream is for a final cosmology, not merely a unified theory of urbanization. If Weinberg succeeds in finding the former, there would be little need for any elaboration of the latter. However, for the time being we content ourselves with less abstract matters, such as the future of the world's greatest rainforest, now not only a humanized place but also very much an urbanized place. While physicists may be on the verge of a final theory, urbanists would be well advised to seek pluralism, not hegemony, in urban theory.

TOWARD CONCEPTUAL PLURALITY

Ferré (1988:94) deserves repeating: "The postmodern images of reality that come from ecology portray the world as an endlessly complex network of organic and inorganic systems locked in constant interaction." This is also our characterization of the Amazonian frontiers. Such interactions result in a myriad of hybridized microsocial formations, each with its own unique physical scaffolds (settlements and their systems). On the surface these settlements appear much the same to the researcher laboring under a single master principle. But below the surface each is different in many significant ways: local history, social composition, institutional affiliation, and subregional and regional economic functions. This wide range of local nuances cannot be summarily swept away by the big brush of globalism; the evidence is simply lacking. Although we would not suggest a return to the simple biological metaphors of the Chicago school of human ecology, contemporary urbanists would do well to give the same attention to the finer points of unifying theories as good ecologists do to the details of nature.

Our attempts to integrate conceptual pluralism with systems ecology lead back to the basic principles that constitute our theory of disarticulated urbanization as presented in chapters 1 and 4.

> 1. The Amazon is a heterogeneous social space. Constituted by diverse ethnic and social groups, Brazilian Amazonia is a difficult region to characterize. Old historical constructs of the rubber boom and the extractive mercantile mode of production in Amazonia, (e.g., the Madeira-Mamoré Railroad, Baroque opera houses in Parisian style, dilapidated mansions in downtown Belém and Manaus) stand side by side in an awkward juxtaposition to contemporary forms of development and spatial organization (such as the

new inland interstate highways, the Manaus International Airport, the Tucuruí dam and associated power grid, the Carajás–São Luis do Maranhão industrial railroad, and a satellite communications system that is arguably unrivaled anywhere in the developing world). The confluence of different social groups originating from processes occurring at different sociospatial levels of the global space economy during overlapping historical periods renders the Amazonian landscape a "mosaic within mosaics," something that is incomprehensible through the optic of any single conceptual lens or master principle. Amazonia is a heterogeneous social space.

2. The configuration of settlement systems in Amazonia is irregular and polymorphous, disarticulated from any single master principle of spatial organization. We conceive of the frontier as a dynamic sociospatial continuum along the extensive margin of the national space on which different social and economic groups emerge. Their socioeconomic interactions, often mediated by institutional alliances, produce emergent properties that engender different spatial points on the continuum with prevalent characters. For example, the rubber boom inscribed a distinctly mercantile orientation into the economic life in the Amazon, leaving as its legacy remnants of the *barracão* settlement system. The Amazon mahogany boom (1975–1985) introduced a highly advanced extractive mode of production in the *terra firme* rainforest far beyond the reaches of any market economy and thus engendered an expeditionary settlement system deep in the forest periphery. The Amazon gold rush of the 1980s created countless provisional bivouacs, some perched in the shadows of corporate megaprojects, some becoming permanent settlements and *municípios* over time. The mosaic of organizational logics represented by each one of these highly diverse urban formations in the frontier defies crude agglomeration into a unified master theory. Not surprisingly, we find rainforest city systems in Brazilian Amazonia to be irregular and polymorphous across the region, with functions articulated differentially within the national economy.

3. Urbanization in Brazilian Amazonia is functionally disarticulated from both regional agricultural development and industrialization. Whereas agricultural expansion and industrial growth are limited to specific locations in Amazonia, urban population growth is generalized throughout the region, suggesting that the process of urbanization is disarticulated from local development processes. True, some urban centers support both agricultural populations

and factories, but the generalized growth in the urban population across the region contrasts with the spatially selective emergence of industrial poles and agricultural fronts. While resource extraction is a powerful *force motris* behind urban growth in certain areas of the Amazon, in other urban systems employment in extractive sectors is negligible. One of the few convergent patterns in the Amazon is the growth of the urban service sector, both public and private, indicating the important role of government spending and inter-governmental fiscal transfers in holding the region's economy together. In the relative absence of productive activity, the expanding urban service sector that feeds upon government spending constitutes a veiled form of extensive unemployment in the region.

4. Urbanization in Brazilian Amazonia is variously linked to global economic forces but is not subordinated to the world economic system. The sectoral disarticulation of economic development widely found in the Amazon suggests some degree of capitalist penetration and external influence upon spatial structure in the region. But even in those specific locations where export-enclave settlements have become generative hubs of economic activity, it cannot be said that capitalism as a system of market forces and wage relations has become established. Rather, we find that selected zones in Amazonia become more closely integrated into the postindustrial technological society emerging in the Global North. For example, Manaus, located in the center of the Amazon's biologically rich *hylea* rainforest, is likely to become an important center for the testing and processing of germ plasm used in pharmaceuticals and biotechnologies. Other zones with great precious and industrial mineral wealth (e.g., Serra Pelada and Carajás) continue to maintain close "high-tech" connections to the global economy. But these two zones are constituted by vastly different socioeconomic interests, populist and corporatist respectively. These examples show how the economic rationales of frontier urbanization extend beyond the borders of the frontier itself but are differentially articulated to the external world.

5. The contemporary Amazonian urban frontier is largely a geopolitical creation but remains politically disarticulated within the central state. Despite its central role in fomenting processes of rapid and often predictable change, the Brazilian government has not been a faithful ally of any one social class or of any single ideological vision of national development in the Amazon region.

Rather, the federal bureaucracy during the military authoritarian period simultaneously (and alternately) represented diverse, even opposing social interests, reinforcing the fragmented social orientation of the region. With the restoration of civilian control (beginning in 1985), political devolution and local political processes have further fragmented the social spaces of Amazonia. Far from being a unifying force in the civil society of Amazonia, the ambiguous and shifting role of the state has contributed to the further socioeconomic differentiation of the region. Urbanization in Amazonia is disarticulated from any coherent set of development policies and from any unified conception of the state and civil society.

6. Established dichotomous categories of rural and urban become problematic when applied to Amazonia. Urban Amazonians are highly mobile people. Most of the urban migrant population in Amazonia came from other urban areas outside the region. In addition, there is considerable fluidity of residence between rural and urban areas. Rural-urban migration within Amazonia is a significant but largely secondary migratory trajectory. In many cases, what appears to be a rural-urban household move is actually the last segment of a longer urban-rural-urban move. Overall, diversity prevails. In Rondônia we find a higher rates of rural-to-urban migration than in Pará. For many among the Rondônia migrants, the first urbanization experience occurs in Amazonia. Our study suggests that categories of rural and urban, as functionally discrete sectors, represent a false dichotomy. The fluidity of movements of household members between rural and urban places indicates a considerable degree of functional integration of space. However, across the Amazon local migration patterns have widely ranging implications for occupational and socioeconomic mobility.

Among urban households in both Pará and Rondônia, interregional migration to Amazonia directly led to significant changes in occupation and socioeconomic position. For unclear reasons, intraregional movements did not have the same effect. In Rondônia a greater proportion of migrants who located in more remote settlements were able to accumulate capital than their counterparts who settled in the larger, more central urban hub and who disproportionately tended to lose capital following migration. In Pará interregional migration to the contested corporatist frontier led to higher rates of occupational mobility across the board than in Rondônia, but predictable linear spatial patterns in the distribution of these changes were absent.

7. Environmental change in Amazonia is increasingly mediated by urban-based interests. The dynamics of environmental change in the Amazon, ranging from tropical deforestation to river contamination, have long been thought to arise from social processes within various groups of autonomous rural-based workers (e.g., farmers, miners, increasingly rubber tappers). Our studies suggest that environmental change is increasingly linked to the urbanization of Amazonia. A growing proportion of the urban population owns rural property, and these rural assets are managed in ways that enhance diverse urban accumulation strategies, not necessarily with sustained forest use in mind. The expansion of the region's hydroelectric capacity to energize its rapidly growing urban centers and the transformation of rainforest lands into gold-mining settlements are urban forces driving environmental change in the Amazon. The growing proportion of the Amazon's total population living in squalid peri-urban shantytowns further attests to the critical and widely neglected urban dimension of environmental change in the Amazon.

These conclusions to our study are also questions for future research. Frontier urbanization in Amazonia is both a pressing policy concern and a novel research topic. The topic raises significant empirical and theoretical challenges for urban and regional studies. Rainforest cities on the world's last great settlement frontier will remain on the research frontier of urban planning and geography for years to come. We encourage those who follow us to shed any limiting blinders of master theories that may burden their inquiry and to enter the region with a mind open to the pluralism of explanatory possibilities.

Notes

1. We include in our synthesis here central place theory, mercantile theory, intersectoral articulation theory, capitalist penetration theory, and world systems theory, excluding diffusion theory because it is conceptualized in terms of abstract space, is unspecific in key aspects of application to urbanization, and borrows heavily from central place theory.

2. As discussed in chapters 3 and 6, among the justifications for both the Transamazon settlement program and the Northwest Region Development Program in Rondônia was the expressed concern about growing rural poverty in the Northeast and South of Brazil, respectively, and the need to move the landless poor from these pockets of potential social unrest.

3. For example, *Movimento Sem Terra* (Landless Movement), the *Federac<ata>o de Trabalhadores Rurais* (Federation of Rural Workers), and numerous allied *Sindicatos de Trabalhadorres Rurais* (rural workers' syndicates).

4. Some, namely those parent companies that are Brazilian commercial banks, however, have investments of various types throughout the world.

Technical Appendix

STUDY SITE SELECTION

With initial seed money support from our respective institutions, the principal investigators traveled to the proposed Amazon study areas in 1988 and 1989 to identify possible research sites. We sought urban settlements representing a transect of the hypothesized local city-system hierarchies, which we expected to find emerging as Amazon towns moved from a reliance on natural resources to commerce and industry. As an urban planner and an urban geographer, we anticipated finding settlement systems based on central place principles of spatial economics (see chapter 2).

In the case of southern Pará, the task of site selection was straightforward. Previous research indicated a continuing dynamic of in-migration and frontier expansion in the Xinguara region, midway between Marabá and Conceição do Araguaia on the PA-150 state highway (fig. 7.1). Xinguara, founded in the late 1970s, subsequently emerged as one of the region's leading urban centers with a population of about 25,000 in 1990 (table 7.2). The town was a logical starting place for a regional study: Xinguara's growth resulted largely from its location at the crossroads of the PA-150 and the PA- 279, the latter leading westward into the active pioneer zones of the Xingu region. Two other urban settlements stood out on the PA-279: Ourilândia do Norte and Tucumã (fig. 7.3). Along with São Félix do Xingu, a smaller town at the end of the road, these were virtually the only urban centers linked directly by road to Xinguara in the west. The absence of competing local central places gave a certain hollow character to the urban system that is typical of southern Pará: after initial spurts of growth, once promising towns often declined in importance, leaving only a few established municipal centers in a vast area, even where abundant natural resources continued to attract migrants. Overall, we decided that the Xinguara–Ourilândia do Norte–Tucumã corridor segment would provide a good case study of evolving urban-system dynamics in eastern Amazonia.

In Rondônia, the task of site selection was more complicated in various ways. The city of Rolim de Moura served as the principal central place and gateway to four different migration corridors radiating south and west, linking eighteen different known settlements (fig. 6.3). After an extended reconnaissance visit, we selected the Rolim de Moura–Alto Alegre corridor, which proceeded in a southerly direction toward the active frontier along the Guaporé Valley. We might have selected one other corridor, Linha 25, the straight westward extension of Rolim de Moura's main street, linking the BR 364 at Pimenta Bueno with Rondônia State Road 429 at São Miguel do Guaporé. We decided against this route because we believed it represented a truncated frontier, influenced at either end by interregional commerce. Linha Norte-Sul, by contrast, terminated in the forest at the cutting edge of the frontier some 15 kilometers south of Alto Alegre in July 1990 (plate 10.2).

We ultimately selected for our comparative study two different settlement corridors, one in southern Pará and one in Rondônia, each featuring three towns of approximately the same ages. Age cohort comparability among the settlements controlled for possible local history biases. Our study areas were both still relatively remote but active frontier environments in which we could observe the processes of urbanization. Since our study site selections in both Rondônia and Pará were determined by the same criteria, it is unlikely that the variations observed in their urbanization patterns can be attributed to idiosyncrasies of site selection protocols.

QUESTIONNAIRE DEVELOPMENT

The prototype questionnaire reflected each author's extensive previous experience with survey research in Brazil. Between them, Browder and Godfrey had personally directed fifteen questionnaire surveys in the country by 1990, dating back to master's or doctoral research many years before. Questionnaires used by other Amazon researchers (Marianne Schmink, Charles Wood, Timmons Roberts, and Donald Sawyer) were also consulted. An initial prototype questionnaire was developed by the authors in the United States, revised in Belém, and pretested in both Rolim de Moura and Xinguara (thirty pretest interviews were conducted); the final version emerged in consultation with Donald Sawyer in Belo Horizonte.

The final survey instrument was divided into ten parts: household composition, primary and secondary economic activities (both current and previous) and their locations, household-level health, family characteristics (race, literacy, religious affiliation), migration and occupational history, physical characteristics of dwelling units, financial attributes, property ownership and agricultural activities, material possessions, and family attitudes about life in the Amazon. Urban households were defined to include those individuals normally dwelling in a common living unit; residents who had arrived during the last six months, or after January 1, 1990, were eliminated from the survey in order to ensure permanent Amazon residence among the respondents. The household questionnaire was developed from the start in the Portuguese language, and copies of it may be obtained from the authors by written request.

SAMPLE SELECTION

In both Rondônia and Pará advance teams employing a total of seven trained Brazilian researchers visited all six study sites to prepare for the general survey. These advance teams carried out several important preparatory tasks: collecting the most recent local town building maps (provided by the federal malaria control agency, SUCAM, in both cases), arranging the logistics of transportation for the research team and its accommodations in the field, and pretesting the urban household questionnaire. In addition, in each settlement the household populations of several blocks were counted to estimate the average number of houses and residential densities and to ensure an even distribution of interviews.

In selecting the households to visit for interviews, urban blocks were drawn at random from the SUCAM maps in all six study sites, using published random number tables (RAND Corporation 1955). Within each randomly selected block, every other building was systematically selected for visitation and interview. We did not a priori eliminate nonresidential zones (e.g., industrial districts, commercial districts, and municipal dumps) since we correctly assumed that people resided in these areas as well. Any nonresidential space within the sample frame of randomly selected blocks was replaced by the next adjoining building on the same block until an occupied building space was found. If no occupants were found on the entire block, then it was replaced by another randomly selected block. Unattended residential buildings were revisited three times before being declared unoccupied and replaced in a similar systematic fashion.

INTERVIEWER SELECTION AND TRAINING

The Brazilian field research ran from June 17 to August 10, 1990. The project utilized a team of twenty-three researchers, including the authors, all but five of whom were Brazilian nationals. Eleven of the researchers were undergraduate students from the Federal University of Minas Gerais recruited for the Project by the Centro de Desenvolvimento e Planejamento Regional (CEDEPLAR) with which Donald Sawyer was associated at the time. Five other Brazilian researchers were recruited in Porto Velho through the Federal University of Rondônia, two of whom were assigned to the urban project and three to the rural project (described below).

Prior to the commencement of the field research, all researchers received two days of classroom training in interview techniques, including interview simulations and critiques, problems of bias, and orientation to study areas. Then, in the field sites, the researchers spent one additional day doing supervised trial interviews of households located on a test block that had been set aside from the sample for training purposes.

SURVEY EXECUTION

From July 13 to August 3, 1990, our two teams of researchers, eleven in Pará and ten in Rondônia, plus the two authors completed a total of 1,617 household interviews: 904 in Pará and 713 in Rondônia (see tables 6.2 and 7.2 for study site descriptions). The typical interview lasted approximately forty-five minutes. Commonly, interviewers returned with four to eight completed questionnaires per day. All completed questionnaires were checked by the authors each night to identify uncertainties. About 30 percent of all completed questionnaires were returned to the interviewers the following day for clarification, before being approved for tabulation. In this manner considerably less than 5 percent of all completed questionnaires were ultimately rejected as unreliable.

To minimize the possibility of a systematic gender bias in our survey practice, interviewers were trained to ask for "entire household" focus groups and to solicit responses from male and female household members alike to the extent practicable. In most cases interviewers noted which household member responded to key questions and what other distractions occurred during the interview event—generally associated with the child care activities that necessarily arise from this genderwide research strategy. Where there was a gender bias, it was slightly oriented toward the female household cohead who was more frequently found at home during weekday daytime periods. Our interview schedule included both weekend days (to find complete households close to home) and three and one-half to four weekdays. Teams enjoyed one and one-half days of rest per week over the seventeen-day interview period in Rondônia and two days of rest over the twenty-day interview period in Pará (not including two days spent being transported between sites). In Rondônia four of the ten interviewers were female. In Pará three women and six men worked as field interviewers, not including the principal investigator and an undergraduate research assistant.

Several additional key sector surveys were undertaken in both study areas to explore structural aspects of the urban economy not obvious from the household level survey. In Rondônia surveys included the agricultural grain processing sector (seven firms surveyed), financial sector (two banks), retail food products sector (seven supermarkets and thirteen small, family-owned grocery shops), and the industrial woods sector (twenty-one wood processing firms). In all cases, the owner or manager of each firm was interviewed using both standardized questionnaires and open-ended interviews. In addition, vehicular movements entering and leaving Rolim de Moura through the two major entry points over a two-day period were classified and counted. Since Browder had completed a similar traffic count in 1984, a comparison of the changes in freight composition over a ten-year period was possible.

Besides the urban project's six interviewers, the Rondônia research included a rural study with three interviewers. The rural project sought, in part, to identify any significant local urban-rural interactions (e.g., local migration, urban employment patterns, income transfers, etc.) in the study area. A separate survey team applied a different questionnaire to 115 rural households in Rolim de Moura, Santa Luzia, and Alta Floresta *municípios*. Forty-five of these rural farmsteads had been surveyed initially by Browder in 1985; the other seventy farms were randomly selected. Selected findings of this rural survey were published separately (Browder 1990a). At the conclusion of the field research, Brazilian team members provided a public presentation of selected preliminary findings of the research in Rolim de Moura, which was attended by about fifty people. A similar debriefing was provided to government officials of the state of Rondônia in Porto Velho at the conclusion of the field research.

In the Xingu corridor of southern Pará the research teams also examined key economic sectors in addition to carrying out the household interviews. During the pretest, ten timber mills were visited in the region, and questionnaires on wood extraction, mill employment, and regional commerce were administered. Managers at the major banks in the regional centers and in all the local study sites were visited to discuss regional financial flows and economic change. In each of the three towns studied, all commercial establishments were counted, classified, and mapped by an American undergraduate and a local Brazilian helper. In addition, these two workers also surveyed the highway traffic by visual recognizance from posts on access routes on the PA-150 and PA-279 highways around Xinguara, during eight sessions in the morning, afternoon, and early evening from July 14 to 17, 1990. Our research teams visited Xinguara's two supermarkets, and Tucumã's sole supermarket, along with numerous small grocery stores, and the popular markets in all the study sites during June and July of 1990. Standard questionnaires on product origins were administered. In addition, our teams monitored communications data at local telephone posts, frequency of bus transport between settlements, and other linkages between settlements. Because the Pará team was forced to travel longer distances between sites and had to take up residence in two separate places (Xinguara and Tucumã), it was not feasible to hold a final public debriefing on the research. At all three of the principal study sites, however, elected municipal officials and other leaders were repeatedly consulted; their approval ensured local support for the household surveys and other team research efforts.

Instead of a carrying out a rural component to the main urban household surveys, as in Rondônia, the southern Pará teams decided to examine the sizable population involved in the regional gold-mining economy. Due to the importance of *garimpos de ouro* in the economy of southern Pará, our teams visited several gold-mining settlements near Tucumã and Ourilândia do Norte. On visits to the settlements at Cuca, Serrinha, and Vila Bananal, the four members of the pretest team informally interviewed residents about migration history, occupational change, and the gold-mining economy. Our full eleven-person team visited a new *garimpo* at Rio Branco in the municipality of Tucumã, during July of 1990. Our team completed seventy questionnaires on migration and socioeconomic status among the hundreds of miners working at Rio Branco, as reported in Godfrey (1992).

SURVEY DATA ANALYSIS

The Rondônia and Pará data collection activities were carefully synchronized to ensure reliability. Identical survey instruments were used in both study areas. Interviewers (all college students or graduates) received the same training. The authors maintained frequent contact by telephone during the field research to resolve periodic problems in the use of the questionnaire. The close collaborative nature of the project enhanced the comparability of the survey findings.

The field work ultimately generated an enormous data base of some 454,377 data points (281 variables x 1,617 questionnaires), the preliminary descriptive analysis of which took over a year to complete. The household questionnaire contained 106 discrete variables and an average of 175 tabular variables (35 discrete variables each multiplied by the number of household members). One of the Rondônia team researchers (Isidoro) returned to Virginia Tech with Browder to lead a team of U.S.-based graduate student assistants in the coding, tabulation, and descriptive statistical analysis of the data set. Author Godfrey joined the Blacksburg-based data analysis team in October of 1990 to confer on the coding and interpretation of the survey data. Subsequently, the authors remained in close contact during the extended process of writing up the results of field surveys.

References

Abers, Rebecca Neaera. 1992. Urbanization and city-ward migration on a resource frontier: The Amazon gold rush and the case of Boa Vista, Roraima. Master's thesis, University of California, Los Angeles.

Abu-Lughod, Janet. 1991. Going beyond the global babble. In Anthony D. King, ed., *Culture, Globalization, and the World System*, pp. 131–137. London: Macmillan Education.

Adalberto da Prússia, Príncipe. 1977. *Brasil: Amazonas / Xingu*. Translated by Eduardo Lima Castro. São Paulo: Editora da Universidade de São Paulo.

Adam, John A. 1988. Plundering the Amazon for power. *Washington Post*, November 27, p. D3.

Amnesty International. 1988. *Brazil: Authorized Violence in Rural Areas*. London.

Amin, S. 1974. *Accumulation on a World Scale: A Critique of the Theory of Underdevelopment*, 2 vols. New York: Monthly Review Press.

Aramburu, Carlos E. 1984. Expansion of the agrarian and demographic frontier in the Peruvian Selva. In Marianne Schmink and Charles Wood, eds., *Frontier Expansion in Amazônia*, pp. 153–179. Gainesville: University of Florida Press.

Armstrong, Warwick and T. G. McGee. 1985. *Theatres of Accumulation: Studies in Asian and Latin American Urbanization*. London and New York: Methuen.

Arruda, Hélio Palma de. 1977. *Colonização oficial e particular*. Brasilia: Instituto Nacionalo de Colonização e Reforma Agrária.

Asselin, V. 1983. *Grilagem: Corrupção e Violência em Terras do Carajás*. São Paulo: Editora Voces.

Aufderheide, Pat and Bruce M. Rich. 1985. Debacle in the Amazon. *Defenders*, March/April, pp. 20–32.

Baer, Werner. 1989. *The Brazilian Economy: Growth and Development*, 3d ed. New York: Praeger.

Bhaskar, Roy. 1975. *A Realist Theory of Science*. Leeds, UK: Leeds Books.

———. 1986. *Scientific Realism and Human Emancipation*. London: Verso.

Bakx, Keith. 1987. Planning agrarian reform: Amazônian settlement projects, 1970–86. *Development and Change* 18:533–555.

Barreiros da Silva, Ana Cristina. 1993. A produção do espaço em Porto Velho, Rondônia: O papel de um agente múltiplo. Master's thesis, Universidade Federal do Rio de Janeiro.

Barth, Gunther. 1988. *Instant Cities: Urbanization and the Rise of San Francisco and Denver*. Albuquerque: University of New Mexico Press.

Bartra, Roger. 1974. *Estrutura agraria y clases sociales en México*. Mexico City: Ediciones Era.

Bastos, João. 1984. Falta política para ocupação. *O Liberal*, April 16, p. 2.

Becker, Bertha K. 1985. Fronteira e urbanização repensadas. *Revista Brasileira de Geografia* 47 (3/4): 357–371.

———. 1987. Estratégia do estado e povoamento espontâneo na expansão de fronteira agrícola em Rondônia: Interação e conflito. In Gerd Kohlhepp and Achim Schrader, eds., *Homen e Natureza na Amazônia*, 237–252. Tübingen: Geographisches Institut, Universität Tübingen.

———. 1990. *Amazônia*. São Paulo: Atica.

———. 1992. Gestão do território e territorialidade na Amazônia. In Philippe Léna and Aldélia Engrácia de Oliveira, eds., *Amazônia: A fronteira agrícola 20 Anos depois*, 333–350. Belém, Pará: Editora CEJUP.

Becker, Bertha K., M. Miranda, and L. Machado. 1990. *Frontiera amazônica: Questoes sobre a gestão do território*. Brasília: Editoria Universidade de Brasília.

Benchimol, Samuel. 1977. *Amazônia: Un pouco-antes e alem depois*. Manaus: Editora Umberto Calderaro.

Bendix, J. and C. Liebler. 1991. Environmental degradation in Brazilian Amazônia: Perspectives of the U.S. news media. *Professional Geographer* 43 (4): 474–485.

Bendix, Reinhard and Guenther Roth. 1971. *Scholarship and Partisanship: Essays on Max Weber*. Berkeley: University of California Press.

Berry, Brian J. L. 1967. *Geography of Market Centers and Retail Distribution*. Englewood Cliffs, N. J.: Prentice-Hall.

———. 1972. Hierarchical diffusion: The basis of development filtering and spread in a system of growth centers. In Niles M. Hansen, ed., *Growth Centers in Regional Economic Development*. New York: The Free Press.

Bienen, Henry. 1984. Urbanization and third world stability. *World Development* 12 (7): 661–691.

Binswanger, Hans P. 1991. Brazilian policies that encourage deforestation in the Amazon. *World Development* 19 (7): 821–829.

Boischio, Ana Amelia P. and Antonio Barbosa. 1993. Exposição ao mercúrio orgânico em populaçães riberirinhas do Alto Madeira, Rondônia, 1991: Resultados preliminares. *Caderno de Saúde Pública*, 9 (2, April-June): 1–5.

Boudeville, J. R. 1961. *Les pôles de croissance brésiliens: La sidérugie du Minas Gerais.* Cahiers de l'isea, série L, number 9. Paris.

——. 1966. *Problems of Regional Economic Planning.* Part I. Edinburgh: Edinburgh University Press.

Bowman, Isaiah. 1931. *The Pioneer Fringe.* New York: American Geographical Society.

Boxer, C. R. 1962. *The Golden Age of Brazil, 1695–1750.* Berkeley and Los Angeles: University of California Press.

Bridges, Tyler. 1988. The rain forest's road to ruin. *Washington Post*, July 24.

Brooke, James. 1990a. Brazilian moves to rescue tribe: Gold miners routed in effort to save stone-age people. *New York Times*, March 27, pp. 1, 6.

——. 1990b. Brazil blows up miners' airstrip, pressing its drive to save Indians. *New York Times*, May 3.

——. 1990c. Is gold worth this? Amazon is being poisoned. *New York Times*, August 2.

Browder, John O. 1986. Logging the rain forest: A political economy of timber extraction and unequal exchange in the Brazilian Amazon. Ph.D. diss., University of Pennsylvania.

——. 1987. Brazil's export promotion policy (1980–1984): Impacts on the Amazon's industrial wood sector. *Journal of Developing Areas* 21 (April): 285–301.

——. 1988a. Public policy and deforestation in the Brazilian Amazon. In Robert Repetto and Malcolm Gillis, eds., *Public Policies and the Misuse of Forest Resources*, pp. 247–298. Cambridge: Cambridge University Press.

——. 1988b. The social costs of rain forest destruction: A critique and economic analysis of the "hamburger debate." *Interciencia* 13 (3, May-June): 115–120.

——. 1989a. Lumber production and economic development in the Brazilian Amazon: Regional trends and a case study. *Journal of World Forest Resource Management* 4:1–19.

Browder, John O, ed. 1989b. *Fragile Lands of Latin America: Strategies for Sustainable Development.* Boulder, Colo.: Westview Press.

Browder, John O. 1990a. *Desenvolvimento Regional e Conservação de Florestas Tropicais em Rondônia.* Technical report. Office of Rural and Institutional Development, Agency for International Development, Washington, D.C.

——. 1990b. Tropical deforestation and the productivity of farm labor in a Brazilian Amazon colonization project. Manuscript.

——. 1995. Surviving in Rondônia: The dynamics of colonist farming strategies in Brazil's Northwest frontier. *Studies in Comparative Economic Development* 29 (3): 45–69.

Browder, John O. and José Antonio Borello. 1992. The state and the crisis of planning in Latin America. *Journal of Planning Literature* 6 (4): 369–377.

Browder, John O. and Brian J. Godfrey. 1990. Frontier urbanization in the Brazilian Amazon: A theoretical framework for urban transition. *Yearbook of the Conference of Latin Americanist Geographers* 16:56–66.

Browder, John O. and Gilvandro B. Pinheiro, Joana D. M. Silva do Amaral, Benedita do Nascimento Pereira, and Carlos Macedo Dias. 1994. Levantamento sócioeconômico da população da periferia urbana de Porto Velho. Technical report. Porto Velho: Centro de Educação e Assessoria Popular and Blacksburg, VA: Virginia Polytechnic Institute and State University (Center for Urban and Regional Studies).

Brown, Lawrence. 1981. *Innovation Diffusion: A New Perspective*. London and New York: Methuen.

Brown, Lawrence A. and Rodrigo Sierra, Scott Digiacinto, and W. Randy Smith. 1994. Urban system evolution in frontier settings. *The Geographical Review* 84 (3): 249–265.

Brum, Argemiro Jacob. 1988. *Modernização da Agricultura: Trigo e Soja*. Petrópolis: Editora Vozes.

Bunker, Stephen G. 1985. *Underdeveloping the Amazon: Extraction, Unequal Exchange, and the Failure of the Modern State*. Urbana: University of Illinois Press.

Bunyard, Peter. 1987. Dam building in the tropics: Some environmental and social consequences. In Robert E. Dickenson, ed., *The Geophysiology of Amazonia*. New York: Wiley.

Burns, E. Bradford. 1993. *A History of Brazil*. New York: Columbia University Press.

Butland, C. J. 1966. Frontiers of settlement in South America. *Revista Geografica* 65:93–107.

Butler, J. R. 1985. Land, gold, and farmers: Agricultural colonization and frontier expansion in the Brazilian Amazon. Ph.D. diss., University of Florida, Gainesville.

Cardoso, Fernando Henrique and Geraldo Muller. 1977. *Amazônia: Expansão do capitalismo*. São Paulo: Editora Brasiliense.

Carnasciali, Carlos H., Dimas Floriani, Eron José Maranho, Rodolfo José Angulo, Rossana Ribeiro Ciminelli, and Venessa Fleischfresser. 1987. Consequências sociais das transformações tecnológicas na agricultura do Paraná. In George Martine and R. C. Garcia, eds., *Os impactos sociais da modernizacão agrícola*, pp. 125–167. São Paulo: Caetés.

Castells, Manuel. 1977. *The Urban Question: A Marxist Approach*. London: Edward Arnold.

Castells, Manuel and Peter Hall. 1994. *Technopoles of the World: The Making of Twenty-First Century Industrial Complexes*. London: Routledge.

Caufield, Catherine. 1983. Dam the Amazon, full steam ahead. *Natural History* July: 60–67.

Cawson, A. 1986. *Corporatism and Political Theory*. Oxford and New York: Blackwell.

CEM (Centro de Estudos Migratórios). 1986. *Migrações no Brasil: O peregrinar de um povo sem terra*. São Paulo: Edições Paulinas.

CEPA/RO (Comissão Estadual de Planejamento Agricola de Rondônia). 1983. Prognóstico agropecuário de Rondônia. Technical report. Porto Velho.

Chase-Dunn, Christopher. 1983. Urbanization in the world-system: New directions for research. *Comparative Urban Research* 9 (2): 41–47.

Chisholm, M. 1962. *Rural Settlement and Land Use*. London: Hutchinson.

Christaller, Walter. 1966 (1933). *Central Places in Southern Germany*. Translated by Carlisle W. Baskin. Englewood Cliffs, N. J.: Prentice-Hall.

Clay, Jason W. 1979. The articulation of noncapitalist agricultural production systems with capitalist exchange systems: The case of Garanhuns, Brazil, 1845–1977. Ph.D. diss., Cornell University.

Cleary, David. 1990 *Anatomy of the Amazon Gold Rush*. Iowa City: University of Iowa Press.

———. 1993. After the frontier: Problems with political economy in the modern Brazilian Amazon. *Journal of Latin American Studies* 25:331–349.

Cloke, Paul, Chris Philo, and David Sadler. 1991. *Approaching Human Geography: An Introduction to Contemporary Theoretical Debates*. New York and London: Guilford Press.

CODEMA (Centro de Desenvolvimento, Pesquisa e Tecnologia do Estado do Amazonas). 1984. *IV pesquisa sócio-econômica da cidade de Manaus*. Manaus, Amazonas.

Collins, Jane. 1986. Smallholder settlement of tropical South America: The social causes of ecological destruction. *Human Organization* 45 (1): 1–10.

———. 1987. Labor scarcity and ecological change. In Peter D. Little and Michael M. Horowitz, eds., *Lands at Risk in the Third World*, pp. 19–37. Boulder, Colo.: Westview Press.

Comissão Brandão Pró-Emancipação. 1993. *Estado de Carajás: Dividir para Administrar e Desenvolver*. Marabá, Pará.

CONSAG (Construtora Andrade Gutierrez). 1978. *Projeto Tucumã: Colonização por Múltiplas Atividades en São Félix do Xingu, Pará*. 3 Vols. Belém, Pará. Mimeographed.

Conzen, Michael P. 1975. A transport interpretation of the growth of urban regions: An American example. *Journal of Historical Geography* 1 (4): 361–382.

Correa, Roberto Lobato. 1987. A periodização da rede urbana da Amazônia. *Revista Brasileira de Geografia* 49 (3): 39–68.

———. 1991. A organização urbana. *Geografia do Brasil, Região Norte*, 3: 255–272. Rio de Janeiro, Fundação Instituto Brasileiro de Geografia e Estatística.

Costa, Manoel Augusto, Douglas H. Graham, et al. 1971. *Migrações Internas no Brasil*. Monografia No. 5. Rio de Janeiro: Instituto de Planejamento Econômico e Social.

Couto, R. D. de Sena, V. M. Câmara, and P. C. Sabroza. 1988. Intoxicação mercurial: Resultados preliminares em duas áreas garimpeiras no Estrado do Pará. *Pará Desenvolvimento* 23 (Jan/Jun.): 63–77.

Coudreau, Henri. 1897. *Voyage au Tocantins-Araguaia: 31 décembre 1896–23 mai 1897*. Paris: A. Lahure.

——. 1977. *Voyage ao Xingu*. São Paulo: Editora da Universidade de São Paulo.

Coy, Martin. 1987. Rondônia: Frente pioneira e programa POLONOROESTE. O processo de diferenciação sócioeconômica na periferia e os limites do planejamento público. In Gerd Kohlhepp and Achim Schrader, eds., *Homen e natureza na Amazônia*, pp. 253–270. Tübingen: Geographisches Institut, Universität Tübingen.

Cunha Camargo, José Geraldo da. 1973. Rural urbanization. Brasília: Ministério da Agricultura. Summary of a paper presented to the Third International Housing Conference held in Rio de Janeiro, March 28, 1971.

Dahl, R. A. 1956. *A Preface to Democratic Theory*. Chicago: University of Chicago Press.

de Silva, L. F., R. Carvalho Filho, and M. B. M. Santana. 1973. Solos do Projeto Ouro Preto. CEPLAC *Boletim Tecnica* 23.

Dean, Robert D., William H. Leahy, and David L. McKee. 1970. *Spatial Economy Theory*. New York: Free Press.

Despres, Leo A. 1991. *Manaus: Social Life and Work in Brazil's Free Trade Zone*. Albany, N.Y.: State University of New York Press.

di Tella, Torcuato S. 1989. *Latin American Politics: A Theoretical Approach*. Austin: University of Texas Press.

Dickens, Peter. 1992. *Society and Nature: Towards a Green Social Theory*. Philadelphia: Temple University Press.

Dickenson, John. 1982. *Brazil*. London and New York: Longman.

D'Incao, Maria Conceição. 1975. *O "bóia-fria": Acumulação e miséria*. Petrópolis: Editora Vozes.

Dogan, Mattei and John D. Kasarda. 1988. Introduction: How giant cities will multiply and grow. In M. Dogan and J. Kasarda, eds., *The Metropolis Era: A World of Giant Cities*, pp. 12–29. Newbury Park, California: Sage Publications.

Dourojeanni, Marc J. 1984. Brazil: Report of the forest and environment components of Projects I, II, III of the Northwest Region Development Program. Memorandum to M. Asseo. World Bank.

Drake, Durrant, Arthur O. Lovejoy, James B. Pratt, Arthur K. Rogers, Goerge Santayana, Roy W. Sellars, and C. A. Strong. 1968 [1920]. *Essays in Critical Realism: A Co-operative Study of the Problem of Knowledge*. New York: Gordian Press.

Dryzek, John S. 1987. *Rational Ecology: Environment and Political Economy*. Oxford, UK, and Cambridge, Mass.: Blackwell.

Eicher, Carl, Thomas Zalla, James Kocher, and Fred Winch. 1970. *Employment Generation in Agriculture*. East Lansing: Institute of International Agriculture, Michigan State University.

EMBRAPA (Empresa Brasileira de Pesquisa Agropecuária). 1975. Mapa esquemático dos solos das regiões norte, meio-norte, e centro-oeste do Brasil. Rio de Janeiro: Centro de Pesquisas Pedológicas, *Boletim Técnica* 17.

Emmanuel, Arghiri. 1972. *Unequal Exchange: A Study of the Imperialism of Trade.* New York: Monthly Review Press.

Escobar, Gabriel. 1995a. Brazil's Indians organize. *Washington Post*, August 12, p. A-15.

———. 1995b. Indians tugged by town try going halfway. *Washington Post*, August 15, p. A-12.

Evans, P. 1979. *Dependent Development: The Alliance of Multinational, State, and Local Capital in Brazil.* Princeton: Princeton University Press.

Falesi, I. 1974. O solo da Amazônia e sua relação com a definição de sistemas de produção agrícola. In *Reunião do grupo interdisciplinar de trabalho sobre directrizes de pesquisa agrícola para Amazônia.* Documento 2. Brasília: Empresa Brasileira de Pesquisa Agropecuária.

Faris, Ralph M. 1991. *Corporate Networks and Corporate Control: The Case of the Delaware Valley.* New York: Greenwood Press.

Fearnside, Philip M. 1978. Estimation of the carrying capacity for human populations in a part of the Transamazon Highway colonization area of Brazil. Ph.D. diss., University of Michigan.

———. 1984. Brazil's Amazon settlement schemes: Conflicting objectives and human carrying capacity. *Habitat International* 8 (1): 45–61.

———. 1986. Settlement in Rondônia and the token role of science and technology in Brazil's Amazônian development planning. *Interciencia* 11 (5): 229–236.

———. 1989. Brazil's Balbina Dam: Environment versus the legacy of the pharaohs in Amazonia. *Environmental Management* 13 (4): 401–423.

———. 1990. Environmental destruction in the Brazilian Amazon. In D. Goodman and A. Hall, eds., *The Future of Amazonia: Destruction or Sustainable Development,* pp. 179–225. New York: St. Martin's Press.

Feijao, A. J. and J. A. Pinto. 1990. *The Amazon Gold Rush.* Amapá: União dos Sindicatos e Associações de Garimpeiros da Amazônia Legal.

Ferré, Frederick. 1988. Religious World Modeling and Postmodern Science. In David Ray Griffin, ed., *The Reenchantment of Science,* pp. 87–98. Albany, N.Y.: State University of New York Press.

Ferreira, Ignez Costa Barbosa. 1987. Expansão da fronteira agrícola e urbanização. In Lena Lavinas, ed., *A Urbanização da fronteira,* 2:1–26. Rio de Janeiro: Publipur, Universidade Federal do Rio de Janeiro.

FGV (Fundação Getúlio Vargas). var. yrs. *Preços médios de arrendamentos, vendas de terrenos, salários, empreita do transporte.* Rio de Janeiro: FGV.

FIERO (Federação das Indústrias do Estado de Rondônia). ca. 1993. *A indústria de Rondônia: Proposta para o desenvolvimento industrial.* Porto Velho: FIERO.

Fifer, Valerie J. 1982. The Search for a series of small successes: Frontiers of settlement in eastern Bolivia. *Journal of Latin American Studies* 14 (2): 408–432.

FIPE (Fundação Instituto de Pesquisas Economica de São Paulo). 1983. Estudo sócioeconômico de populações assentadas nos projetos de colonização do Estado de Rondônia. Survey table of the Jí-Paraná. PIC. Porto Velho: SEPLAN.

FJP (Fundação João Pinheiro). 1975. *Levantamento de reconhecimento de solos da aptidão agropastoril, das formas vegetais e do uso da terra em área do território federal de Rondônia*. Belo Horizonte: Centro de Recursos Naturais.

Foin, Theodore C., Jr. 1976. *Ecological Systems and the Environment*. Boston: Houghton Mifflin.

Foresta, Ronald. 1992. Amazônia and the politics of geopolitics. *The Geographical Review* 82 (2): 128–142.

Foweraker, Joe. 1981. *The Struggle for Land: A Political Economy of the Pioneer Frontier in Brazil from 1930 to the Present Day*. Cambridge, UK: Cambridge University Press.

Frank, Andre G., ed. 1969. *Latin America: Underdevelopment or Revolution*. New York: Monthly Review Press.

———. 1978. *Dependent Accumulation and Underdevelopment*. New York: Monthly Review Press.

Friedmann, John. 1986. The world city hypothesis. *Development and Change* 17 (1): 69–84.

Friedmann, John and Goetz Wolff. 1982. World city formation: An agenda for research and action. *International Journal of Urban and Regional Research* 6 (3): 309–344.

Furley, Peter. 1980. Development planning in Rondônia based on naturally renewable resource surveys. In Francoise Barbira-Scazzocchio, ed., *Land, People, and Planning in Contemporary Amazônia*, pp. 37–45. Occasional Publication No. 3. Cambridge, UK: Centre of Latin American Studies.

———. 1986. Radar surveys for resource evaluation in Brazil: An illustration from Rondônia. In M. J. Eden and J. T. Parry, ed., *Remote Sensing and Tropical Land Management*, pp. 79–99. London: Wiley.

Garcia, Afrânio Raul, Jr. 1990. *O sul: Caminho do roçado*. São Paulo and Brasília: Editora Marco Zero and Editora da Universidade de Brasília.

Gazeta Mercantil. 1978. Um potencial de 13.5 milhões de metros cúbicos para exportação. October 5. São Paulo.

Geiger, P. P. and F. R. Davidovich. 1986. The spatial strategies of the state in the political-economic development of Brazil. In A. J. Scott and M. Storper, eds., *Production, Work, Territory: The Geographical Anatomy of Industrial Capitalism*, pp. 281–300. Boston: Allen and Unwin.

Gentil, Janete M. L. 1988. A noção do urbano em áreas de fronteira: Uma revisão teórica. Paper presented at the seminar amazônia: A fronteira agrícola 20 anos depois. Belém: Museu Paraense Emilio Goeldi, December 5–7.

Giddens, Anthony. 1979. *Central Problems in Social Theory*. London: Macmillan.

———. 1984. *The Constitution of Society: Outline of the Theory of Structuration*. Cambridge, U.K.: Polity Press.

———. 1990. *The Consequences of Modernity*. Cambridge, U.K.: Polity Press.

Ginsburg, N., B. Koppel, and T. G. McGee, eds. 1991. *The Extended Metropolis: Settlement Transition in Asia*. Hawaii: University of Hawaii Press.

Godfrey, Brian J. 1978. Brazil's Transamazon Highway: Migration and agricultural production on the Amazon frontier. Paper presented at the Conference of Latin Americanist Geographers, Sonoma State College, California, June 8–9.

———. 1979. Road to the Xingu: Frontier settlement in southern Pará, Brazil. Master's thesis, University of California, Berkeley.

———. 1982. Xingu Junction: Rural migration and land conflict in the Brazilian Amazon. *Proceedings of the Pacific Coast Council on Latin American Studies* 9:71–82.

———. 1988. Frentes de expansão na Amazônia: Uma perspectiva geográfico-histórica. *Geosul* 6:7–20. Florianópolis: Universidade Federal de Santa Catarina.

———. 1990. Boom towns of the Amazon. *Geographical Review* 80 (2): 103–117.

———. 1991. Modernizing the Brazilian city. *Geographical Review* 81 (1): 18–34

———. 1992. Migration to the gold-mining frontier in Brazilian Amazônia. *Geographical Review* 82 (4): 458–469.

———. 1993. Regional depiction in contemporary film. *Geographical Review* 83 (4): 428–440.

Gomes da Silva, Amizael. 1984. *No rastro dos pioneiros: Um pouco da história Rondôniana.* Porto Velho: SEDUC.

———. 1991. *Amazonia Porto Velho: Pequena história de Porto Velho.* Porto Velho: Palmares.

Goodland, Robert. 1985. Brazil's environmental progress in Amazonian development. In John Hemming, ed., *Change in the Amazon Basin: Man's Impact on Forests and Rivers.* Manchester, U.K.: Manchester University Press.

Goodman, David, Bernardo Sorj, and John Wilkinson. 1984. Agro-industry, state policy, and rural social structures: Recent analyses of proletarianisation in Brazilian agriculture. In B. Munslow and H. Finch, eds., *Proletarianisation in the Third World: Studies in the Creation of a Labour Force Under Dependent Capitalism.* London: Croom Helm.

Gordon, D. M. 1978. Capitalist development and the history of American cities. In W. K. Tabb and L. Sawers, eds., *Marxism and the Metropolis*, pp. 21–53. New York: Oxford University Press.

Gottmann, Jean. 1961. *Megalopolis: The Urbanized Northeastern Seaboard of the United States.* New York: Twentieth Century Fund.

Gould, Stephen Jay. 1995. Spin doctoring Darwin. *Natural History* July: 6–71.

Governo do Estado do Pará. 1976. *Rodovia PA-150: Relatório do anteprojeto de engenharia.* Belém, Pará.

Graham, Douglas,. 1986. Land use in the Amazon: Soil erosion, sedimentation, and the Samuel Dam. *Latinamericanist* 22 (1): 1–5. Gainesville, Florida: Center for Latin American Studies, University of Florida.

Graziano da Silva, José. 1982. *A modernização dolorosa: Estrutura agrária, Fronteira agrícola, e Trabalhadores rurais no Brasil.* Rio de Janeiro: Zahar Editores.

Greenwood, Davydd J. 1973. The political economy of peasant family farming: Some anthropological perspectives on rationality and adaptation. Occasional Paper no. 2. Rural Development Committee. Ithaca, N.Y.: Cornell University.

Gregory, Derek. 1986. Humanistic geography, realism, and structuration theory. In R. J. Johnston, D. Gregory, and D. M. Smith, eds., *The Dictionary of Human Geography*, pp. 207–10, 387–90. 2d ed. Oxford, U.K.: Blackwell.

Gulley, J. L. M. 1959. The Turnerian frontier: A study in the migration of ideas. *Tijidschrift Voor Economische en Sociale Geografie* 3 (March): 65–71.

Hägerstrand, T. 1952. The propagation of innovation waves. *Lund Studies in Geography*, series B, Human Geography. New York: St. Martin's Press.

Haggett, Peter. 1966. *Locational Analysis in Human Geography*. New York: St. Martin's Press.

Hall, Anthony L. 1989. *Developing Amazonia: Deforestation and Social Conflict in Brazil's Carajás Programme*. Manchester, U.K., and New York: Manchester University Press.

Hall, P. and D. Hay. 1980. *Growth Centres in the European Urban System* London: Heinemann.

Hamer, Andrew Marshall. 1985. Decentralized urban development and industrial location behavior in São Paulo, Brazil. World Bank Staff Working Papers, No. 732. Washington, D.C.: The World Bank.

Hardoy, Jorge E. and David Satterthwaite, eds. 1986. *Small and Intermediate Urban Centres: Their Role in National and Regional Development in the Third World*. London: Hodder and Stoughton.

Hardman, Francisco Foot. 1988. *Trem fantasma: A modernidade na selva*. São Paulo: Compania das Letras.

Harré, R. 1986. *Varieties of Realism*. Oxford, U.K.: Blackwell.

Harvey, David. 1976. Labor, capital, and class struggle around the built environment in advanced capitalist societies. *Politics and Society* 6:265–295.

———. 1978. The urban process under capitalism: A framework for analysis. *International Journal of Urban and Regional Research* 2:101–131.

———. 1985. *The Urbanization of Capital*. Baltimore: Johns Hopkins University Press.

———. 1987. Three myths in search of a reality in urban studies. *Environment and Planning D: Society and Space* 5:367–76.

Hathaway, Dale E. 1964. Urban industrial development and income differentials between occupations. *Journal of Farm Economics* 46 (February): 55–66.

Hays, Samuel P. 1987. *Beauty, Health, and Permanence: Environmental Politics in the United States, 1955–1985*. New York: Cambridge University Press.

Hebette, Jean and Rosa Acevedo. 1979. *Colonização para quem?* Belém: Núcleo de altos estudos amazônicos, Universidade Federal do Pará.

Hecht, Susanna B. 1985. Environment, Development, and Politics: Capital Accumulation and the Livestock Sector in Eastern Amazônia. *World Development* 13 (6): 663–684.

———. 1986. Development and deforestation in the Amazon: Current and future policies, investment, and impact on forest conversion. Washington, D.C.: Unpublished report to the World Resources Institute.

——. 1986. Regional development: Some comments on the discourse in Latin America. *Environment and Planning D: Society and Space* 4:201–209.

Hecht, Susanna and A. Cockburn. 1989. *The Fate of the Forest: Developers, Destroyers, and Defenders of the Amazon*. London and New York: Verso.

Helmsing, Bert. 1982. Agricultural production in the periphery: Settlement schemes reconsidered. *Development and Change* 13:401–419.

Hennessy, Alistair. 1978. *The Frontier in Latin American History*. Albuquerque: University of New Mexico Press.

Henriques, Maria Helena F. T. 1980. *A dinâmica demográfica de uma área de fronteira: Rondônia*. Technical Report. Brasília: Instituto Brasileiro de Geografia e Estatística.

Herbert, David T. and Colin J. Thomas. 1990. *Cities in Space: City as Place*. Savage, Maryland: Barnes & Noble.

——. 1988. The colonization experience in Brazil. In A. S. Oberai ed., *Land Settlement Policies and Population Redistribution in Developing Countries: Achievements, Problems, and Prospects*, pp. 317–354. New York: Praeger.

Higgins, Benjamin. 1963. Requirements for rapid development in Latin America: The view of an economist. In Egbert de Vries and José Medina Echavarria, eds., *Social Aspects of Economic Development in Latin America*, vol. 1. Paris: United Nations Educational, Scientific, and Cultural Organization.

——. 1968. *Economic Development: Principles, Problems, and Policies*. Revised edition. New York: Norton.

Hinderink, J. and M. J. Titus. 1988. Paradigms of regional development and the role of small centres. *Development and Change* 19 (3): 401–424.

Hiraoka, Mario. 1980. Settlement and development of the Upper Amazon: The East Bolivian example. *The Journal of Developing Areas* 14 (April): 327–347.

Hirschman, Albert. 1958. *The Strategy of Economic Development*. New Haven: Yale University Press.

Ho, Mae-Won and Peter T. Saunders. 1984. Pluralism and convergence in evolutionary theory. In Mae-Won Ho and Peter T. Saunders, eds., *Beyond New-Darwinism: An Introduction to a New Evolutionary Paradigm*, pp. 1–12. London: Academic Press.

Hoffmann, Ronald. 1994. Winning gold. *American Scientist* 82 (January-February): 15–17.

Hoornaert, Eduardo. 1992. *História da igreja na Amazônia*. Comissão de Estudos da História da Igreja na América Latina. Petrópolis: Editora Vozes.

Hoover, Edgar M. 1948. *The Location of Economic Activity*. New York: McGraw-Hill.

Human Rights Watch. 1991. *Rural Violence in Brazil: An Americas Watch Report*. New York and Washington, D.C.: Human Rights Watch.

Ianni, Octavio. 1979. *Colonização e contra-reforma agrária na Amazônia*. Petrópolis: Editora Vozes.

IBGE (Instituto Brasileiro de Geografia e Estatística). 1977. *Geografia do Brasil, Região Norte*, vol. 1. (also vols. 2–5).

———. 1979a. *Atividade de simulação na área econômico-demográfica no* IBGE, serie estudos e pesquisa, 3.

———. 1979b. *Indicadores sociais*. Rio de Janeiro: IBGE.

———. 1980a. *Censo agropecuário, Rondônia*. Rio de Janeiro: IBGE.

———. 1980b. *Censo agropecuário, Brasil*. Rio de Janeiro: IBGE.

———. 1985. *Anuário estatístico do Brasil*. Rio de Janeiro: IBGE.

———. 1986. *Produção agrícola municipal, Regiões Norte e Nordeste*. Rio de Janeiro: IBGE.

———. 1987. *Regiões de influência das cidades*. Rio de Janeiro: IBGE.

———. 1989. *Anuário estatístico do Brasil*. Rio de Janeiro: IBGE.

———. 1990. População residente em 1980 e estimativa em 01.07.90. Belém, Pará, mimeographed.

———. 1991. *Sinopse preliminar do censo demográfico, 1991*. Rio de Janeiro.

———. 1992. *Anuário estatístico do Brasil*. Rio de Janeiro: IBGE.

IDESP (Instituto do Desenvolvimento Econômico-Social do Pará). 1990a. *Pará Agrário: Informativo da Situação Fundiária*. Belém.

———. 1990b. *Municípios paraenses: Ourilândia do Norte*. Belém.

———. 1990c. *Municípios paraenses: Tucumã*. Belém.

———. 1991. *Zoneamento ecológico-econômico do estado do Pará*. Belém.

INCRA (Instituto Nacional de Colonização e Reforma Agrária). 1972. *A new March Northward: Altamira, Model of Amazon Colonization*. Brasília: Ministério da Agricultura.

———. 1973. *Summary of INCRA's Operations in the Amazon Region*. Translated by Winia J. Farberow. Brasília: Ministério da Agricultura.

———. 1975. Annual report. Belém, Pará: Pan American Health Organization, mimeographed.

———. 1982. Doze anos de colonização. Technical report. Porto Velho: INCRA.

Ingram, Gregory K. and Alan Carroll. 1981. The spatial structure of Latin American cities. *Journal of Urban Economics* 9:257–273.

Innis, Harold A. 1956 (1930). *The Fur Trade in Canada: An Introduction to Canadian Economic History*. Revised edition. Toronto: University of Toronto Press.

James, Preston. 1938. Changing patterns of population in São Paulo state, Brazil. *Geographical Review* 28:353–362.

Janvry, Alain de. 1981. *The Agrarian Question and Reformism in Latin America*. Baltimore and London: Johns Hopkins University Press.

Jardim, Antonio de Ponto. 1987. Aspectos do processo de urbanização recente na região centro-oeste. In Lena Lavinas, ed., *Urbanização da Fronteira*, vol. 2. Rio de Janeiro: Publipur, Universidade Federal do Rio de Janeiro.

Jasperson, Frederick Z. 1968. *The economic impact of the Venezuelan agrarian reform*. Ph.D. diss., Indiana University, Bloomington.

Johnson, E. A. J. 1970. *The Organization of Space in Developing Countries*. Cambridge, Mass.: Harvard University Press.

Junk, W. J. 1984. Ecology, fisheries, and fish culture in Amazonia. In H. Sioli, ed., *The Amazon: Limnology and Landscape Ecology of a Mighty Tropical River and Its Basin*. Hingham, Mass.: Kluwer Academic Publisher.

Kammen, Michael. 1991. *Mystic Chords of Memory: The Transformation of Tradition in American Culture*. New York: Knopf.

Katzman, Martin T. 1976. Paradoxes of Amazonian development in a "resource-starved" world. *Journal of Developing Areas* 10 (July): 445–459.

———. 1977. *Cities and Frontiers in Brazil*. Cambridge, Mass.: Harvard University Press.

Keller, Francisca Isabel Viera. 1975. O homem da frente da expansão: Permanência, mudança e conflito. *Revista da História* 51 (102): 665–709.

Kelly, Brian and Mark London. 1983. *Amazon*. New York: Harcourt, Brace, Jovanovich.

King, A. D., ed. 1991. *Culture, Globalization, and the World-System*. New York: Macmillan.

Klaassen, L. H., W. T. Mille, and J. H. P. Paelnick. 1981. *Dynamics of Urban Development*. Farnborough, U.K.: Gower.

Knox, Paul L. 1994. *World Cities and the Organization of Global Space*. Working Paper 94–1. Center for Urban and Regional Studies. Blacksburg, VA: Virginia Polytechnic Institute and State University.

Knox, Paul L. and Peter J. Taylor. 1995. *World Cities in a World System*. Cambridge, U.K.: Cambridge University Press.

Kuper, Leo and M. G. Smith, eds. 1969. *Pluralism in Africa*. Berkeley, Calif.: University of California Press.

Kuznets, Simon. 1966. *Modern Economic Growth: Rate, Structure, and Spread*. New Haven: Yale University Press.

Lacerda de Melo, Mário and Hélio A. de Moura. 1990. *Migrações para Manaus*. Fundação Joaquim Nabuco. Recife: Editora Massangana.

Laraia, Roque de Barros and Roberto da Mata. 1967. *Indios e Castanheiros: A empresa extrativa e os Indios no Médio Tocantins*. São Paulo: Difuszto Européia do Livro.

Lefebvre, Henri. 1970. *Du rural a l'urbain*. Paris: Anthropos.

Leff, Nathaniel H. 1968. *Economic Policy-Making and Development in Brazil, 1947–1964*. New York: Wiley.

Leighly, John, ed. 1963. *Land and Life: A Selection from the Writings of Carl Ortwin Sauer*. Berkeley and Los Angeles: University of California Press.

Lemos de Sá, Rosa M. 1992. A view of hydroelectric dams in the Amazon, with emphasis on the Samuel Dam, Rondonia. *Tropical Conservation and Development Program Newsletter* 25 (April): 1–3. Gainesville, Florida: University of Florida.

Levin, Jonathan V. 1960. *The Export Economies: Their Pattern in Historical Perspective*. Cambridge, Mass.: Harvard University Press.

Lewis, Arthur W. 1954. Economic development with unlimited supplies of labour. *Manchester School of Economics and Social Studies* 22 (May): 139–191.

———. 1955. *The Theory of Economic Growth*. London: Allen and Unwin.

———. 1967. *Reflections on Nigeria's Economic Growth*. Paris: Organization for Economic Cooperation and Development.

Lewis, Pierce F. 1979. Axioms for reading the landscape. In D. W. Meinig, ed., *Interpretation of Ordinary Landscapes*. Oxford: Oxford University Press.

Liberal, O. 1984. Peão denuncia trabalho em regime escravo em Xinguara. 16 de junho, p. 19.

Lisansky, Judith. 1990. *Migrants to Amazônia: Spontaneous Colonization in the Brazilian Frontier.* Boulder, Colo.: Westview Press.

Lobo, Susan. 1982. *A House of My Own: Social Organization in the Squatter Settlements of Lima, Peru.* Tucson: University of Arizona Press.

Logan, John R. and Harvey Molotch. 1987. *Urban Fortunes: The Political Economy of Place.* Berkeley, Calif.: University of California Press.

Lopes, João Batista. 1989. *Rolim de Moura: Seus Pioneiros e Desbravadores.* Privately published by the author. Rolim de Moura, Rondônia.

Lösch, August. 1954. *The Economics of Location.* 2d rev. ed. Translated by Wolfgang F. Stolper. New Haven: Yale University Press.

Loy, Jane. 1973. *The Llanos in Colombian History: Some Implications of a Static Frontier.* Occasional Papers Series, number 2. Amherst: University of Massachusetts.

Mahar, Dennis. 1979. *Frontier Development Policy in Brazil: A Study of Amazonia.* New York: Praeger.

—— 1989. *Government Policies and Deforestation in Brazil's Amazon Region.* World Bank monograph. Washington, D.C.: The World Bank.

Malloy, J., ed. 1977. *Authoritarianism and Corporatism in Latin America.* Pittsburgh: University of Pittsburgh Press.

Margolis, Maxine L. 1973. *The Moving Frontier: Social and Economic Change in a Southern Brazilian Community.* Gainesville, Florida: University of Florida Press.

Marques, A. C. 1986. Síntese histórica do combate da malária no Brasil e situação atual: Evolução da luta antipalúdica em Rondônia. Porto Velho: *Jornada de Malariologia,* n.n.

Martine, George. 1988. Changes in agricultural production and rural migration. Manuscript. Brasília: Instituto de Planejamento Econômico e Social.

——. 1987. Exodo rural, concentração urbana e fronteira agrícola. In George Martine and Ronaldo Coutinho Garcia, eds., *Os impactos sociais da modernização agrícola,* pp. 59–80. São Paulo: Editora Caetés.

——. 1990. Rondônia and the fate of small producers. In David Goodman and Anthony Hall, eds., *The Future of Amazônia: Destruction or Sustainable Development,* pp. 23–48. New York: St. Martin's Press.

——. 1992. *Processos recentes de concentração e desconcentração urbana no Brasil: Determinantes e implicações.* Brasilia: Documento de trabalho No. 11, Instituto Sociedade, População e Natureza.

Mattos, Gen. M. 1980. *Uma Geopolítica Pan-Amazônica.* Rio de Janeiro: Biblioteca do Exército.

McMillan, Gordon. 1993. Gold mining and land use change in the Brazilian Amazon. Ph.D. diss., University of Edinburgh.

Mendes, Chico. 1989. *Fight for the Forest: Chico Mendes in His Own Words.* London: Latin American Bureau.

Menezes, Marilda A. and Alfredo J. Gonçalves. 1986. *Migrações no Brasil: O peregrinas de um povo sem terra.* Centro de Estudos Migratórios. São Paulo: Editora Paulinas.

Merrick, Thomas W. and Douglas H. Graham. 1979. *Population and Economic Development in Brazil, 1800 to the Present*. Baltimore: Johns Hopkins University Press.

Meyer, Alfred H. 1956. Circulation and settlement patterns of the Calumet region of northwest Indiana and northeast Illinois (the second stage of occupance-pioneer settler and subsistence economy, 1830–1850). *Annals, Association of American Geographers* 46:312–356.

Meyer, David. R. 1980. A dynamic model of the integration of frontier urban places into the United States system of cities. *Economic Geography* 56 (2): 120–140.

———. 1986. System of cities dynamics in newly industrializing nations. *Studies in Comparative International Development* 21 (1): 3–22.

Miller, Darrell. 1985. Replacement of traditional elites: An Amazon case study. In J. Hemming, ed., *Change in the Amazon Basin*, 2:158–171. Manchester University Press.

Miller, James C. 1979. *Regional Development: A Review of the State of the Art*. Office of Urban Development. Washington, D.C.: U.S. Agency for International Development.

Millikan, Brent H. 1984. *Diagnóstico de dez núcleos urbanos de apoio rural*. Avaliação do PDRIRO/POLONOROESTE, Relatório Final. São Paulo: Fundação Instituto de Pesquisas Econômicas.

———. 1988. The dialectics of devastation: Tropical deforestation, land degradation, and society in Rondônia, Brazil. Master's thesis, University of California, Berkeley.

Mintz, Sidney. 1976. On the concept of a Third World. *Dialectical Anthropology* 1:377–382.

Mitschein, T. A., H. R. Miranda, and M. C. Paraense. 1989. *Urbanização selvagem e proletarização passiva na Amazônia: O caso de Belém*. Belém: Edições CEJUP.

Mohan, Rakesh. 1984. "The city study": Understanding the developing metropolis. The World Bank. *Research News* 5 (3): 3–14.

Molotch, Harvey. 1976. The city as a growth machine: Toward a political economy of place. *American Journal of Sociology* 82 (2): 309–337.

Monte-Mor, Roberto Luis de M. 1980a. Espaço e planejamento urbano: Considerações sobre o caso de Rondônia. Master's thesis, Universidade Federal do Rio de Janeiro.

———. 1980b. Transformação de colonização em Rondônia: A questão do planejamento urbano. Paper presented at the Seminário NAEA/CEDEPLAR Urbanização na Amazônia. July 3–4, 1980. Belém.

Moog, Viana. 1964. *Bandeirantes and Pioneers*. New York: Braziller.

Moran, Emilio F. 1975. Pioneer farmers of the Transamazon Highway: Adaptation and agricultural production in the lowland tropics. Ph.D. diss., University of Florida, Gainesville.

———. 1976. *Agricultural Development in the Transamazon Highway*. Latin American Studies Working Papers. Bloomington: Indiana University.

——. 1981. *Developing the Amazon: The Social and Ecological Consequences of Government-Directed Colonization Along Brazil's Transamazon Highway.* Bloomington: Indiana University Press.

——. 1984a. Colonization in the Transamazon and Rondônia. In Marianne Schmink and Charles Wood, eds., *Frontier Expansion in Amazônia*, pp. 285–306. Gainesville: University of Florida Press.

——. 1984b. Amazon Basin colonization. *Interciencia* 9 (6): 377–385.

Moreira, Ruy. 1985. *O movimento operário e a questão cidade-campo no Brasil* Petrópolis: Vozes.

Morrill, Richard L. 1965. Waves of spatial diffusion. *Journal of Regional Science* 8:1–18.

——. 1974. *The Spatial Organization of Society.* 2d ed. Belmont, Calif.: Wadsworth.

Mougeot, Luc J. A. 1990. Future hydroelectric development in Brazilian Amazonia: Towards comprehensive population resettlement. In David Goodman and Anthony Hall, eds., *The Future of Amazonia: Destruction or Sustainable Development*, pp. 90–129. New York: St. Martin's Press.

Mousasticoshvily, Jr., Igor. 1994. Diagnóstico da indústria madeireira na Amazônia ocidental: O caso das serrarias no eixo da estrada PA-279. Report. São Paulo: Centro Ecumênico de Documentação e Informação.

Mueller, Charles C. 1980. Recent frontier expansion in Brazil: The case of Rondônia. In Françoise Barbira-Scazzocchio, ed., *Land, People, and Planning in Contemporary Amazônia*, pp. 141–153. Cambridge, U.K.: Cambridge University Press.

Muller, Edward K. 1976. Selective urban growth in the middle Ohio Valley, 1800–1860. *Geographical Review* 66 (2): 178–199.

——. 1977. Regional urbanization and the selective growth of towns in North American regions. *Journal of Historical Geography* 3 (1): 21–39.

Muller, Geraldo. 1982. *Complexo agroindustrial e modernização agrária.* São Paulo: Editores Hucitec.

Nascimento, Cristo and Alfredo Homma. 1984. *Amazônia: Meio ambiente e tecnologia agrícola.* Belém: Empresa Brasileira de Pesquisa Agropecuária.

Nelson, Michael. 1973. *The Development of Tropical Lands: Policy Issues in Latin America.* Baltimore: Johns Hopkins University Press.

Nicholls, D. 1975. *The Pluralist State.* New York: St. Martin's Press.

Nicholls, William H. 1961. Industrialization, factor markets, and agricultural development. *Journal of Political Economy* 69 (August): 319–340.

Norgaard, Richard B. 1994. *Development Betrayed: The End of Progress and a Coevolutionary Revisioning of the Future.* London: Routledge.

North, Douglass C. 1955. Location theory and regional growth. *Journal of Political Economy* 63 (3): 243–258.

——. 1956. A reply (to Charles M. Tiebout). *Journal of Political Economy* 64 (2): 165–168.

Odum, Eugene P. 1983. *Basic Ecology.* Philadelphia: Saunders College Publications.

Oelschlaeger, Max. 1991. *The Idea of Wilderness.* New Haven: Yale University Press.

Oliveira, F. de. 1972. A economia brasileira: Crítica a razão dualista. *Estudos Cebrap* no. 2:5–82.

Ortiz, Sutti. 1984. Colonization in the Colombian Amazon. In Marianne Schmink and Charles Wood, eds., *Frontier Expansion in Amazônia*, pp. 204–230. Gainesville: University of Florida Press.

Ostrom, Elinor. 1986. An agenda for the study of institutions. *Public Choice* 48:3–25.

——. 1991. *Governing the Commons: The Evolution of Institutions for Collective Action.* Cambridge and New York: Cambridge University Press.

Outhwaite, William and Tom Bottomore, eds. 1993. *Twentieth Century Social Thought.* Oxford, U.K.: Blackwell.

Ozorio de Almeida, Anna Luiza. 1987. Os comerciantes da fronteira. In Lena Lavinas, ed. *Urbanização da fronteira*, vol. 1. Rio de Janeiro: Publipur, Universidade Federal do Rio de Janeiro.

Pace, Richard Brown. 1987. Economic and political change in the Amazonian community of Ita, Brazil. Ph.D. diss., University of Florida, Gainesville.

Painter, Michael. 1987. Unequal exchange: The dynamics of settler impoverishment and environmental destruction in lowland Bolivia. In P. D. Little and M. M. Horowitz, eds., *Lands at Risk in the Third World: Local-Level Perspectives*, pp. 164–192. Boulder, Colo.: Westview Press.

Parker, Eugene P. 1985. *The Amazon Caboclo: Historical and Contemporary Perspectives.* Studies in Third World Societies, No. 32. Williamsburg: Department of Anthropology, College of William and Mary.

Parr, John B. 1973. Growth poles, regional development, and central place theory. *Papers of the Regional Science Association* 31:173–212.

Peasants: Capital penetration and class structure in Rural Latin America. In *Latin American Perspectives*, vol. 27, parts 1–3, 1980.

Pedersen, Poul Ove. 1970. Innovation diffusion within and between national urban systems. *Geographical Analysis* 2 (July): 203–254.

Peluso de Oliveira, Marília Luiza. 1987. Cidade e gestão na fronteira agrícola. In Lena Lavinas, ed., *Urbanização da fronteira*, 2:27–72. Rio de Janeiro: Publipur, Universidade Federal do Rio de Janeiro.

Penteado, Antonio Rocha. 1968. *Belém do Pará: Estudo de geografia urbana.* Belém: Universidade Federal do Pará.

Perdigão, Francinete and Luiz Bassegio. 1992. *Migrantes Amazônicos: Rondônia, a trajetória da ilusao.* São Paulo: Edições Loyola.

Pereira, Luiz Bresser. 1984. *Development and Crisis in Brazil, 1930–1983.* Boulder, Colo.: Westview Press.

Perlman, Janice E. 1976. *The Myth of Marginality: Urban Poverty and Politics in Rio de Janeiro.* Berkeley and Los Angeles: University of California Press.

Perloff, Harvey. 1963. How a region grows: Area development in the U.S. economy. Government document. New York: Committee for Economic Development.

Perroux, Francios. 1950. Economic space: Theory and application. *Quarterly Journal of Economics* 64:89–104.

——. 1955. Note sur la notion de "pole de croissance." *Économie Appliquée* 64:307–320.

Pinheiro, Gil. 1994. O CEAP e os atores urbanos: Quem é quem na reforma urbana. *Boletim Informativo de* ANSUR, no. 16 (março), Porto Velho.

Pinheiro, Sebastião. 1984. Apenas a ponta de um iceberg. *O Liberal*, April 16, p. 2.

Pinto, Lúcio Flávio. 1978. Informe amazônico: No faroeste amazônico. *O Liberal*, July 31, p. 1.

——. 1984. Novas pistas. *O Liberal*, April 7, p. 6.

Pompermayer, Malori. 1979. The state and the frontier in Brazil: A case study of the Amazon. Ph.D. diss., Stanford University.

——. 1982. *Estratégias do grande capital na fronteira amazônica brasileira*. Programa de Estudos Comparativos Latinoamericanos. Departamento de Ciencias Politicas. Belo Horizonte: Universidade Federal de Minas Gerais.

Portes, A. and J. Walton. 1981. *Labor, Class, and the International System*. New York: Academic Press.

Pred, Allan. 1983. Structuration and place: On the becoming of sense of place and structure of feeling. *Journal for the Theory of Social Behaviour* 13:45–68.

——. 1984. Place as historically contingent process: Structuration and the time-geography of becoming places. *Annals of the Association of American Geographers* 74:279–97.

——. 1985. The social becomes the spatial, the spatial becomes the social: Enclosures, social change, and the becoming of place in the Swedish province of Skane. In D. Gregory and J. Urry, eds., *Social Relations and Spatial Structures*, pp. 337–365. London: Macmillan.

Prefeitura Municipal de Porto Velho. 1990. *Plano diretor de Porto Velho*. Porto Velho, Rondônia.

Prefeitura Municipal de Tucumã. 1990. Perfil de potencialidades do município de Tucumã. Tucumã, Pará. Mimeograph.

Presidência da República, Brasíl. n.d. *Programa grande carajás*. Brasília: Secretaria de Planejamento (official published brochure).

Provincia do Pará, A. 1984. Peões são escravizados na fazenda. June 16.

——. 1984. Dioxina empregada em Tucuruí. March 24.

Quandt. C. 1986. Technical and social changes in an export-oriented agricultural region: Coffee and soy in northern Paraná, 1935–1985. Manuscript. Graduate School of Architecture and Urban Planning, University of California, Los Angeles.

RADAM (Radar of Amazônia). 1974. Folha SB.22 Araguaia e SC.22 Tocantins. Rio de Janeiro.

Rand Corporation. 1955. *A Million Random Digits*. New York: Free Press.

Reiner, Thomas A. 1972. The region as a unit of social analysis and its place in development planning. In Manfred Stanley, ed., *Social Development: Critical Perspectives*, pp. 170–202. New York: Basic Books.

Reis, A. C. F. 1982. *A Amazônia e a cobiça internacional*. Rio de Janeiro: Editora Civilização Brasileira.

Reps, J. 1980. *Town Planning in Frontier America.* Columbia and London: University of Missouri Press.

Revkin, Andrew. 1990. *The Burning Season: The Murder of Chico Mendes and the Fight for the Amazon Rain Forest.* Boston: Houghton Mifflin.

Ricardo, Cassiano. 1970. *Marcha para oeste.* São Paulo: Editora da Universidade de São Paulo.

Rich, Bruce. 1989. The "greening" of the development banks: Rhetoric and reality. *Ecologist* 19 (2): 44–52.

Richardson, H. W. 1972. Optimality in city size, systems of cities, and urban policy: A skeptic's view. *Urban Studies* 9 (February): 29–48.

———. 1980. Polarization reversal in developing countries. *Papers of the Regional Science Association* 45:67–85.

Rivière, P. 1972. *The Forgotten Frontier: Ranchers of Northern Brazil.* New York: Holt, Rinehart and Winston.

Roberts, Bryan. 1978. *Cities of Peasants: The Political Economy of Urbanization in the Third World.* London: Edward Arnold.

Roberts, J. Timmons. 1991. Forging development, fragmenting labor: Subcontracting and local response in an Amazon boomtown. Ph.D. diss., Johns Hopkins University.

———. 1992. Squatters and Amazônian urban growth. *Geographical Review* 82 (4): 441–457.

———. 1995. Expansion of television in eastern Amazonia. *Geographical Review* 85 (1): 41–49.

Rocque, C. 1973. *Antônio Lemos e sua época: História política do Pará.* Belém, Pará: Edições Culturais.

Rodriguez Pereira, Alvaro. 1975. Transamazônica: Colonos ricos e pobres. *Veja,* July 16, pp. 26–27.

Rolim de Moura. 1989. *Dados estatísticos.* Prefeitura, Rolim de Moura, Rondônia.

Romeiro, A. R. 1987. Alternative developments in Brazil. In Bernhard Glaeser, ed., *The Green Revolution Revisited.* London: Oxford University Press.

Rondinelli, Dennis A. 1983. *Secondary Cities in Developing Countries: Policies for Diffusing Urbanization.* London and Beverly Hills: Sage.

———1986. The urban transition and agricultural development: Implications for international assistance policy. *Development and Change* 17:231–263.

——— 1987. Cities as agricultural markets. *Geographical Review* 77 (4): 408–420.

Rozman, Gilbert. 1978. Urban networks and historical stages. *Journal of Interdisciplinary History* 9 (1): 65–91.

Rudel, Thomas K. with Bruce Horowitz. 1993. *Tropical Deforestation: Small Farmers and Land Clearing in the Ecuadorian Amazon.* New York: Columbia University Press.

Ruttan, V. W. 1959. The impact of urban-industrial development on agriculture in the Tennessee Valley and Southeast. *Journal of Farm Economics* 37 (February): 38–56.

Salazar, João Pinheiro. 1992. O novo proletariado industrial de Manaus e as suas transformações possíveis (Estudo de um grupo de operários). Ph.D. diss., Universidade de São Paulo.

Sampaio, Aidil and Risalva Vasconcelos Rocha. 1989. *Tendências das migrações no Nordeste, 1940–1980.* Recife: Superintêndencia do Desenvolvimento do Nordeste.

Sampaio, Plínio. 1980. *Capital estranjeiro e agricultura no Brasil.* Petrópolis: Vozes.

Santos, M. 1979. *The Shared Space: The Two Circuits of the Urban Economy in Underdeveloped Countries.* London and New York: Methuen.

Santos, Stael Starling Moreira dos, Ciléa Souza de Silva, and Nelly Lamarao Câmara. 1991. Saneamento básico e problemas ambientais na região metropolitana de Belém. IBGE. Mimeograph.

Sawyer, Donald. 1984. Frontier expansion and retraction in Brazil. In Marianne Schmink and Charles Wood, eds., *Frontier Expansion in Amazônia*, pp. 180–203. Gainesville: University of Florida Press.

——. 1987. Urbanização da fronteira agrícola no Brasil. In Lena Lavinas, ed., *Urbanização da Fronteira*, 1:43–57. Rio de Janeiro: Publipur, Universidade Federal do Rio de Janeiro.

——. 1993. População e meio ambiente na amazônia brasileira. In George Martine, ed., *População, meio ambiente, e desenvolvimento: Verdades e contradições*, pp. 149–165. Campinas, SP: Editora da UNICAMP.

Sawyer, Donald, H. da Gama Torres, A. Lourenço Pereira, and R. Abers. 1990. Fronteiras na Amazônia: Significado e perspectivas. Mimeograph. Belo Horizonte: CEDEPLAR.

Sawyer, Donald R. and Diana R. T. O. Sawyer. 1987. Malaria on the Amazon frontier: Economic and social aspects of transmission and control. Research report. Belo Horizonte: Centro de Desenvolvimento e Planejamento Regional.

Sayer, Andrew. 1984. *Method in Social Science: A Realist Approach.* London: Hutchinson.

——. 1989. On the dialogue between humanism and historical materialism in geography. In A. Kobayashi and S. Mackenzie, eds., *Remaking Human Geography*. London: Unwin Hyman.

Scherl, Ilana and José S. Netto. 1979. O outro lado de uma obra grandiosa. *Isto É*, July 3, pp. 24–27.

Schmink, Marianne. 1982. Land conflicts in Amazônia. *American Ethnologist* 9 (2): 341–357.

——. 1987. The rationality of forest destruction. In J. C. Figueroa Colón, ed., *Management of the Forests of Tropical America: Prospects and Technologies*, pp. 11–30. Rio Pedras: Institute of Tropical Forestry, U.S. Department of Agriculture.

Schmink, Marianne and Charles Wood, eds. 1984. *Frontier Expansion in Amazônia.* Gainesville: University of Florida Press.

——. 1992. *Contested Frontiers in Amazônia.* New York: Columbia University Press.

Schneider, R. 1991. *"Order and Progress": A Political History of Brazil.* Boulder, Colo.: Westview Press.

Schultz, T. W. 1953. *The Economic Organization of Agriculture.* New York: McGraw-Hill.

Sellars, Roy Wood. 1969 [1916]. *Critical Realism: A Study of the Nature and Conditions of Knowledge.* New York: Russel and Russell.

Sennett, Richard, ed. 1969. *Classic Essays on the Culture of Cities*. Englewood Cliffs, N.J.: Prentice Hall.

SEPLAN (Secretaria de Planejamento e Coordenação Geral). 1976. *Zona franca desenvolvimento*. Estado de Amazonas. Manaus, Amazonas.

———. 1983 (and various years). *Anuário estatístico de Rondônia*. Porto Velho, Rondônia.

———. 1985. *Projeção populacional*. Porto Velho, Rondônia.

———. 1988a. *Perfil sócioeconômico de Santa Luzia*. Porto Velho, Rondônia.

———. 1988b. *Perfil sócioeconômico de Rolim de Moura*. Porto Velho, Rondônia.

———. 1988c. *Perfil sócioeconômico de Alta Floresta do Oeste*. Porto Velho, Rondônia.

———. 1988d. *Informativo sócioeconômico dos municípios*. Porto Velho, Rondônia.

———. 1990. *Levantamento sistemático da produção agrícola: Rondônia*. Porto Velho, Rondônia.

———. 1992. *Rondonia: Indicadores municipais*. Brasília: SEPLAN.

SET (Secretaria Estadual do Trabalho, Rondônia). 1983. *A oferta de mao-de-obra migrante e o mercado de trabalho na area urbana de Jí-Paraná*. Porto Velho.

Sheldrake, Rupert. 1988. The laws of nature as habits: A postmodern basis for science. In David Ray Griffin, ed., *The Reenchantment of Science*, pp. 79–86. Albany, N.Y.: State University of New York Press.

Silva, Sergio. 1979. Formas de acumulação e desenvolvimento do capitalism no campo. In Jaime Pinsky, ed., *Capital e Trabalho no Campo*, pp. 7–24. São Paulo: Editora Hucitec,

Simon, David. 1992. *Cities, Capital, and Development: African Cities in the World Economy*. London: Belhaven.

———. 1995. The world city hypothesis: Reflections from the periphery. In P. L. Knox and P. J. Taylor, eds., *World Cities in a World-System*, pp. 132–155. Cambridge: Cambridge University Press.

Singer, Paul. 1979. Capital e trabalho no campo. In Jaime Pinsky, ed., *Capital e Trabalho no Campo*, pp. 1–7. São Paulo: Editora Hucitec.

Sisler, Daniel G. 1959. Regional differences in the impact of urban-industrial development on farm and nonfarm income. *Journal of Farm Economics* 41:1100–1112.

Skillings, Robert F. 1985. Economic development of the Brazilian Amazon: Opportunities and constraints. In John Hemming, ed., *Change in the Amazon Basin*. Vol. 1: *Man's Impact on Forests and Rivers*, pp. 36–43. Manchester: Manchester University Press.

Smith, Michael G. ed. 1969. *Pluralism in Africa*. Berkeley: University of California Press.

Smith, M. P. 1980. *The City and Social Theory*. Oxford: Blackwell.

Smith, Neil. 1993. Homeless/global: Scaling places. In Jon Bird, Barry Curtis, Tim Putnam, George Robertson, and Lisa Turner, eds., *Mapping the Futures: Local Cultures, Global Change*, pp. 87–119. London: Routledge.

Smith, Nigel J. H. 1976. Brazil's Transamazon Highway settlement scheme: Agrovilas, agropoli, and ruropoli. *Association of American Geographers Proceedings* 8:129–132.

———. 1981. Colonization lesson from a tropical forest. *Science* 214 (13): 755–761.

———. 1982. *Rainforest Corridors: The Transamazon Colonization Scheme.* Berkeley: University of California Press.

Smolowe, Jim and Mac Margolis. 1984. Poison in the jungle? *Newsweek,* January 9, p. 11.

Snethlage, Emilia. 1910. A travessia entre o Xingu e o Tapajoz. *Boletim do Museu Paraense "Emilio Goeldi"* 7:49–99.

Snyder, Gary. 1990. *The Practice of the Wild.* San Francisco: North Point Press.

Socorro Pessoa, Maria do. 1988. *Rolim de Moura: Um ponto de vista.* Brasilia: CEGRAF.

Sorj, Bernardo. 1980. *Estado e classes sociais na agricultura brasileira.* Rio de Janeiro: Zahar Editores.

Souza, Itamar de. 1980. *Migrações internas no Brasil.* Petrópolis: Vozes and Fundação José Augusto (Natal, Rio Grande do Norte).

Spencer, Herbert. 1972. *On Social Evolution: Selected Writings.* Edited with an introduction by J. D. Y. Peel. Chicago: University of Chicago Press.

Stearman, Allyn MacLean. 1984. Colonization in Santa Cruz, Bolivia: A comparative study of the Yapacani and San Julian projects. In Marianne Schmink and Charles Wood, eds., *Frontier Expansion in Amazônia,* pp. 231–260. Gainesville, Florida: University of Florida Press.

Steinen, Karl von den. 1940. *Entre os aborigenes do Brasil central.* São Paulo: Separata renumerada da *Revista do Arquivo.* Nos. 34–58:713.

Sternberg, H. O. 1975. *The Amazon River of Brazil.* Wiesbaden: Franz Steiner Verlag.

———. 1987. "Manifest destiny" and the Brazilian Amazon: A backdrop to contemporary security and development issues. *Conference of Latin Americanist Geographers Yearbook* 13:25–35

Stewart, I. 1989. *Does God Play Dice?* London: Penguin.

Storper, Michael. 1984. Who benefits from industrial decentralization? Social power in the labour market, income distribution, and spatial policy in Brazil. *Regional Studies* 18:143–164.

SUDAM (Superintendência do Desenvolvimento da Amazônia). 1968. *Acre e Rondônia: Diagnóstico econômico.* Published statement of SUDAM President José Marcelino Monteiro da Costa. Belém, Pará: SUDAM.

———. 1976. *Programa de pólos agropecuários e agrominerais da Amazônia: Polamazônia, Carajás.* Belém.

Swimme, Brian. 1988. The cosmic creation story. In David Ray Griffin, ed., *The Reenchantment of Science,* pp. 47–56. Albany, N.Y.: State University of New York Press.

Syvrud, Donald E. 1974. *Foundations of Brazilian Economic Growth.* Stanford, Calif.: Hoover Institution Press.

Thompson, Stephen I. 1973. Pioneer colonization: A cross-cultural view. *Addison-Wesley Module in Anthropology* 33:23.

Thornton, Ian, David Cleary, and Simon Worthington. 1990. Mercury contamination in the Brazilian Amazon. Report to the Commission of the European Communities. Contract reference-article B946/90.

Thrift, Nigel. 1983. On the determination of social action in space and time. *Environment and Planning D: Society and Space* 1:23–57.

———. 1985. Bear and mouse or bear and tree? Anthony Giddens's reconstitution of social theory. *Sociology* 17:609–623.

Thunen, Johann H. von. 1966. *Von Thunen's Isolated State.* Translated by Carla M. Wartenburg. London: Oxford University Press.

Time. 1976. Ludwig's wild Amazon kingdom. November 15, pp. 79–80.

Time. 1979. Billionaire Ludwig's Brazilian gamble. September 10, pp. 76–79.

Todaro, Michael P. 1969. A model of labor migration and urban unemployment in less developed countries. *American Economic Review* 59 (1): 138–148.

———. 1980. Internal migration, urban population growth, and unemployment in developing nations: Issues and controversies. *Journal of Economic Development* 5 (1): 7–23.

Todaro, Michael P. and John R. Harris. 1970. Migration, unemployment, and development: A two-sector analysis. *American Economic Review* 60 (1): 126–142.

Tonnies, Ferdinand. 1887 (1963). *Community and Society.* New York: Harper & Row.

Torres, Haroldo da Gama. 1988. A urbanização e o migrante de origem urbana na Amazônia. Manuscript. Brasília: Instituto de Sociedade, População, e Natureza.

Townroe, Peter M. and David Keen. 1983. Polarization reversal in the state of São Paulo. *Regional Studies* 18 (1): 45–54.

Turner, Frederick Jackson. 1920. *The Frontier in American History.* New York: Holt.

UFRRJ (Universidade Federal Rural do Rio de Janeiro). 1985a. *Contribuição do mercado madeireiro no desenvolvimento regional: Rondônia.* Seropédica: Instituto de Florestas.

———. 1985b. *Proposta para o plano de manejo da floresta nacional do Jamari.* Seropédica: Instituto de Florestas.

Uquillas, Jorge. 1984. Colonization and spontaneous settlement in the Ecuadoran Amazon. In Marianne Schmink and Charles Wood, eds., *Frontier Expansion in Amazônia,* pp. 261–284. Gainesville, Florida: University of Florida Press.

Vance, J. E., Jr. 1970. *The Merchant's World: The Geography of Wholesaling.* Englewood Cliffs, N. J.: Prentice Hall.

———. 1982. *Location in a System of Global Extent: A Social Model of Settlement.* Geographical Papers. Reading: University of Reading.

———. 1990. *The Continuing City: Urban Morphology in Western Civilization.* Baltimore: Johns Hopkins University Press.

Veja. 1978. O passado ainda vivo: No tenso sul do Pará, a vigilancia do Exército é ainda considerado indispensável. Sept. 6, pp. 61–62.

Velho, Otávio Guilherme. 1972. *Frentes de expansão e estrutura agrária: Estudo do processo de penetração numa area da Transamazônica.* Rio de Janeiro: Zahar Editores.

———. 1976. *Capitalismo autoritário e campesinato: Um estudo comparativo a partir da fronteira em movimento.* São Paulo: DIFEL.

Vining, D. R. 1984. Industrial decentralization in Brazil: A query. *Regional Studies* 19 (2): 163–164.

Vining, D. R. and T. Kontuly. 1978. Population dispersal from a major metropolis: The case of São Paulo. *Regional Science Review* 3:49–73.

Volbeda, Sjouke. 1986. Pioneer towns in the jungle: Urbanization at an agricultural colonization frontier in the Brazilian Amazon. *Revista Geográfica* 104 (July-December): 115–140.

Wagley, Charles. 1976 [1953]. *Amazon Town: A Study of Man in the Tropics.* London: Oxford University Press.

Waibel, Leo. 1950. European colonization in southern Brazil. *Geographical Review* 40 (4): 529–547.

Wallerstein, Immanuel. 1974a. The rise and future demise of the world capitalist system. *Comparative Studies in History and Society* 16 (4): 387–415.

———. 1974b. *The Modern World System.* New York: Academic Press.

———. 1991. The national and the universal: Can there be such a thing as world culture? In Anthony D. King, ed., *Culture, Globalization, and the World System,* pp. 91–105. London: Macmillan.

Walton, J. 1976a. Political economy of world urban systems: Directions for comparative research. In J. Walton and L. H. Massoti, eds., *The City in Comparative Perspective: Crossnational Research and New Directions in Theory,* pp. 301–314. New York: Wiley.

———. 1976b. Urban hierarchies and patterns of dependence in Latin America: Theoretical bases for a new research agenda. In A. Portes and H. L. Browning, eds., *Current Perspectives in Latin American Urban Research.* Austin, Tex.: University of Texas Press.

Watkins, Melvin H. 1967. A staple theory of economic growth. In W. T. Easterbrook and M. H. Watkins, eds., *Approaches to Canadian Economic History,* pp. 49–73. Toronto: McClelland and Steward.

Webb, Walter Prescott. 1964. *The Great Frontier.* Austin, Tex.: University of Texas Press.

Weinberg, Steven. 1992. *Dreams of a Final Theory: The Scientist's Search for the Ultimate Laws of Nature.* New York: Vintage Books.

Weinstein, Barbara. 1983. *The Amazon Rubber Boom, 1850–1920.* Stanford, Calif.: Stanford University Press.

Wells, Robin F. 1973. Frontier systems as a sociocultural type. In Graydon H. Doolitter and Christopher Lintz, eds., *Papers in Anthropology* 14 (1): 6–15. Norman: University of Oklahoma.

Wenz, Peter. 1988. *Environmental Justice.* Albany, N.Y.: SUNY Press.

Wesche, Rolf. 1974. Planned rain forest family-farming on Brazil's Transamazon Highway. *Revista Geográfica* 81 (December): 105–113.

Willems, Emilio. 1972. The rise of a rural middle class in frontier society. In Riordan Roett, ed., *Brazil in the Sixties,* pp. 325–344. Nashville: Vanderbilt University Press.

Wirth, John D., Edson de Oliveira Nunes, and Thomas E. Bogenschild. 1987. *State and Society in Brazil: Continuity and Change.* Boulder, Colo.: Westview Press.

Wood, Charles H. and Marianne Schmink. 1979. Blaming the victim: Small farmer production in an Amazon colonization project. *Studies in Third World Societies* 7:77–93.

Wood, Charles H. and John Wilson. 1984. The magnitude of migration to the Brazilian frontier. In Marianne Schmink and Charles Wood, eds., *Frontier Expansion in Amazônia*, pp. 142–152. Gainesville, Florida: University of Florida Press.

Wood, Charles and J. A. Mongo de Carvalho. 1988. *The Demography of Inequality in Brazil.* Cambridge: Cambridge University Press.

Wood, William B. 1986. Intermediate cities on a resource frontier. *Geographical Review* 76 (2): 149–159.

World Bank. 1981a. Brazil: Northwest regional development program, first phase, agricultural development and environmental protection project. Staff Appraisal Report No. 3512b-BR.

——. 1981b. Brazil: Integrated development of the northwest frontier. Monograph.

——. 1983. Brazil: Northwest region development program, phase III: New settlements project. Staff Appraisal Report.

——. 1987. Brazil: Northwest I, II, and III technical review. Final Report.

——. 1989. *World Development Report.*

——. 1991. Brazil: An analysis of environmental problems in the Amazon. Vols. 1 and 2. Report No. 9104-BR. Technical Report. Washington, D.C.: World Bank.

——. 1992. *World Development Report.* Oxford, U.K., and New York: Oxford University Press.

——. 1993. *World Development Report.* Oxford, U.K., and New York: Oxford University Press.

World Resources Institute (WRI). 1987. *World Resources.* Washington, D.C.: World Resources Institute.

Yuill, R. S. 1965. A simulation study of barrier effects in spatial diffusion problems. *Michigan Inter-University Community of Mathematical Geographers, Discussion Papers* 5. Ann Arbor: University of Michigan.

Zavala, Silvio. 1957. The frontiers of Hispanic America. In Walker D. Wyman and Clifton B. Kroeber, eds., *The Frontier in Perspective.* Madison: University of Wisconsin Press.

Author Index

Subject Index

Printed in the USA
CPSIA information can be obtained
at www.ICGtesting.com
JSHW011519221024
72172JS00014B/112

9 780231 106559